Pest Management and Residual Analysis in Horticultural Crops : An Integrated Approach

NIPA® GENX ELECTRONIC RESOURCES & SOLUTIONS P. LTD.
New Delhi-110 034

www.nipaers.com
Browse, Search, Read & Buy...
Print Books
eChapters
eBooks
Publishers
Forthcoming
Subject Catalogues
New Titles
Online Resources on:
✓ Current Affairs
✓ Reasoning, Logic & Aptitude
✓ Competitive Examinations
✓ English Language Lab
✓ Writing & Pronunciation Tools
✓ Personality Development
✓ Online Programmes for Professionals
✓ Interview Preparation
Pay Using
PhonePe
Google Pay
Paytm
PayPal
UPI
CCAvenue

Pest Management and Residual Analysis in Horticultural Crops : An Integrated Approach

Edited by

Rachna Gulati

and

Beena Kumari

NIPA® GENX ELECTRONIC RESOURCES & SOLUTIONS P. LTD.

New Delhi-110 034

NIPA® GENX ELECTRONIC RESOURCES & SOLUTIONS P. LTD.

101,103, Vikas Surya Plaza, CU Block
L.S.C. Market, Pitam Pura, New Delhi-110 034
Ph : +91 11 27341616, 27341717, 27341718
E-mail: newindiapublishingagency@gmail.com
web: www.nipabooks.com

For customer assistance, please contact
Phone: + 91-11-27 34 17 17
Fax: + 91-11- 27 34 16 16
E-Mail: feedbacks@nipabooks.com

ISBN: 978-81-19254-86-6

Composed and Designed by NIPA®.

Preface

The horticultural crops are important component of diversified agriculture which contribute to nutritional and livelihood security. Today, India ranks second in the world in the production of fruits. Horticultural crops are preferred because of their high returns compared with other field crops. More than 50,000 plant species are meeting the food (calories) needs of human world wide. Horticultural crops encompass fruit crops, vegetables, ornamentals, plantation crops, spices, aromatic and medicinal plants, tuber crops and mushrooms.

In the recent years, major thrust is being given to the production of high value cash crops such as vegetables, ornamentals and mushrooms. However, these are susceptible to a large number of pests and diseases causing reduction in yield. Likewise, reliance on the use of synthetic organic pesticides has resulted in aggravated resistance, residues, human risk, effect on other non target organisms and environmental problems. These are crucial factors limiting the spectacular development of the horticulture industry. Higher yields can be harnessed through better pest and disease management strategies.

The book is intended to provide a clear overview on the management of pests and diseases of horticulture crops, associated soil and beneficial fauna, residue status of pesticides and their estimation techniques. It is divided in four parts: **Part I** explain the practices followed in the pest management of horticulture crops. Chapters include pest status of insects, mites and rodents in fruits, vegetables, ornamentals, spices and mushrooms and their management. Different aspects of biological, cultural, and mechanical controls are also highlighted. Harmful and beneficial soil fauna associated with horticulture crops are dealt in **Part II.** Keeping in view the potential of beneficial organisms, the effects of pesticides on predators and parasites compatibility of microbial pesticides with synthetic have also been discussed in this section. The recent scientific developments related to diseases and residue status in vegetables, fruits and spices are provided in **Part III. Part IV** includes the residue estimation techniques of various pesticides. This book is hoped to serve as a useful source of information to teachers, postgraduate students and researchers.

Editors

Contents

Part III

Part IV

Part - I

Chapter-1

Acarine Pests and their Management in Vegetable Crops

Manmeet Brar Bhullar and Sunita Yadav
AINP on Agricultural Acarology,
Department of Entomology, Punjab Agricultural University, Ludhiana-141004
Department of Entomology, CCS Haryana Agricultural University, Hisar - 125004

Chemicals which are used specifically for the control of mite pests are known as acaricides, though there are many insecticides, which show acaricidal properties as well. In the last few decades, there has been tremendous progress in the development of broad-spectrum acaricides *i.e.* the compounds that will knock down different species of mites at the same time. Prior to the commercial introduction of broad-spectrum insecticides like DDT and other organochlorine insecticides, mites did not commercially cause wide-spread damage under natural or semi-natural environments which were little influenced by man. This might be due to the natural processes of prey-predator relationship in undisturbed situations. Contrastingly in situations where broad-spectrum pesticides are commonly used, the natural enemies are more adversely affected than the mites, and an outbreak of the latter occurs. Besides becoming generally intensified with the introduction of modern insecticides, the mite problem in some cases has subsequently become even more serious due to development of resistance against these materials.

Many spider mite species, which were rarely found in pest form in some situations, appeared commonly as major pests with the persistent use of pesticides. In Indian literature, spider mites were never reported in pest form prior to 1940 but with the intensive use of pesticides, every year new mite problems are appearing and becoming troublesome to control.

Two main hypotheses concerning mite-outbreaks have been found in literature:

i) The general upsurge of spider mites following World War II due to improved cultural methods or stimulative effects of the pesticide used (including use of fertilizers) promoting increased plant growth and better spider mite nutrition. In this view, it is emphasized that more nutritious foliage promotes increased fecundity so that the outbreak of mites occur independent of the action of their natural enemies.

ii) The more general hypothesis is that the shift to modern insecticides adversely affected the natural enemies of spider mites thus producing more frequent outbreaks of species formerly troublesome and rise to pest status of ones previously innocuous. The adverse effects of chlorinated hydrocarbon, organophosphate, carbamate and various pyrethroid pesticides, is well known (Dhooria and Mann, 1989). Resurgences of spider mites have commonly been reported with the application of several types of pesticides with different modes of action in a wide variety of climates, indicates that chemical control in these cases upsets their population dynamics.

1. ACARINE FAUNA OF VEGETABLE CROPS

Phytophagous mites have long been considered as potentially serious pests of a wide range of food and fiber crops, and ornamentals. They are covered under subclass Acari of class Arachnida and phylum Arthropoda. Mites unlike insects are eight legged creatures and lack compound eyes, antennae and wings. Plant feeding mites measure from 200μ to 600μ. Majority of the plant feeding mites belong to suborder Prostigmata, and phytophagous mites belong to family Tetranychidae (spider mites); Eriophyidae (bud or gall mites); Tenuipalpidae (false spider mites); Tarsonemidae (broad mites) and Tuckerellidae, in the decreasing order of their importance. Of these, the first three are exclusively phytophagous while some tarsonemids are phytophagous (Zhang, 2003).

1.1 Major Mite Pests of Vegetables

Majority of the vegetable crops like brinjal, okra, cucurbits, beans, tomato, chillies , potato *etc.* are attacked by large number of mite pests which greatly affects the productivity (Gupta , 1991). The major mites belonging to different families are:

1.1.1 Family: Tetranychidae

Mites belonging to this family are commonly called as spider mites because of their web forming nature. These are soft-bodied, variously colored, found on both surfaces of the leaves, colony forming, have four pairs of legs and size varies from 0.5 - 0.6mm. The spider mites, *Tetranychus urticae* (syn. *T. cinnabarinus*), *T. neocaledonicus, T. ludeni* and *T. macfarlanei*, severely infest brinjal, okra, cucurbits, cowpea crops throughout the country and the mite problem becomes more severe during summer months particularly during April to June. The four species of spider mites mentioned above are similar in appearance, behaviour, and type of damage they cause and many times mixed populations of these are found. The infested leaves develop yellowish-white speckling, dry and later fall off. Plants remain covered with dense webs harbouring sometimes thousands of mites and the heavily infested plants die.

1.1.2 Family: Eriophyidae

These are found on leaves, and also buds, galls, may form blisters, erineum etc. They have only two pairs of legs and their size varies from 0.2-0.25mm. *Aceria lycopersici* attacks potato, brinjal, tomato crops and causes crinkling, curling of leaves, erineum formation in leaves of brinjal and 'tambera' disease in potato.

1.1.3 Family: Tenuipalpidae

These mites inhabit mostly lower surface of leaves/ twigs, fruits. They are commonly known as false spider mites because they don't form webs like spider mites. They are small (150µ - 250µ), pear-shaped , flat mites with dorso-ventrally flattened body, mostly brightly coloured and colony forming. Mites of *Brevipalpus*, particularly *B. deleonis*, have been seen infesting Sweet potato crop in Tamil Nadu causing browning of leaves.

1.1.4 Family: Tarsonemidae

These are tiny, yellow, broader anteriorly, fast moving, have 4 pairs of legs and size varies from 0.25-0.30mm. They occur mostly in young leaves. A very characteristic feature in these mites is that often in colonies of these mites, males are seen carrying female deutonymphs, for mating with them as soon as they emerge as adults. A tarsonemid mite, *Polyphagotarsonemus latus*, causes severe damage to chilli crop in Tamil Nadu, West Bengal and several other parts of the country. Leaves show crinkling, downward curling, flower and fruit formation is affected and in turn yield is reduced considerably. It causes 'Murda' disease of chilli. Besides chilli this mite also attacks tomato, potato, French beans, etc.

1.1.5 Family: Tuckerellidae

They look similar to spider mites and are rare. These are brightly coloured having fan-like dorsal body setae and long whip like caudal setae. They are mostly associated with fruit crops.

Table 1 : Crop wise Status of Mite Pests in India

Crops	Total mite species	Total mite pest	Major pest	Minor pest	Yield losses
Vegetables	**54**	**6**	**4**	**2**	
All vegetables			*Tetranychus urticae*		25-30% in okra, brinjal
Brinjal			*T. urticae*, *T. neocaledonicus*		
Cowpea			*T. ludeni*		
Chilli, potato, tomato			*Polyphagotar -sonemus latus*		27-39% in chillies
Cucurbits			*T. urticae*	*T. macfarlanei*	

Table 2 : Phytophagous Mites Associated with vegetable crops

Family	Mites	Host range
Tetranychidae		
	T. cinnabarinus (Boisduval) (Syn. *T. urticae, T. telarius*)	Cabbage^{+}, beans^{+}, lettuce^{+}, okra^{++}, brinjal^{++}, fenugreek^{+}, spinach^{+}, tomato^{+}, capsicum^{+}, mint
	T. cucurbitae Rahman & Sapra	Beans, brinjal, cucurbits, cabbage, tomato
	T. ludeni Zacher	Beans^{+}, cowpea^{++}, brinjal, potato, pumpkin,
	T. macfarlanei Baker & Pritchard	Okra^{+}
	T. neocaledonicus Andre	Cauliflower^{+}, french bean^{++}, brinjal, okra

Contd.

Eriophyidae		
	A. hibisci (Nalepa)	Okra
	A. lycopersici (Wolff)	Brinjal+, tomato+
	Calacarus capsicae Chak & Mond.	Capsicum
Tenuipalpidae		
	Brevipalpus obovatus (Donn.)	
	B. californicus (Banks)	Brinjal
	B. deleonis	Sweet potato
	B. phoenicis (Geijskes)	Brinjal, tomato
Tarsonemidae		
	Polyphagotarsonemus latus (Banks)	Brinjal, chilli++, potato++

++ Major pest, + Minor pest

1.2 Nature and Extent of Damage

Spider mites damage the stomata and spongy parenchyma of the leaf surface, and some even inject toxic substances into the leaf surfaces which interferes with the vital processes of the infested plants. Phytophagous mites attack almost all the cultivated crops like vegetables, fruits, cereals, ornamentals and even wild plants and in turn reduce their yield. They also serve as vectors for many plant diseases. From India, losses in yield due to mite damage vary from 10-15 percent in vegetables due to spider mites (Gupta, 2001) and in some cases losses may also go up to as high as 25-50 percent due to spider mites on brinjal and yellow mites on chilli (Singh and Singh, 2004). Although crop loss estimates are not available for many crops but the damage must be considerable because of strong host-plant relationships of mites.

1.3 Biology of Mites

The life-cycle of mites consist of four stages i.e. egg, larva, nymph (protonymph and deutonymph) and adult. The three immature stages (larva and the two nymphal stages) are each followed by a quiescent stage known respectively as nymphochrysalis, deutochrysalis and teleiochrysalis. Males mature earlier than females and copulation takes place almost immediately after emergence of the young female. The eggs are laid singly or in clusters. Females lay round, whitish, pink, red or light yellow eggs on the lower surface of the leaves. The young larva with six legs emerges out of the eggs after an incubation period of 4-5 days. After 2-3 days, the larva changes into an eight-legged nymph. There are 1-2 nymphal stages in different species which are completed in 3-4 days. In most of the tetranychids and tenuipalpids, the above pattern is followed and life-cycle is completed in 7-10 days in Tetranychidae while in Tenuipalpidae the time taken is longer (15-20 days). In tarsonemids, only two life-stages are seen i.e. larva and adult. In eriophyids, there are two active stages, i.e. first stage nymph and second stage nymph (Gupta, 1985). The development is influenced by temperature as life-cycle is longer in winter months as compared to summer months (Banerjee, 1988; Singh and Singh, 2004).

2. ACARICIDES USED IN VEGETABLE CROPS

Although mites are different from insects yet their control is within the realm of the entomologists and the chemicals used for this purpose are called acaricides. Ovicides, which kill eggs, are also important since the life span of the adult mite is short and a large proportion of the population is always in the egg stage. When nonspecific insecticides are used, natural mite predators are often eliminated without impairing mites. Their short developmental cycle tends to rapidly develop the resistant populations by removing unfavourable genes for their existence. The history of miticide or acaricide development indicates the treadmill between chemicals and mites and many types of compounds should have been developed (Perry *et al.*, 1998). In the early part of the 20th century inorganic acaricides like sulphur were used, which were followed by the use of oils and then synthetic organic acaricides having long residual action.

2.1 Acaricide Usage in India

Acaricides are widely used in crops like tea, apple, roses, chillies, and many vegetables. Currently there are approximately 204 pesticides registered for use and production has gone up to 85,000 metric tones. As the cropping pattern is becoming more intensive, use of these pesticides is also increasing. Consumption of insecticides in agriculture has been increased more than 100% from 1971 to 1994-95, e.g. insecticide consumption in India, which was to the tune of 22013 tonnes, has increased to 51755 tonnes by 1994-95. Consumption of all these pesticides in same duration has increased more than two times, that is from 24305 tonnes to 61357 tonnes. However, in 2003-04 and 2004-05 the consumption has come down to 41, 3504 MT (Tech grade).

World-wide acaricide market is around 700 million USD with 21% consumption in USA followed by Japan (15%), Brazil (12.5%), Spain (7.5%), France and Itlay (4.5%) and the rest of the world 35%. Cropwise consumption of acaricides lead by citrus (38%), cotton (22%), orchards including tea (18%), vine (8%), vegetables (5%), corn (3%) and other crops (6%).

In India currently the acaricide consumption is on the increasing trend in the last few years due to menace of phyophagous mites. Sudden outbreaks of mites in various crops are mainly due to changes in the agro-climatical conditions and agronomic practices. In 4000million rupees Indian Pesticide market, acaricide growth is around 2% during 2005-06 from the existing 5% acaricide market. Acaricides are widely used in crops like tea, apple, roses, chillies and many vegetable crops. Every year one or two new acaricides are registered in India. Approach should be to acaricides having minimum or no impact on natural enemies, no resurgence of pests, new chemistry/mode of action which will not lead to resistance of mites (Pandey and Selvasundaram, 2007).

2.2 Acaricide Groups

Presently available acaricides can be classified into five major groups:

i) Inorganic acaricides

ii) Oil based acaricides

iii) Synthetic organic acaricides

iv) Microbial acaricides

v) Growth regulators and Juvenile hormones

The characteristics of some of the major acaricides used for the control of mite pests are discussed below:

2.2.1 Inorganic Acaricides

Sulphur is an inorganic acaricide and in earlier days was the only chemical used to control different mite pests. Majority of tetranychids, eriophyids, and tenuipalpids are susceptible to sulphur preparations like lime sulphur and sulphur dust or even finely grounded colloidal suspension. However, sulphur shows some selectivity as some mites of genus *Tetranychus* and some tarsonemids are not easily controlled by sulphur preparations.

Sulphur is non-systemic but targets the mite directly by exposing the vapours that are toxic to them. Due to this reason, sulphur is many times used to control mites living within the buds which are otherwise difficult to control. Sulphur is generally non-toxic to plants, but it is not advisable to use it under polyhouse conditions. It is relatively safer to insect predators (Puttaswamy and ChannaBasvanna, 1976).

2.2.2 Oil Based Acaricides

Petroleum based oils were used basically to control the eggs of red spider mites on twigs and foliage of fruit trees. Oil preparations consist of mostly aliphatic hydrocarbons, both saturated and unsaturated. Their greatest disadvantage is their phytotoxic effect on the plants i.e. scorching and damage to leaves. However, in some oil preparations like summer oils, phytotoxicity has been greatly reduced by chemical modifications and refinements. Horticultural mineral oils are used to control tetranychid mites like *Panonychus ulmi* on apple, *Tetranychus* and *Bryobia* on fruit trees (Jeppson *et al.*, 1975; Bharadwaj and Bharadwaj, 2005). These mineral oils have both ovicidal as well as adulticidal properties and considered as good substitutes for acaricides/ insecticides. One of the main advantage of mineral or petroleum oil is that mites do not develop resistance easily to them. However, their application under tropical conditions may cause serious phytotoxicity especially when applied under strong strong sun.

2.2.3 Synthetic Organic Acaricides

Many acaricides have been developed in the last few decades and some of them are still being used (Anonymous, 2005; 2007; Hill, 1987; Matsumura, 1975). Their characterictics, mode of action and efficacy against mites are given below:

1. Chlorinated hydrocarbons : They are commonly referred to as organochlorines and they are a broad-spectrum and very persistent group which usually kill both by contact and as stomach poisons. There are three major kinds of chlorinated hydrocarbon insecticides i.e. DDT analogues, cyclodiene compounds and benzene hexachloride isomers.

DDT analogues (bridged diphenyl acaricides) :

i) Chlorobenzilate: It is a non-systemic acaricide and gives an effective kill of all stages of red spider mites, but has a low ovicidal property. It has some marginal effect on eriophyids like bud mites (Das and Sengupta, 1962). It causes some phytoxicity effects on pear, plum and some apple varieties.

ii) Dicofol (Kelthane) : It is a representative of diphenyl carbinols, a detoxication product of DDT, and is a non-systemic acaricide recommended for the control of mites on a wide range of crops. It is effective on all stages of mites, causing gradual paralysis and eventual death. It has a good knockdown effect and is effective against many families of mites which includes tetranychids, tenuipalpids, tarsonemids and eriophyids. It effectively controls mites resistant to organophosphorous compounds. Dicofol resistance is not so high and the adaptation of resistant mites to the environment is often too low to survive. Dicofol may act on the octopamine receptor or inhibit cAMP synthesis by acting on the GTP-binding protein (Perry *et al.*, 1998). However, it is toxic to predatory mites, some insect parasites and some predators.

iii) DMC (Dimite) : It is effective against both eggs and adult stages of many mite species . It is not easily manufactured and is therefore very expensive. It has a moderate residual effect, produces a semiparalysis of the mite and kills it slowly.

Cyclodiene compounds :

iv) Endosulfan (Thiodan) : It is a non-systemic contact and stomach insecticide with acaricidal properties as well. It is a mixture of two isomers. It is effective against most of the crop mites.

2. Organophosphorous compounds : These were discovered and developed during second world war and are amongst the most toxic substances known to man. These compounds have phosphorous chemically bonded to the carbon atoms of organic radicals and are effective as both contact and systemic insecticides and acaricides. Majority of these compounds are toxic to mammals and so are to be handled with care. The mode of action is inhibition of acetyl cholinesterase or similar enzymes.

Phosphates :

i) TEPP(Tetraethyl pyrophosphate) : A non-systemic acaricide of brief persistence but very high mammalian toxicity. It is very effective against many tetranychid mites. It was used to provide immediate reduction of active mites under field conditions, these compounds kill very quickly and cause mites to shrivel and dry rapidly.

ii) Phosphamidon : A systemic acaricide, rapidly absorbed by plants, but only little contact action. It is non-tainting and non-phytotoxic and effective against many phytophagous mites.

iii) Mevinphos : It is a contact and systemic acaricide and insecticide of short persistence and is effective against sap-feeding mites. It is non-phytotoxic.

Phosphorothionates :

iv) Parathion : A non-systemic, contact and stomach insecticide and acaricide and effective against most of the mite species.

v) Diazinon : A non-systemic insecticide with some acaricidal action and used against some mite species.

Phosphorothiolates :

vi) Demeton – S- methyl (Metasystox, Demetox) : A systemic and contact insecticide with acaricidal properties, metabolized in the plant to the sulphoxide and sulphone, rapid in action and moderate in persistence. It is effective against all stages of red spider mites. It was used to control two-spotted spider mite and European red mites on decidous fruit trees and also on vegetables, field and ornamental crops.

vii) Azinphos-methyl : It is a non-systemic, broad-spectrum, insecticide with acaricidal properties and of relatively long persistence. It is effective against tetranychids.

Phosphorodithioates :

viii) Dimethoate (Rogor) : A systemic and contact insecticide with acaricidal action, used mainly against red spider mites.

ix) Malathion : A wide-spectrum insecticide non-systemic insecticide and acaricide, of brief to moderate persistence and effective against red spider mites.

x) Phosmet : A non-systemic acaricide and insecticide used at concentrations safe for a variety of predators of mites and thus useful for integrated control programmes. It is used mainly against spider mites.

Phosphoramidates :

xi) Schradan : A systemic acaricide and insecticide with little contact effect and effective against red spider mites on a variety of crops.

Miscellaneous :

xii) Bromopropylate : A contact acaricide with residual action and effective against mites resistant to organophosphorous compounds. It is effective against all phytophagous mites on field crops and is recommended for use on fruits , vegetables, cotton, ornamentals.

xiii) Carbophenothion : A non-systemic acaricide with insecticidal properties as well. It is phytotoxic at higher concentrations. It is mainly used on decidous fruits, in combination with petroleum oil, as a dormant spray for the control of overwintering mites. It is particularly effective on all stages of *Panonychus* mites, mainly on eggs before hatching.

xiv) Ethion (Fosmite) : It is a non-systemic insecticide with acaricidal properties, used mainly in combination with petroleum oils on dormant fruits as an ovicide. It is effective against red spider mites, brevipalpid mites, and some eriophyid mites. It is not phytotoxic. It has a good ovicidal property and acts on both developing stages and adults.

xv) Formothion : A contact and systemic insecticide and acaricide effective against red spider mites

xvi) Phenisobromolate : A contact acaricide with long residual activity and of promising use against mites on many pome and stone fruits, citrus, hops, cotton, beans, cucurbits and many ornamentals.

xvii) Phosalone : A non-systemic acaricide with insecticidal properties as well, and used on decidous fruit trees, field and garden crops. It persists on plants for about two weeks and is very effective against red spider mites

xviii) Quinomethionate : A selective, non-systemic quinoxaline acaricide and effectively controls all stages of red spider and other phytophagous mites. Hydrolysis gives a dithiol compound which inhibits SH enzymes.

xix) Triazophos (Hostathion) : A broad-spectrum incesticide/ acaricide used for either foliar application. It can penetrate plant tissues generally but has no systemic action. It is however very toxic to bees.

xx) Profenofos : It shows excellent larvicidal and good ovicidal activity and is used for control of some mite species on chillies, citrus, vegetables

3. Carbamates - Oximes

i) **Aldicarb (Temik) :** It is a systemic insecticide/acaricide which is absorbed into the root system of the plants and has a long residual life, it is effective against some phytophagous spider mites.

ii) **Methomyl :** A broad spectrum, systemic and contact insecticide/acaricide and its foliar spray is effective against many red spider mites.

4. Dinitrophenols : These were the early acaricides which were first used in the 1930's. They still have some usefulness today as they are incorporated in dormant oil sprays during winters and kill mite eggs .

i) **Dinitrocrescol :** It is an acaricide with ovicidal properties and used mainly as a dormant spray for orchards. It is highly toxic.

ii) **Binapacryl :** It is a non-systemic acaricide mainly used against red spider mites. It is effective against all the stages of red spider mites and even some of the eriophyids (Mukerjee, 1966). Both the larval stages and adults of eriophyid mites, particularly *Acaphylla* and the tenuipalpid mite, *Brevipalpus* sp. are effectively controlled by this acaricide. Due to its solubility in lipids binapacryl is easily penetrated in the integument of the mites, where it precipitates the protoplast and kills the cells. Due to its special mode of action, binapacryl is fully effective against mites which have developed resistance against phosphoric acid esters and chlorinated hydrocarbons

5. Pyrethroids : Many pyrethroids cause resurgence of mites, but some pyrethroids are effective on mites, though having no ovicidal activity. Fluvalinate, flucythrinate, fenpropathrin and bifenthrin are some of the examples. Fenpropathrin is highly effective against various species of spider mites like *Tetranychus, Panonychus* and *Eotetranychus* infesting fruits, vegetables, flowers and field crops, but is less effective to rust mites. Though the primary mode of action is strong repellency which is unique for this acaricide, contact killing activity against nymphs, adults and ovicidal activity are also seen.

6. Sulfones, sulfonates, sulfides, sulfite esters : They are those group of acaricidal compounds that normally contain two chlorinated benzene rings and are either sulphones, sullfonates, sulfite esters or sulfides but never sulphates.

i) **CPCBS and Tetradifon (Tedion) :** CPCBS is a selective miticide effective on eggs and larvae. Tetradifon is a systemic acaricide which is toxic to the eggs and all stages (larvae, nymphs) of phytophagous mites. The female adults treated with tetradifon bear infertile eggs. Eggs treated with CPCBS develop until hatching and then die, while eggs treated with tetradifon die at an early stage. There is no cross-resistance between the two compounds. CPAS and diphenylsulphide are converted to tetradifon. CPCBS inhibits olygomymycin-sensitive Mg^{2+-} ATPase, but its significance is not clear (Perry *et al.,* 1998). Tetradifon is non-phytotoxic at acaricidal concentrations and is recommended for application to top fruit, citrus, tea, cotton, grapes, vegetables, ornamentals and nursery stock. It is relatively safer to natural enemies. It is a very good acaricide against tetranychids and persists for about three to four weeks even during rains and can therefore kill the eggs and larvae during this period even after one application.

ii) **Sulfite esters – BPPS :** BPPS acts on mites, but it is not ovicidal. The mode of action is not clear, though it inhibits olygomycin-sensitive Mg^{2+}ATPase. Propargite is another sulfite ester which is highly effective against many mite families mainly Tetranychidae, Tarsonemidae, Eriophyidae and Tenuipalpidae on a wide range of vegetable, fruit and field crops. It's primary activity is on the larval and adult stages of mites with limited ovicidal activity. It acts primarily by contact, necessitating through coverage of the foliage. After application of propargite, female mites stop egg-laying and mortality of motile stages occurs after 48 to 96 hours. It controls the hatching larvae for a longer period because of its residual action and effectively controls mites for more than 2 weeks depending upon application rate, climate, crop stages and population pressure. It is a specific acaricide and can be successfully used in IPM programmes. It is relatively safer to predatory mites especially phytoseiids and other insect predators.

iii) **Tetrasul :** A non-systemic acaricide, highly toxic to eggs and all stages of phytophagous mites except adults. Being highly selective it does not pose any hazard to beneficial insects. It is recommended for use on fruits at the time when the winter eggs are hatching.

iv) **Amitraz :** An acaricide effective against a wide range of phytophagous mites, especially red spider mites on fruit trees, eriophyid mites, tenuipalpids and tarsonemids on a wide range of crops. It is effective on mite strains resistant to other acaricides, gives good knock down and residual control of all species of mites and is also safer to the crop. All the stages of mites are susceptible, except winter eggs of *Panonychus ulmi*. But it is relatively non-toxic to predaceous insects and bees.

v) **Aramite :** It is used commonly for the control of mite pests of shade trees, ornamental plants and vegetables. It is particularly effective against red spider mites, tetranychids, but has virtually no effect on eriophyids except for a few species (Das *et al.*,1962), tarsonemids, and brevipalpids. It acts against only the adults of tetranychids and has a fairly long residual action which enables it to kill mites emerging from eggs. Aramite is translocated through the leaves and can therefore kill the mites on the undersurface of the leaf as well. It can however, taint consumable plant parts and tea. It has little influence on predators and parasites of mites.

vi) **Ovex :** Unlike Aramite, Ovex is basically an ovicide i.e it kills the eggs and to some extent the larval stages of tetranychid mites. It has a long residual action and remains active for several weeks after application and has a bad odour. Ovex is a highly persistent acaricide that remains unaffected by rains or high temperature. However, mite populations have been reported to become resistant to ovex after repeated applications (Jeppson and Jesser, 1962).

vii) **Quinomethionate :** It is a slow acting, selective non-systemic acaricide effective against adult red spider mites, some brevipalpid mites and some of the eriophyids. It may cause phytotoxicity if used at high temperatures. The acaricide remains active for about 3-4 weeks and slowly kills the adult and newly emergent mites when they come in contact with the chemical.

viii) **Sulphenone :** It is a sulfone acaricide and ovicide. Because of its low phytotoxicity to otherwise susceptible crops (e.g. apple, pear), it is used as a substitute acaricide. It is a short-life acaricide.

ix) **Tricyclohexyltin (Plictran) :** This acaricide gives good control of active stages of tetranychids but has no effect on the eggs. It is used for application on fruit trees.

7. Organotin compounds : Fenbutatin oxide and cyhexatin are effective on all motile stages except eggs. They inhibit olygomycin-sensitive Mg^{2+}ATPase as well as Na^{+}, K^{+}-ATPase. Fenbutatin oxide is not toxic to predator (*Phytoseiulus*) or *Encarsia* parasites and is ideal for use in IPM programmes. These are highly effective against two-spotted spider mites in glass house vegetables, ornamentals and tunnel grown strawberries and act by contact and stomach poisoning.

Diafenthiuron : It has translaminar and vapour action and is specific to arthropod ATPase. It is broad-spectrum but selective in some crops as stomach acting. It is used on vegetables, fruits and ornamentals. It is non-systemic with excellent contact and stomach action and some ovicidal activity. It has good translaminar activity and is selective on beneficial insects and predatory mites.

8. Tetronic acid derivatives : e.g. Spiromesifen, Spirodiclofen,. These were introduced in 2002 and are highly effective against Tetranychids particularly *Tetranychus* spp. Spiromesifen is an excellent ovicide and acts on juvenile stages and interferes with lipid biosynthesis. Spirodiclofen is relatively safer to predatory mites, *Typhlodromus pyri*, even under field conditions (Oberwalder, 1998).

9. Mite growth regulators (Hexythiazox, Clofentizine) and Chitin synthesis inhibitors : Clofentizine (tetrazine acaricide) is a specific acaricide of novel structure which gives excellent residual control of spider mites on a wide range of crops. It is a contact acaricide acting primarily as an ovicide with some effect on young motile stages and with long residual activity. It is particularly effective against mite eggs including winter eggs of the European red mite. It is relatively safer to phytoseiid mites, coccinellid predaors. Hexythiazox (thiazolidine acaricide) and clofentizine affect all stages of mites except adults. Both are different in chemical structure, but often cross-resistance occurs. The mites die without moulting by abnormality of cuticle formation. Mode of action is not clear, but cross-resistance occurs between clofentizine and benzoylphenyl urea-type miticides such as flufenoxuron and flucycloxuron. Benzoylphenyl ureas and buprofezin are both insecticides and acaricides. Etoxazole, a 2,4-diphenyloxazoline derivative, has excellent activities on eggs, larvae and nymphs of mites, but not adult mites. Inhibition of a moulting process similar to hexythiazox occurs, but it is effective on the hexythiazox-resistant mites (Ishida *et al.*, 1994).

A. Uncouplers : Binapacryl is converted to phenol to affect all stages of mites. Fluzinam acts on eggs and larvae, but not on adults. Chlorfenapyr is a broad-spectrum pyrrole acaricide/insecticide, but also effective on *Tetranychus* mites.

10. Inhibitors of electron transport system (Complex I) : These inhibit the mitochondrial electron transport of complex I and have assumed new toxicological significance due to appearance of many miticides (Hollingsworth *et al.*, 1994). Pyridaben, Fenazaquin (Quinazalin acaricide), Fenpyroximate (Pyrazole acaricide), tebufenpyrad and pyrimidifen are effective on all stages of mites. Pyridaben shows excellent activity against a wide range of phytophagous mites including Tetranychidae, Eriophyidae and Tarsonemidae (Hirata *et al.*, 1995). Fenpyroximate shows high activity against tetranychids, eriophyids, tarsonemids and tenuipalpids, but the activity on predaceous mites is relatively low (Hamaguchi *et al.*, 1995), whereas Tebufenpyrad is effective on eggs and adults of *Tetranychus* and *Panonychus* mites.

11. Microbial pesticides : These include antibiotic acaricides (nikkomycins, thuringiensin), macrocyclic lactone acaricides (tetranactin), avermectin (abamectin, doramectin, eprinomectin, ivermectin, selamectin) and milbemycin acaricides (milbemectin, milbemycin oxime, moxidectin). These miticidal antibiotics are supposed to inhibit GABA receptor function. Abamectin , a mixture of avermectin-B_{1a} and $-B_{1b}$, is a macrolide disaccharide from *Streptomyces avermitilis.* Milbemectin, a mixture of milbemycin A_3 and A_4, is from *Streptomyces hygroscopicus* subsp. *Aureolacrimosus* and the structure is closely related to abamectin but without disaccharide moiety. Both are effective on all stages of various mites by acting as GABA antagonists. It has long lasting effect against target pests, has translaminar and contact action, pests become paralyzed soon after contact or ingestion, is non-phytotoxic and excellently fits into IPM systems. It is highly effective against spider mites, *Tetranychus urticae,* on a wide range of vegetables, *Panonychus ulmi* on pome fruits, *Panonychus citri, Polyphagotarsonemus latus* and *Phyllocoptruta oleivora* on citrus. Polynactin is also a miticidal antibiotic, produced by *Streptomyces aureus,* which consist of tetranactin, trinactin and dinactin. The major component is tetranactin.

12. Insect Growth Regulators (IGR) and Juvenile Hormones (JH) : In the search for effective and specific pesticides having minimal disruptive effects on the environment and not encouraging resistance build-up in the pests, insect hormones have shown great promise. They fall into two categories : The juvenile hormones, if applied to full -grown larvae disturb the process of metamorphosis to the extent that the insect dies as a deformed larva/pupa. These hormones are species specific. On the other hand Insect growth regulators often act as pesticides by interfering with cuticle formation at the time of ecdysis, killing the moulting larvae. These chemicals are not species specific.

i) **Diflubenzuron (Dimilin) :** An insect growth regulator, produced in 1974, which operates by ingestion and interference with the deposition of new cuticle at ecdysis. It is effective against some rust mite species on apple and citrus .

ii) **Flufenoxuron and Lufenuron :** These are chemically Benzoylureas and act by contact and stomach action. These are used against some mite species on fruit and vegetable crops.

Specificity of acaricides : Tetranychid mites are susceptible to at least initial application of many OPs and sulfoxide, sulfone, chlorbenside and ovex. False spider mites are generally not susceptible to OPs and moderately susceptible to other pesticides mentioned above. But both groups of acaricides have proved more effective against *Polyphagotarsonemus latus*. Members of phyoseiid mites are highly susceptible to DDT and related compounds, and synthetic pyrethroids. These compounds are completely ineffective in the control of spider mite pests. So their use should be discouraged in situations where spider mites are a problem (Dhooria *et al.,* 2001).

Stage specificites : Acaricides toxic to adult mites and whose residues degrade rapidly after application are usually ineffective in controlling mites under many field situations. Such acaricides may be the only means of chemical control on food crops infested with mites near the time of harvest. Acaricides which are toxic in vapour phase have been successfully used for mite control in glass houses and under field situations where frequent applications of acaricides having short residual life fail to provide adequate control of mite populations in the field. Acaricides like dicofol, which is effective against eggs and mobile stages of mite, and have more persistent residues are essential for economical mite control when the cost prohibits frequent treatments (Dhooria *et al.,* 2001).

Resistance to acaricides : The first report of resistance in mites to acaricides was by Crompton and Kearns in the two-spotted spider mite to ammonium potassium selenosilfide (SelecideR) in 1937. With the introduction of OP's in 1947, virtual elimination of mites in greenhouses was made by first TEPP and later parathion. In 1949 resistance to parathion and TEPP was observed and by 1950 resistant mites were present in a large percentage of rose houses in the eastern USA (Jeppson *et al.,* 1975; Halle and Sabelis, 1985). In Europe, resistance to acaricides occurred in several countries by 1950 and 1960's, development of resistance in mites to insecticides and acaricides, except petroleum oil, had become a world wide problem. Some of the acaricides resistance reported in major mite pests are : Mcdaniel mite, *Tetranychus mcdanieli* to dicofol in 1958 (Jeppson *et al.,* 1975); European red mite, *Panonychus ulmi,* to OP's and tetradifon (Nomura, 1973 and Asada, 1978); Carmine mite, *Tetranychus cinnabarinus,* on ornamental plants to methyl demeton (Nomura and Nakagaki, 1959); Kanzawa spider mite, *T. kanzawai,* on tea to Phencapton (Osakabe, 1971); *Panonychus citri* to benzomate (Benzoxamate) (Yamada *et al.,* 1986).

Table 3 : Different Acaricides/ Chemicals Tested against Phytophagous Mites on vegetable crops

Sr. No.	Mite species	Crop	Acaricides
1.	*Tetranychus urticae*	Okra,Pointed gourd	Wettable sulphur (0.5%), Dicofol (0.04%), fenazaquin (0.02%), fenpropathrin (2ml/L), Sulphur (4g/L) Dioxathion (0.1%), Replin (1.5%), malathion (400 ml/ acre)
		Brinjal	Sulphur (0.2%), Dicofol (0.09%), tetradifon (0.03%), endosulfan (0.075%), monocrotophos (0.04%), malathion (400 ml/ acre)
		Beans	Dicofol (0.04%), ethion (0.1%), Neemark (0.5%), Replin (1.5%)
		Cucumber	Malathion (250 ml/ acre)
2.	*T. cucurbitae*	Brinjal	Plictran (0.03%), wettable sulphur (0.15%)
3.	*T. ludeni*	Cowpea	Dicofol (0.05%), ethion (0.1%)
4.	*T. macfarlanei*	Brinjal	Sevisulf (Carbaryl+ sulphur) (0.1%)
		Okra	Dicofol (0.04%), monocrotophos (0.04%)
5.	*T. neocaledonicus*	Brinjal	Plictran (0.03%), sulphur (0.15%), dicofol (0.04%), tetradifon (0.03%), endosulfan (0.075%)
6.	*T. telarius*	Brinjal	Dimethoate , methyl oxydemeton
7.	*Polyphagotarsonemus latus*	Chilli	Triazophos, phosalone (0.1%), propargite (2l/ ha), ethion (0.05%), wettable sulphur (0.5%), fenpropathrin (0.01%)
	Capsicum	Dicofol (0.025%)	

References

Anonymous. 2005. *Insecticide and Acaricides* : 2. Non-neuroactive and novel Biological and Chemical Control Methods. : L20 RPB : Insecticide_2:v1.1

Anonymous 2007. IRAC Mode of Action Classification (Revised), July 2007 *Version: 5.3* www.irac-online.org.

Asada, M. 1978. Genetics and biochemical mechanisms of acaricide resistance in phytophagous mites. *J. Pestic. Sci.*, **3** : 61.

Banerjee, B. 1988. *An introduction to Agricultural Acarology, Biology and Control of Mite Pests in the tropics.* Associated Publishing Co. New Delhi. 118pp.

Bhardwaj, S.P. and Bhardwaj, S. 2005. Evaluation of horticultural mineral oils against European red mite, *Panonychus ulmi* (Koch) on apple in Himachal Pradesh. p. **209**. In : *Proceedings of 1st Congress on Insect Science, Dec. 15-17, 2005, PAU, Ludhiana.*

Crompton, C.C. and Kearns, W.W. 1937. Improved control of red spider on greenhouse crops with sulphur and cyclohexylamine derivatives. *J. Econ. Entomol.*, **30** : 512.

Das, G.M. and Sengupta, N.S. 1962. Biology and control of the purple mite, *Calcarus carinatus* (Green), a pest of tea in North-Esat India. *J.Zool. Soc. India*, **14** : 64-72.

Dhooria, M.S., Kapur-Ghai, J. and Bhullar, M.B. 2001. Chemical control of mite pests. pp. 137-140. In : *Mites, Their Identification and Management* (Eds. P.R. Yadav, R. Chauhan, B.N. Putatunda and B.S. Chillar, Dept. of Entomology, CCSHAU, Hisar).

Dhooria, M.S.and Mann ,G.S.1989. Role of insecticides in inducing mite-outbreaks . pp.45-62. In : '*Soil Pollution and Soil Organism*'. Ashish Publishing Co, New Delhi.

Gupta, S.K. 1985. *Handbook. Plant Mites of India.* ZSI pub. 520pp.

Gupta, S.K. 1991. *Mites of Agricultural Importance in India and their Management.* Tech. Bull. No. 1. AICRP on Agricultural Acarology, UAS, Bangalore.25pp.

Gupta, S.K. 2001. Mite pests of agricultural crops in India, their management and identification. pp. 48-61. In : *Mites, Their Identification and Management* (Eds. P.R. Yadav, R. Chauhan, B.N. Putatunda and B.S. Chillar, Dept. of Entomology, CCSHAU, Hisar).

Halle, W. and Sabelis, M.W. 1985. *Spider Mites : Their Biology, Natural Enemies and Control,* Vol. **1B**. Elsevier, Amsterdam, Tokyo.

Hamaguchi, H.; Kajihara, O. and Katoh, M. 1995. Development of a new acaricide, fenpyroximate. *J. Pestic. Sci.*, **20** : 173 -175, 203-212.

Hill, D.S. 1987. *Agricultural Insect pests of the Tropics and their control.* (Second edition). Cambridge University Press, Cambridge.

Hirata, K.; Kawamura, Y.; Kudo, M. and Igarashi, H. 1995. Development of a new acaricide, pyridaben. *J. Pestic. Sci.*, **20** : 213-221.

Ishida, T; Suzuki, J and Tsukidate, Y. 1994. YI-5301, a novel oxazoline acaricide. In : *Brighton Crop Protection Conference, Pests and diseases,* Vol. **1**. British Crop Protection Council, Farnham, pp 37-44.

Jeppson, L.R. and Jesser, M.J. 1962. Laboratory studies on resistance to the pacific spider mites to acaricides. *J. Econ. Entomol.*, **55** : 78-82.

Jeppson, L.R., Keifer, H.H. and Baker, E.W. 1975. *Mites Injurious to Economic Plants.* University of California Press, Berkeley, 614pp.

Matsumura, F. 1975. *Toxicology of Insecticides.* Plenum Press, New York and London503pp.

Mukherjee, T.D. 1966. A new acaricide for the control of scarlet and pink mites. *Two and a bud,* **13** : 94-100.

Nomura, K. 1973. review of acaricide resistance in red spider mites in Japan. *Rev. Plant Protect. Res.*, **6** : 44.

Nomura, K. and Nakagaki, S. 1959. On resistance of red spider mite, *Tetranychus cinnabarinus*, to methyl demeton (Metasystox). *Tech. Bull. Fac. Hort. Chiba Univ.*, **7** : 39.

Oberwalder, C. 1998. A field study to evaluate the effects of a BAJ 2740SC240 containing 240g/L ai against the predatory mite, *Typhlodromus pyri* in vines (one location in Germany). In : Wolf, C. and Schnorbach, H.-J. (2002). *Ecobiological profile of the acaricide spirodiclofen.* Planzenschutz-Nachrichten Bayer , **55(3)** : 177-196.

Osakabe, M. 1971. Studies on insecticide resistance of the kanzawa spider mite, *Tetranychus kanzawai* Kishida, I-III, *Rev. Plant Protect. Res.*, **4** : 132.

Pandey, B.K. and Selvasundaram, R. 2007. Acaricide usage in India. P 24. In : *Souveneir : Acari : Globalisation & Climate Change,* A National Symposium in Acarology, Nov. 24-26, 2007, UAS, Bangalore.

Perry, A.S.; Yamamoto, I; Ishaaya, I. and Perry, R. 1998. *Insecticides in Agriculture and Environment. Retrospects and prospects.* 261pp. Springer-Verlag Berlin Hiedelberg.

Puttaswamy and ChannaBasvanna, G.P. 1976. Effect of some important acaricides on the predator *Stethorus pauperculus* Wise (Coleoptera : Coccinellidae) and its host *Tetranychus cucurbitae* Rahman and Sapra. Mysore. *J. Agric. Sci.*, **10**: 636-641.

Seki, M. 1958. Control of the citrus red mite, *Panonychus citri,* Record of II symposium. *Jap. Soc. Appl. Zool.*, **59** : 62.

Singh, J. and Singh, R.N. 2004. *Major harmful mites of vegetables and other crops and their control.* Publ. of Dhanuka Group, 20pp.

Yamada, T., Hiromi, Y. and Mitsou, A. 1986. Resistance to benzomate in mites. In : *Pest Resistance to Pesticides.* (Eds. G.P. Georghio and T. Saito) Plenum Publishing Corporation, New York pp 445-451.

Zhang, Z-Q. 2003. *Mites of Green Houses. Identification, Biology and Control.* CABI Publishing, Wallingford, UK.

Chapter-2

Insect-Pests and their Management in Vegetable Crops

P.C. Sharma and R.S. Chandel
Department of Entomology
CSK Himachal Pradesh Krishi Vishvavidyalaya, Palampur-176 062, Himachal Pradesh.

The vegetables form an essential component of the human diet especially in case of India and some South Asian countries where sizeable population is vegetarian. India ranks second in the production of vegetables after China and occupy 5.993 million hectares area and the total annual production is around 90.83 metric tonnes (Gopalakrishnan, 2007). There is a need of around 5-6 million tones of food to feed 1.3 billion India' population by 2020 AD. It is an established fact that any attempt at improvement of agricultural production by using new cultivars with the recommended package of practices has invariably resulted in the increase in activity of damage by pests and diseases. Vegetables are more prone to insect-pests and diseases mainly due to their succulence as compared to other crops. The pest problems on vegetables can be more serious because of the favourable conditions which are provided for multiplication by the present methods of cultivation. The major constraints in vegetable production include the extensive crop losses due to increased pest infestation directly or due to viral diseases vectored by insects. The extent of crop losses in vegetables varies with the plant type, location, damage potential of the pest involved and cropping season. The crop losses to the tune of 40 per cent have observed in vegetable crops. The increased demand and limited productivity of local cultivars, led to introduction of high yielding varieties and hybrids. As a result dramatic changes in pest scenario occurred and minor pests became major pests. Indiscriminate use of pesticides has led to severe ecological consequences like destruction of natural enemies, effect on non-target organisms, pesticide residues in vegetables and resistance of the pests to pesticides. The modern pest management is aimed at educating the farmers on various issues of pest management like identification of the pests, natural enemies, and judicious use of pesticides with alternate methods of pest control like cultural, mechanical, biological and behavioural (use of pheromones) management. The strategy for the control of insect-pests on vegetable crops is necessarily to be different from the other crops because of the nature of utilization of vegetables. This is especially true in the case of chemical control. Insecticides which are highly toxic and are known to leave hazardous residues cannot be recommended on vegetables. The vegetable crops have been grouped into summer and winter vegetables depending upon their season of cultivation in majority of areas. The present chapter deals with the insect pests infesting summer and winter vegetables and their possible management options.

I. INSECT-PESTS OF SUMMER VEGETABLES

1. Brinjal : Brinjal is one of the most common vegetables and there are numerous recipes for cooking brinjal. It contains vitamin A, B and C and has got ayurvedic medicinal properties.Brinjal is attacked by a number of insect-pests which are listed in Table 1.

Shoot and fruit borer : The pest is widely distributed in Indian sub-continent. It also damages potato and other solanaceous crops. The pest is active throughout the year in places having moderate climate, but its activity is adversely affected by severe cold.

The damage by this pest starts soon after transplanting of seedlings and continues till harvest of fruits. As a result of feeding, the affected leaves dry up and drop down while in case of shoots, the growing point is killed. Appearance of wilted drooping shoots is the typical symptom indicating damage by this pest. At later stages, the caterpillar bore into flower buds and fruits, entering from under the calyx; they leave no visible sign of infestation, while the caterpillars feed inside. The damaged flower buds are shed without blooming, whereas the fruits show circular exit holes. Such fruits, being partially unfit for human consumption lose their market value considerably. The yield loss due to this pest usually ranges from 54 to 66 per cent which may go upto 80 per cent (Lal *et al.*, 2004).

Table 1 : Insect-pests of brinjal

Common Names	Scientific Names	Order	Family
Shoot and fruit borer	*Leucinodes orbonalis* Guenee	Lepidoptera	Pyralidae
Brinjal stem borer	*Euzophera perticella* Ragonot	Lepidoptera	Pyralidae
Hadda beetle	*Henosepilachna vigintioctopunctata* Fabricius, *Epilachna dodecastigma* Widemann	Coleoptera	Coccinellidae
Brinjal lace-wing bug	*Urentius sentis* Distant	Hemiptera	Tingidae
Jassids	*Amrasca biguttula biguttula* (Ishida)	Hemiptera	Cicadellidae
Leaf roller	*Eublemma olivacea* (Walker)	Lepidoptera	Noctuidae
Aphids	*Aphis gossypii* Glover, *Myzus persicae* (Sulzer)	Homoptera	Aphididae
Mealy bug	*Coccidohystrix insolita* (Green)	Homoptera	Coccidae
Scale insect	*Aonidiella orientalis* (Newstead)	Homoptera	Coccidae
Thrips	*Thrip tabaci* Lindemann, *Scirtothrips dorsalis* Hood	Thysanoptera	Thripidae
Leaf miner	*Phthorimaea operculella* (Zeller)	Lepidoptera	Gelechiidae
Leaf miner	*Liriomyza trifolii*(Burgess)	Diptera	Agromyzidae
Hairy caterpillars	*Spilosoma obliqua* Walker	Lepidoptera	Arctiidae
Black ant	*Camponotus infuscus* Forel	Hymenoptera	Formicidae

The moth is white but has pale brown or black spots on the dorsum of the thorax and abdomen. Its wings are white with a pinkish or bluish tinge and are ringed with small hair along the apical and anal margins. The forewings are ornamented with a number of black, pale and light brown spots. The moth measures about 20-22 mm across the spread wings. The moths appear in March-April and during their life span of 2-5 days, lay 80-120 creamy-white eggs singly or in batches of 2-4 on the underside of leaves, on green stems, flower buds or on the calyx of fruits. The eggs hatch in 3-6 days and the young caterpillars bore into petioles and mid ribs of large leaves and young tender shoots, close the entry point with their excreta and feed within. The larvae grow through five stages and are full-fed in 9-28 days. The mature larvae come out of the feeding tunnels and pupate in tough silken cocoons among the fallen leaves. The pupal stage lasts for 6-17 days and the life cycle is completed in 20-43 days. There are five overlapping generations in a year.

Management : Some of the promising varieties found resistant/ tolerant to this pest viz., Long green, PBR-129-5, Purple Cluster, Pusa Purple Long, Punjab Neelam and Pusa Kranti, Arka Keshav, Punjab Chamkila and Punjab Barsati may be grown. Ratooning of brinjal crop should be avoided. Remove and destroy the infested shoots with borers inside. Natural enemies like *Trathala flavoorbitalis* and *Goryphus nursei* are some of the larval parasitoids which need to be conserved.

Higher doses of nitrogen and phosphorus in general favour high pest incidence whereas potassium application has inverse relation with the incidence of the pest. Avoid closer spacing and continuous planting of brinjal. Release of egg parasitoid, *Trichogramma chilonis* @ 50,000 parasitized eggs/ ha six times from the shoot formation stage at weekly intervals will help in checking the pest population.

Spray 0.1 % carbaryl or 0.05 % endosulphan or 0.01% cypermethrin or 0.004 % lambdacyhalothrin as soon as the attack is noticed on foliage. Repeat after 15 days in case the attack is still there. Observe 7 days waiting period for picking of fruits. Nimbecidine 0.03% also gives good control of brinjal fruit borer.

Brinjal stem borer : It is another destructive pest of brinjal which also attacks chilli, tomato and even potato. It is widely distributed all over the Indian sub-continent. The caterpillars feed exclusively in the main stem. The attacked plants wither and wilt, the growth is stunted and fruit bearing capacity is adversely affected.

Eggs are laid singly or in clusters on young leaves, petioles or even tender shoots. A single female may lay 104-363 eggs in its life time of about one week. The eggs hatch in 3-10 days. Soon after hatching, the caterpillars bore into the stem and move downwards. They pass through 4 or 5 stages and are full fed in 26-58 days. When full grown, they make silken cocoons within the feeding galleries or in the cracks and crevices in the soil. After pupation, they transform into adults in 6-8 days. The life cycle is completed in 35-76 days and 5-6 overlapping generations are completed in a year.

It is desirable to check the infestation in the initial stage by mechanical methods and ratoon crop should be avoided. Uproot and destroy the infested plants. Ratooning of brinjal plants should be avoided. Spray 0.05 % malathion or 0.05% dichlorvos, if the attack is serious.

Hadda beetles : Being polyphagous, hadda beetles have a wide host range including brinjal, cucurbits, potato and tomato. Both the adults and grubs of this beetle feed by scrapping chlorophyll from the epidermal layers of leaves. The leaves present a lace-like appearance, turn brown, thus resulting in skeletonization and drying.

The beetle passes the winter as a hibernating adult in heaps of dry plants or in cracks and crevices in soil. It resumes its activity during March-April. The beetle of *H. viginoctopunctata* is deep red in colour and usually has 7-14 spots on each elytron with a pointed tip. The beetles of *E. dodecastigma* are deep copper-coloured and have six black spots on each elytron and tip is somewhat rounded. A single female can lay upto 400 eggs in her lifetime and lays yellow cigar-shaped eggs in batches of 5-40 each mostly on the underside of leaves. The eggs hatch in 3-5 days and the grubs feed on the lower epidermis of leaves and are full grown in7-18 days depending upon the temperature. The grubs are about 6 mm long, yellowish in colour and have six rows of long branched spines. The pupae are darker and found fixed on the leaves, stems and mostly at the base of the plants. The pupal period lasts for 5-13 days. Several generations are completed from March to October.

Management : Collect the egg masses, larvae and adults and destroy them in kerosenized water. Hand picking and destruction of eggs, grubs and adult beetles is effective, if the cropped area is small.

Grow resistant varieties like Punjab Chamkila, Arka Shrish and Hisar Sel 1-4. No insecticidal application should be given if the population is less than 2 beetles per plant.

Spraying of 1 per cent extract of *Ageratum*, *Lantana* or *Eupatorium* has been found effective as antifeedant. The insecticides recommended for the management of shoot and fruit borer will take care of this pest also. Foliar application of malathion (0.05%) or carbaryl (0.1%) or endosulfan (0.05%) checks the pest infestation.

Brinjal lace-wing bug : It is distributed throughout the north-western parts of Indian sub-continent and is common in the plains. It is a specific pest of brinjal.

Both the nymphs and adults are destructive. The adults and nymphs suck sap from the leaves and cause yellowish spots which together with the black scale-like excreta deposited by them impart a characteristic mottled appearance to the infested leaves. When the attack is severe, about 50 per cent of the crop may be destroyed.

The pest is active from April to October and hibernates from November-March in cracks and crevices in soil under the brinjal plants as an adult. The bugs live for 30-40 days and lay 35-44 shining white nipple shaped eggs singly in the tissues on the underside of the leaves. The eggs hatch in 3-12 days and the young nymphs feed gregariously on the lower surface of the leaves but the fully developed nymphs feed individually. They grow through five stages in 10-23 days. There are 8 overlapping generations in a year. Apply 0.03 % dimethoate for the suppression of the pest. A waiting period of 15 days should be observed if fruits are ready for picking.

Leaf hopper : It is a polyphagous pest and infests cotton, bhendi, brinjal, beans, castor, cucurbit, hollyhock, potato, sunflower and other malvaceous plants.

Both adults and nymphs suck cell sap from the ventral surface of leaves and inject their toxic saliva into plant tissues. Infested leaves turn yellowish and curl. In case of severe infestation, leaves turn dark brick red, become brittle and crumple.

Pear-shaped, elongated and yellowish-white eggs are laid singly in the tissues of main veins on the undersurface of leaves. Egg period lasts for 4- 10 days. Nymphs are whitish-pale green, wingless and move diagonally. Nymphal period is 7-20 days. Adults are wedge-shaped, pale green in colour with a black dot on posterior portion of each forewing. Pest population appears with the onset of cloudy weather and decreases after heavy monsoon showers.

Management : Spray Neem seed kernel extract (5%) at early stages of these pests. Use yellow sticky traps @ 10/ha for attracting these pests. Spray 0.05% malathion for the suppression of these pests. Use of systemic insecticides like oxy demeton methyl @0.025% can be effective but a waiting period of 15 days be observed.

2. Tomato : Tomato is grown all over tropical and temperate regions of the world. The fruits are rich in vitamin C coupled with vitamins A, B_1 and B_2. It is eaten raw or baked and is also made into syrup, pickles, ketchup, sauce and juice. The major pests reported from India are given in Table 2.

Table 2 : Insect-pests of tomato

Common Names	Scientific Names	Order	Family
Fruit borer	*Helicoverpa armigera* (Hubner)	Lepidoptera	Noctuidae
Tobacco caterpillar	*Spodoptera litura* (Fabricius)	Lepidoptera	Noctuidae
Fruit flies	*Bactrocera* spp.	Diptera	Tephritidae
Greenhouse whitefly	*Trialeurodes vaporariorum* (Westwood)	Homoptera	Aleyrodidae
Whitefly	*Bemisia tabaci* (Gennadius)	Homoptera	Aleyrodidae
Serpentine leaf miner	*Liriomyza trifoli* (Burgess)	Diptera	Agromyzidae
Spotted beetles	*Henosepilachan vigintioctopunctata* Fabricius, *Epilachna dodecastigma* Widemann	Coleoptera	Coccinellidae
Striped mealy bug	*Ferrisia virgata* (Cockerell)	Homoptera	Coccidae
Cutworms	*Agrotis* spp.	Lepidoptera	Noctuidae
Thrips	*Thrips tabaci* Lindemann *Frankliniella schultzei* Schm.	Thysanoptera	Thripidae
Jassids	*Amrasca biguttula biguttula* (Ishida)	Hemiptera	Cicadellidae
Fruit sucking moths	*Othreis fullonica, O. materna*	Lepidoptera	Noctuidae

Fruit borer : The pest is of international importance and is widely distributed in tropical, sub tropical and warmer temperate regions of the world. The main host plant is chickpea, though it also attacks cotton, cowpea, okra, tomato, tobacco, sunflower, linseed, etc.

The newly hatched caterpillars feed on flowers, flower buds and tender fruits and sometimes scrap tender leaves. The young caterpillars cut circular holes through the sepals and petals and reach the ovary to feed on its contents. The flowers wither and fall off. The caterpillars prefer newly formed fruits due to their soft skin although all stages of fruit growth are attacked. A single caterpillar damages several fruits before completion of its development. While feeding, half of the body of the caterpillar remains outside the infested fruit. In India, *H. armigera* is reported to cause 5-80 per cent loss in tomato crop under different agro-climatic situations. Spring and summer planted tomato crop in

Himachal Pradesh is vulnerable to fruit borer with infestation ranging from 15-30 per cent.

The peak period of pest activity occurs from April-May on tomato in Punjab while it is from May-June in Himachal Pradesh. The adult is a medium-sized moth, light brown in colour with a V-shaped mark on the forewings. Hind wings are of dull-grey colour with a black border on the distal end. The female moth, being bigger, has a tuft of hair on the tip of abdomen. The oviposition occurs 4-5 days after mating and the eggs are laid singly, late in the evening. The eggs are spherical in shape with a flat base, giving a dome-shaped appearance. Freshly laid eggs are yellowish-white but change to dark brown near hatching. Incubation period is 3-4 days. The average fecundity of female moth is 500 eggs, though it may range from 1200-1600 eggs. The newly hatched caterpillars are sluggish and white green in colour. The full grown caterpillars are 30-40 mm in length and pale green in colour, however, the colour varies with the food ingested. Dorsal surface bears dark brown stripes. The larval period lasts for 15-24 days. The pupation takes place in earthen cell. The pupae are dark brown in colour, 11-14 mm long and sharp spine at the anal end. The pupal period lasts for 7-14 days. The entire life cycle is completed in about 4-6 weeks with 5-8 generations depending upon the environmental conditions and availability of host plants.

Management : Crop transplanted late in summer (June) escapes the damage of fruit borer. Hand collection of caterpillars during early stages of infestation can be helpful. Deep ploughings after harvesting of crop exposes the pupae to natural mortality. Use of African marigold (*Tagetes erecta*) as a trap crop provides good control of the pest. Grow pest tolerant varieties like Punjab Kesri, Punjab Chuhara, Pant Bahar, Azad and Pusa Hybrid-4.

The economic threshold of 0.3 larvae/ m^2 has been established. Inundative release of egg parasitoid, *Trichogramma chilonis* and *T. brasilensis* @ 2,50,000 adults/ha is advocated for the management of this pest.

Application of *Bacillus thuringiensis* alone (1.0 litre/ha) or in combination with endosulfan (0.035%) at flowering and fruit setting stage can suppress the damage. Spray Ha NPV @ 250 LE with jaggery (10 g/litre), soap water (5 g/litre) and tinopal (1 ml/litre) during evening hours.

The pest can also be suppressed by foliar application of endosulfan (0.07%) or carbaryl (0.1%) or deltamethrin (0.0028%) or acephate (0.05%). However, no two consecutive sprays should be undertaken with insecticides of same group to slow the development of insecticide resistance.

Tobacco caterpillar : It is found throughout the tropical and sub-tropical parts of the world. It is widespread in India. It is a polyphagous pest. Besides tobacco, it feeds on tomato, brinjal, cruciferous crops, sunflower, groundnut etc.

The young larvae feed gregariously for few days on green material of leaf and skeletonize it and then disperse to feed individually. They feed on leaves by making big holes. In other crops, if the leaves are not available for larval feeding, fruits are also bored by this pest. They are voracious feeders and faecal matter can be seen on leaves. This pest is found to cause damage in all stages of crop growth, but green leaves should be present for egg laying.

The moths are active at night when they mate and the female lays about 300 eggs in clusters. The moths are 22 mm long and measure 40 mm across the wings. The fore-wings have beautiful golden and grayish brown patterns. These clusters are covered with brown hairs and they hatch in 3-5 days. The damage is done by the caterpillars which measure 35-40 mm in length. They are velvety-black with yellowish-green dorsal stripes and lateral white bands. The caterpillars feed on the leaves and fresh growth and cause extensive damage particularly at night. They pass through six instars and are full-fed in 15-30 days. The full grown larva enters the soil to pupate and the pupal period lasts for 7-15 days.

Management : Collection and destruction of egg masses and early stage caterpillars will be beneficial. Dichlorvos (0.05%), quinalphos (0.05%), endosulfan (0.05%), malathion (0.05%), deltamethrin (0.028%), cypermethrin (0.0075%) and fenvalerate (0.01%) can be used for effective suppression of caterpillars.

Fruit flies : The fruit flies are widely distributed throughout tropical, sub tropical and temperate areas of the world. There are three species of fruit flies damaging different fruits and vegetables. These are *Bactrocera cucurbitae* (Coquillett), *B. dorsalis* Hendel and *B. tau* (Walker). *B. cucurbitae* is the commonest and most destructive. It is an international pest and is found in Pakistan, Myanmar, China, Malaysia, Japan, East Africa, Australia and the Hawaiian islands. It has been reported to infest melons, tomato, chillies, guava, pear, fig, citrus, etc. *B. tau* is closely related to *B.cucurbitae* and also causes serious damage to tomato and cucurbitaceous crops.

Only the maggots cause damage by feeding on the near ripe fruits. The maggots destroy fruits by feeding on the pulp. The damage caused by this pest is most serious especially after first shower of monsoons.

Management : The regular removal and destruction of the infested fruits help in the suppression of the pest. Ploughing the infested fields after harvest can help in killing the pupae. Stirring the soil nearby plants at regular interval reduces adult emergence.

Growing 2-3 rows of maize as trap crop planted 8-10 meter apart in cucurbit fields act as resting site for the adults of fruit fly and any of the contact insecticides can be applied on maize during evening hours to kill the adults.

Methyl eugenol and Cue-lure have been reported to attract *B. cucurbitae* males from mid July to mid November. Use of *Ocimum sanctum* as a border crop sprayed with protein bait (protein derived from corn, wheat or other source) containing spinosad as a toxicant is also effective against *B.cucurbitae*. Pupae are also parasitized by some chalcids and braconids. Use pheromone traps for monitoring pest population. Application of bait spray containing 50 ml malathion 50 EC + 500 *gur* (molasses) in 50 litres of water per hectare is recommended. It can be repeated after 7 days.

Whitefly : Whitefly is found in most of the countries in tropics and sub tropics. The main host plants are cotton, tobacco and some winter vegetables, including tomato. Though it is a major pest of cotton in many cotton growing countries including India but it also feeds on other plants such as cabbage, cauliflower, sarson, toria, melon, tomato, potato, brinjal, okra and some weeds.

Vitality of the plant is lowered through the loss of cell sap and normal photosynthesis is hampered due to the growth of sooty moulds on the honey dew excreted by the insect. White tiny scale like insects may be seen near the plants or crowding in between the veins on the ventral surface of leaves sucking sap from the infested parts. This whitefly also acts as a vector, transmitting the leaf curl virus.

Eggs are pear-shaped, light yellowish in colour, about 2 mm long placed upright on the leaves being anchored by a tail like appendage. On hatching the nymphs crawl for a while, settle down on a succulent spot on the same leaf and thereafter never move during that stage. Adults are minute insects about 1 mm long, covered completely with a white waxy bloom. They have a pair of reniform compound eyes, a pair of ocelli and seven segmented antenna. Wings are of equal thickness, opaque and milky white in colour, wing expanse 4-5 mm. Reproduction is mostly oviparous though parthenogenesis also occurs. Only male emerge from unfertilized eggs whereas both sexes develop from fertilized eggs. Incubation period ranges from 3-5 days in summer to 33 days in winter. Nymphal development lasts for 9-14 days in summer and pupal period varies from 2-8 days. A life cycle is completed in 14 days in summer and it may be prolonged upto 107 days in winter. Adult longevity is 2-5 days in summer. There are about 12 overlapping generations in a year.

Management : Prophylactic spraying with dimethoate (0.03%) to prevent the population buildup of the pest is suggested. In severe infestation, spray triazophos (0.08%) or oxy demeton methyl (0.025%) can be used. *Propsaltella smithi* Silvestri, *Pteropteryx bemisia* Mani and *Eretmocerus masii* Silvestri have been found to parasitize the whitefly. Yellow sticky traps @ 10/ha for whitefly may be set up in the fields. Use of nylon nets in the nursery beds to avoid entry of whitefly can be helpful. Seed treatment with imidacloprid (2.5 g/kg) provides protection for 25-30 days.

Greenhouse whitefly : *Trialeurodes vaporariorum* infests 44 plant species belonging to 20 families comprising vegetables, field crops, ornamentals, medicinal plants and weeds. Among vegetables, tomato supports the highest population. *T. vaporariorum* remains active from March to October-November on different hosts but it breeds throughout the year on ornamentals like fuschia and salvia.

The nymphs and adults suck large quantities of phloem sap from the under surface of leaves and secrete a sticky liquid, the honeydew, on which dark sooty mould grows under humid conditions. This interferes with the photosynthetic activity of the plants resulting in withering, premature dehiscence and defoliation and finally the death of the plant. The greenhouse whitefly is also reported to act as vector of several economically important viral plant pathogens.

The females deposit eggs singly or in clusters of 5-6 eggs on the lower surface of leaves, whereas on glabrous leaves, a circular pattern of egg laying is observed. The female on an average lays 108 to 373 eggs in different generations (Sood, 2002). The freshly laid eggs are creamish-yellow in colour and turn dark prior to hatching. The eggs are stalked and rounded at the base with apical tapering. The length of the eggs ranges between 0.18 to 0.23 mm and breadth from 0.07 to 0.11 mm. The mean incubation period is 4.7 days (June-July) to 12.7 days (January-February). There are three nymphal instars. The freshly hatched nymphs are oval and light yellow in colour. They possess functional legs and crawl throughout the leaf on which they hatch. They settle in a few hours and insert their proboscis into the phloem tissue for sucking up sap. The second and third instar nymphs are oval in shape, depressed, scale-like and greenish yellow in colour. The setae like projections are prominent in these stages. The total nymphal period ranges from 10-24 days during different generations. The pupae are flattened and translucent initially and become expanded and opaque with the white lateral and dorsal setae in the later stages.

The mean duration of pupal period ranges from 3.0 to 7.7 days. The yellow bodied adults have two pairs of wings covered with white waxy powder. In resting position, the forewings cover the hind wings and are held flat but loosely over the body. They prefer to feed and oviposit on the lower surface of tender leaves. The females are larger than the males. The females measure 0.963 mm in length with mean wing expanse of 2.613 mm whereas, in males the corresponding values are 0.844 mm and 2.174 mm, respectively. Under laboratory conditions, on French bean, it completes 13 overlapping generations in a year and the total developmental period (egg to adult) ranges from 20.7 to 27.3 days. The females live longer than the males and they live for 7.1 to 42.6 days in different generations, whereas the males live for 5.5 to 26.7 days.

Management : Application of dimethoate (0.03%) or imidacloprid (0.05%) or triazophos (0.03%) or acetamiprid (0.004%) on the infested crop is effective. Insect growth regulator, buprafezin (0.02%) has been found effective but with delayed effect on population. Repeated sprays of azadirachtin (0.0045 %) are effective in controlling the whitefly population. *Beauveria bassiana* (@ 7.5 x 10^6 cfu/ml) is also effective under protected environment.

Cutworms : Cutworms are pests of world wide occurrence and are found in America, Europe, North Africa, Syria, Japan, Indonesia, New Zealand, China, Myanmar, Sri Lanka and India. They are prolific polyphagous insects that infest cultivated crops especially during summer season. Five species of cutworms have been reported from India, namely *A.ipsilon, A.flammatra, A.segetum, A.interacta* and *A. spinifera. A. ipsilon* is predominant in tropics while *A. interacta* is common in temperate regions. *Agrotis (Ochropleura) flammatra* is occasionally abundant in India, particularly destructive in the north region. Cutworms attack a wide variety of cultivated plants including maize, tobacco, wheat, cotton, pea, tomato, cabbage, potato, cauliflower, broccoli, brinjal, capsicum, coffee, etc. The larva of *A. ipsilon* is commonly called greasy cutworm while that of *A.segetum* is known as black cutworm.

The young larvae feed on the epidermis of the leaves. As they grow, their habit changes. During the daytime they live in cracks and crevices in the ground and come out during night and cut the plants at ground level. The cut branches are sometimes seen to have been dragged into holes where leaves are eaten. They generally consume a little part of the plant parts and move on to attack other seedlings.

The adult moth is greyish-brown with their forewings dirty brown and hind wings are semi-hyaline white. The moths hide themselves in day time under leaves or in cracks in the soil. The moths come out at dusk and fly about until darkness sets in. The female lays upto 1000 eggs in her life span, either singly or in groups of 2-30 on various parts of the plant or even in soil or in debris. The eggs are creamy-white but become darker later. The incubation period is 8 days. The newly hatched larvae feed on their egg shells and move like semi looper. The larval stage lasts for 30-34 days in February–April. The older larvae may become cannibalistic. The colour of larvae varies from dark brown to pale grey or even black and greasy to touch. The larva passes through six instars and the full grown larva forms earthen cocoon for pupation inside the soil at a depth of 5-7 cm. The pupae are reddish brown in colour and the pupal period lasts for 10-30 days. The life cycle is completed in one to two months.

Management : Use of well decomposed manure helps in reducing the incidence. Collect and destroy the larvae after flooding of fields/ beds. Soil drenching with chlorpyriphos (0.04%) or spraying of cypermethrin (0.0075%) on foliage and soil surface reduces the incidence. Poison bait with wheat bran + molasses + acephate and carbaryl has been found effective in controlling the caterpillars.

Serpentine leaf miner : Serpentine leaf miner is a new introduced pest in India and the incidence is severe in Karnataka, Andhra Pradesh, Maharashtra, Gujarat and Delhi. The pest is a native of southern United States of Amercia and Central America. It has spread to other countries in 1970's alongwith Chrysanthemum. It infests 47 genera in 10 families in America including economically important crops namely chrysanthemum, aster, calendula, gerbera, sugar beet, tomato, potato, cucumber and celery.

The female fly makes feeding and egg laying punctures in the leaves. These are often not seen in the fields due to their minute size. The larva feeds between the two layers of leaf on mesophyll, leaving the papery thin layer intact. It makes narrow serpentine mines that appear whitish when observed from the upper surface; the mines are sharply curled and contorted thus causing blotches and holes. *L. trifoli* is also known to be a vector of many plant viruses.

The adult fly is 1.5-2.0 mm long and greyish-black with yellow spot on the top of the thorax with plum red eyes. It flies quite rapidly and may travel several meters before landing. When left undisturbed, the adults tend to congregate on individual leaves. The length of the life cycle is greatly influenced by temperature and the host. Adult lives 2-4 weeks and may lay 250 eggs. Eggs are translucent, 0.2 mm long and 0.15 mm broad. These are laid singly in small incisions in the leaf. Females make incisions on the plants with their ovipositors and feed on them. Same incisions are used for egg laying. The larva is legless, orange-yellow and about 2 mm long. The full grown larva leaves the mine in 7-18 days, crawls on the leaf or drop to the soil for pupation. The pupal period lasts for 14 days. The total development time is 50 days at 16°C and 14 days at 30°C.

Management : The control of this pest is difficult with conventional insecticides as it has become resistant to most of the commonly used insecticides within 6-8 generations. In glasshouses, limited eradication is possible by a combination of insecticide treatment and destruction of affected plants. Under field conditions, spraying either with oxy demeton methyl (0.025%) or acephate (0.05%) or phosphamidon (0.05%) or neem seed kernel extract (3%) was found effective. Infested leaves should be removed and destroyed.

Striped mealy bug : It is found all over the Indian sub-continent and South-east Asia. Being polyphagous pest, has a wide range of host plants including beans, coffee, citrus, guava, groundnut, sugarcane, and sweet potato. The pest is found throughout the year, though less active during winter. The population normally increases from February onwards and reaches peak in mid April. During rains, the population declines and after rains, the population again increases till November and again decreases.

White, cottony mealy bugs can be easily seen on the leaves and twigs. Crawlers are pale white in colour. These crawlers become active and move swiftly when they find a succulent spot where they puncture the epidermis of the host plant, inject there toxic saliva and suck cell sap. The injury serves as an entry for various disease producing organisms. Both nymphs and adults suck sap from the leaves and twigs of the plants, resulting in stunted growth. The nymphs, second instar onwards secrete honey dew on the sooty mould develops and the normal photosynthetic activity of the plant is hindered.

Eggs are laid in clusters in cottony ovisac remaining concealed under the female body. A single female lays 100-400 eggs, which on hatching give rise to crawlers; those remain together in cottony nest under the mother. Reproduction is both sexual and parthenogenetic. Incubation time ranges from 15 minutes to 4 hours, nymphal period 20-

60 days in males and 19-47 days in females. Longevity of males and females ranges from 1-3 and 507 weeks, respectively.

Management : Remove and destroy mechanically all the affected leaves and twigs when the infestation is first observed. In large areas of infestation, spray dimethoate (0.03%) or oxy demeton methyl (0.025%). Observe a waiting period of 10 days for picking of fruits.

3. Okra : Okra is an important vegetable grown in tropical and sub-tropical areas of the world. Fruits are rich in pectin and mucilage and a fair source of iron and calcium. The abundance of pests varies from region to region and the major species that attack okra crop in India are listed in Table 3.

Table 3 : Insect-pests of okra

Common Names	Scientific Names	Order	Family
Jassid	*Amrasca biguttula biguttula* (Ishida)	Homoptera	Cicadellidae
Shoot and fruit borer	*Earias vitella* (Boisduval) *E. insulana* (Fabricius)	Lepidoptera	Noctuidae
Blister beetles	*Mylabris pustulata* Thunberg, *M. phalerata* Pallas, *M. mecilenta* Marhsall, *M. tifiensis* Bills	Coleoptera	Meloidae
Cotton leaf roller	*Sylepta derogata* (Fabricius)	Lepidoptera	Pyralidae
Okra aphid	*Aphis gossypii* Glover	Homoptera	Aphididae
Red cotton bug	*Dysdercus koengii* (F.)	Homoptera	Pyrrhocoridae
Green semilooper	*Anomis flava* (F.)	Lepidoptera	Noctuidae
Fruit borer	*Helicoverpa armigera* (Hubner)	Lepidoptera	Noctuidae
Whitefly	*Bemisia tabaci* (Gennadius)	Homoptera	Aleyrodidae

Okra shoot and fruit borer: *Earias vittella* is commonly known as spotted boll worm of cotton. It is widely distributed and is recorded from Pakistan, Srilanka, Bangladesh, Myanmar, Indonesia, New Guinea and Fiji. It is an oligophagous pest, though found feeding on a large number of other malvaceous plants, both wild as well as cultivated.

The caterpillars bore into the shoots during vegetative stage and feeds inside as a result of which the shoots droop down and dry-up. In the later stages it infests the capsules (fruits) which become disfigured and show damage holes. Fruits become distorted and are rendered unfit for consumption. The pest can cause up to 90 per cent loss in yield.

The moths appear in April, live for 8-22 days and females lay 200-400 eggs singly on flower buds, brackets, tender leaves, fruits and other parts of plant. The hairy parts of the plants are preferred for oviposition. On hatching the caterpillars make their way inside the plant tissues. In warm weather, the eggs hatch in 3-4 days and the caterpillars pass through 6 stages, becoming full grown in 10-16 days. The larvae are light yellow with black spots. Full grown caterpillars are 18 to 24 mm long, stout spindle-shaped having long stiff setae. They pupate either on the plants or on the ground among fallen leaves and the moths emerge in 4-9 days. Pupae are 13 to 16 mm long and chocolate brown in colour, bluntly rounded and enclosed in inverted boat-shaped cocoons. The life cycle is completed in 17-29 days during the summer.

Earias insulana which is also known as spotted boll worm of cotton is found damaging okra, especially in dry regions. Its immature stages are more or less similar in appearance as those of *E. vittella* except for the size. Full grown caterpillars are 15 to 18 mm long and pupae are 12 to 14 mm long.

Management : Grow resistant/ tolerant varieties. Moths dislike red coloured plants. Narnaul Special, Perkins Long Green, Varsha Uphar, Rashmi Sagar are tolerant to the damage by this pest. Avoid overuse of nitrogenous fertilizers.

Removal and destruction of damaged shoots/ fruits at regular interval helps in reducing the pest damage. Among chemicals, carbaryl (0.1%), endosulfan (0.05%) and fenvalerate (0.01%) are effective. The eggs are parasitized by *Trichogramma chilonis*, which is widely distributed in India and Pakistan.

Blister beetles : Various species of blister beetles are known to cause damage in Africa, Southeast Asia, Bangladesh and Sri Lanka and India. *Mylabris pustulata* Thunberg and *M. phalerata* Pallas are the common species found feeding on okra. These beetles are widely distributed in India from the high altitude regions of Himalaya to plains of north India and southern parts of the subcontinent.

Blister beetles are polyphagous pests and feed on plants belonging to family Malvaceae, Leguminosae, maize etc. Beetle feeds on pollens, petals of flowers and flower buds, as a result, fruit setting is affected. It has three black and three reddish-orange bands running horizontally and alternately on elytra. When disturbed, beetle exudes an acrid yellow fluid which contains cantharidin, which is irritant to touch and causes blisters on human skin-hence the name.

Only the adults of these beetles cause direct damage by feeding voraciously on petals, anthers and pollens, thus affecting the yield considerably. A single beetle can destroy as many as 20-30 flowers per day.

The beetles are 22 to 26 mm or more in length. *M. phalerata* is smaller in size in comparison to *M. pustulata*. The female lays large number of eggs in soils of cultivated fields. Eggs are light yellowish in colour and cylindrical in shape. Incubation period is about three weeks. Young grubs are white in colour. The larvae prey upon the eggs of Orthoptera and Hymenoptera. In the first stage they are campodeiform termed 'triungulins' and actively look for the host. They pupate inside the soil tunnels.

Management : Collect and destroy the beetles in the early morning. Apply 0.1% carbaryl or 0.05% malathion or 0.01% fenvalerate during the severe infestation.

Cotton leaf roller : It is a sporadic pest and is widely distributed in the Orient, Africa and India. Besides, cotton, it has also been recorded on various other malvaceous plants like okra, gulkhaira (*Athalea officinalis*) *Abutilon indicum* and some forest trees.

The young caterpillars feed first on the lower surface of the leaves. The older larvae roll leaves from the edges inwards up to the mid rib and feed on leaf tissue from the inside.

The leaf roller is active from March to October and passes the winter as full grown caterpillar among the plant debris. The hibernating larvae pupate by the end of February and the moths appear during March. The moths are active during night, they mate and female lays 200-300 eggs singly on the underside of the leaves. The moths are yellowish-white with black and brown spots on the head and the thorax. The eggs hatch in 2-6 days. The larvae are greenish-grey or pink. The larva grows through seven stages and is full-fed in 15-35 days. They pupate either on the plants, inside the rolled leaves or among the plant debris in the soil. They emerge as moths in 6-12 days and live for about a week. The life cycle is completed in 23-53 days and the pest passes through 5 or 6 generations.

Management : The destruction of alternate host plants or the burining of plant debris minimizes the incidence of this pest. Deep ploughings by the end of February is also helpful in reducing the carry over of the pest to the next season. *Trichogramma* spp. parasitize the eggs while *Brachymeria tachardiae* Cam., *Elasmus indicus* Rohn., *Goryhus nursei* Cam and *Trichospilus pupivora* Ferrieri are important larval parasitoids of this pest. If the attack is severe, then spray 0.05 % endosulphan or 0.025 % quinalphos or 0.04% triazophos or 0.05 % acephate or 0.1% carabryl.

Leaf hoppers : Leaf hoppers are polyphagous pests causing damage to cotton, okra, eggplant, beans, castor, cucurbits, potato, hollyhock, sunflower as also other malvaceous plants.

Both nymphs and adults suck usually from ventral surface of leaves and while feeding inject their toxic saliva into the plant tissues, affected leaves turn yellowish and curl. Leafhoppers infest the lower surface of the leaves. Commonly, if disturbed, they move very rapidly sideways and hop. Infested leaves curl upwards along the margins. Outer leaf areas appear yellowish or burned. Leaves remain small and show a mosaic pattern of yellowing. Fruit-set may be very low. Severe infestation results in cupping of leaves and reduces photosynthesis and stunted growth of the plants.

A female lays 15-30 eggs. Incubation and nymphal periods last for 4-10 and 7 to 21 days, respectively. The eggs are laid singly in the tissues of main veins on the underside of the leaves. The eggs are pear-shaped, elongated and yellowish white in colour. The nymphs are usually 1–3 mm long, greenish-yellow with slender, tapered bodies, with two distinct black spots on the posterior end of the wings. Legs have rows of sharp spines. Longevity of the adults varies from 5 to 8 weeks.

Management : The pest can be suppressed by foliar application of monocrotophos (0.02%), dimethoate (0.03%), oxy demeton methyl (0.025%) or fenvalerate (0.01%) before flower initiation stage. In endemic areas, seed should be treated with imidacloprid (3 g/kg) which will protect the crop for 40-50 days after sowing. In nature, eggs of *A. biguttula biguttula* are parasitized by *Lymaenon empoascae* Subbhrao and *Erythmelus empoascae* Subbarao.

4. Cucurbits : Cucurbits include a number of crops which mostly possess trailing habits and are extensively grown all over the tropical and sub-tropical countries as summer and rainy season vegetables. In India, cucumbers, melons, pumpkins and various types of gourds are grown. In general, these are good source of vitamin A & C and various vital minerals. The insect-pests attacking cucurbits are shown in Table 4.

Table 4: Insect-pests of cucurbits

Common Names	Scientific Names	Order	Family
Red pumpkin beetle	*Aulacophora foevicollis* (Lucas)	Coleoptera	Chrysomelidae
Melon Fruit fly	*Bactrocera cucurbitae* Coquillett, *B. tau* (Walker)	Diptera	Tephritidae
Hadda beetle	*Henosepilachana viginoctopunctata* (F.)	Coleoptera	Coccinellidae
Melon aphid	*Aphis gossypii* Glover	Homoptera	Aphididae
Pumpkin caterpillar	*Diaphania indica* (Saunders)	Lepidoptera	Pyralidae
Stink bug	*Aspongopus janus* F., *A. observus* F., *A. brunneus* Thunberg	Heteroptera	Pentatomidae
Jassid	*Empoasca binotata* Pruthi	Heteroptera	Cicadellidae
Whitefly	*Bemisia tabaci* (Gennadius)	Homoptera	Aleyrodidae
Serpentine leaf miner	*Liriomyza trifolii* (Burgess)	Diptera	Agromyzidae
Flea beetle	*Phyllotreta cruciferae* (Goeze)	Coleoptera	Chrysomelidae
Gall fly	*Lasioptera falcata* Felt	Dipera	Cecidomyiidae
White grub	*Holotrichia insularis* Bern.	Coleoptera	Scarabaeidae

Red pumpkin beetle : The beetle is widely distributed in Asia, Australia, southern Europe and Africa. It is a serous pest of cucurbits such as ash gourd, pumpkin, tinda, cucumber and melon.

Adult beetles are mainly responsible for the damage of the plant above ground, attacking on the leaves, flowers and fruits. The beetles injure the cotyledons, flowers and foliage by biting holes into them. The grubs live in the soil and damage the plants by boring into the roots and stem of the plant. Fruits and leaves are also damaged when they come in contact with soil. In case of heavy infestation, re-sowing is required to be done.

The beetles are found concealed in groups under dry weeds, bushes and plant remains or in the crevices in the soil and resume activity as soon as the season warms up. The pest is active during March-April. Adults are pale orange-yellow to deep pale brown and measures 5-7 mm in length. The female lays about 300 oval yellow eggs singly or in batches of 8-9 in moist soil, near the base of the plants. They hatch in 6-15 days and the grubs remain below the soil surface feeding on roots. Grubs are slender, elongate, pale creamy-yellow with brown head and legs. Full grown grub measures 12 mm in length and are full grown in 13-25 days which pupate in thick-walled earthen chambers in the soil.

The pupal stage lasts for 7-17 days and the beetles on emergence begin to feed and breed.

Management : Collect and destroy the beetles in early stage of infestation. Preventive measures like burning of old creepers, ploughing and harrowing of fields after harvest of the crops are recommended for the destruction of adult, grubs and pupae.

In cucumber, Cu-All season, Cu-Gy-604, Cu-Improved, Long Green, Cu-Korea Dutch and Cu-Gy-598 have been reported to be tolerant against this pest.

Spraying with 0.05% malathion or dusting with 5% malathion @ 10 kg/ha gives satisfactory control of the pest. Apply carbofuran 3 G @ 7.0 kg/ha 3-4 cm deep in the soil near the base of the plants just after germination and irrigate the fields.

Melon fruit fly : Melon fruit fly, *Bactrocera cucurbitae* is the most common and most destructive fruit fly of musk melon and other cucurbits throughout India. It damages over 81 plant species, but plants belonging to family cucurbitaceae are preferred hosts (Dhillon *et al.*, 2005). Two other allied species common in India are *Bactrocera dorsalis* and *Dacus cilitatus* Loew. The extent of losses varies between 30-100 per cent depending on the cucurbit species and the season. The melon fruit fly has been reported to infest 95 per cent of bitter gourd fruits in Papua (New Guinea) and 90 per cent snake gourd and 60 to 87 per cent pumpkin fruits in Solomon Islands. Singh *et al.* (2000) reported 27 per cent damage in bitter gourd and 28.5 per cent on watermelons in India.

The maggots feed on the pulp of the fruits. Infested fruits can be identified by the presence of brown resinous juice which oozes out of the punctures made by the female fly for oviposition. These punctures also serve as an entry for various bacteria and fungi; as a result the infested fruits start rotting, get distorted and malformed and fall prematurely.

The female fly makes a cavity with the help of sharp ovipositor and thrusts the white elongate cigar-shaped eggs singly or in groups of 4-10 into the flowers and tender fruits. After laying the eggs, the female secretes a gummy secretion on the oviposition puncture and makes the entrance waterproof. During the life span of 15-60 days, a female on an average lays 58-95 eggs. The incubation period is 1-9 days. The freshly hatched maggots bore into the pulp, forming galleries. Depending upon weather, the larval period extends from 3 days in summer to 21 days in winter. The full grown maggots come out of the fruits and drop down to the ground and pupate in the soil at a depth of 1.5 to 15 cm. The pupal period is 5-9 days in summer. The total life cycle is completed in 12-34 days. There are several generations in a year. The pest is active through out the year.

Management : As in tomato.

5. Chilli : Among the vegetables, chilli is a good source of vitamin A and C. It is an important condiment crop, fruits are used green as well as ripe and dried to impart pungency to various foods. It has a tonic and carminative action and is especially useful in atonic dyspepsia. Over 25 insect species have been recored from South-east Asia infesting chilli leaves and fruits and the important ones are given in Table 5.

Table 5 : Insect-pests of chillies

Common Names	Scientific Names	Order	Family
Thrips	*Scirtothrips dorsalis* Hood, *Caliothrips indicus* (Bagnall), *Franklineilla sulphurea* Schm.	Thysanoptera	Thripidae
Aphids	*Aphis gossypii* Glover, *Myzus persicae* (Sulzer)	Homoptera	Aphididae
Termites	*Odontotermes* sp.	Isoptera	Termitidae
Cotton whitefly	*Bemisia tabaci*(Gennadius)	Homoptera	Aleyrodidae
White grubs	*Holotrichia consanguinea* Blanchard, *H. insularis* (Bren.), *Anomala bengalensis* Bl.	Coleoptera	Scarabaeidae
Stem borer	*Euzophera perticella* Ragonot	Lepidoptera	Pyralidae
Fruit borer	*Spodoptera exigua* (Hubner), *S. litura* (Fab.), *Helicoverpa armigera* Hubner	Lepidoptera	Noctuidae
Root grub	*Arthrodeis* spp.	Coleoptera	Scarabaeidae
Leaf defoliators	*Mythimna loreyi* (Duponchel), *Spodoptera exigua* (Hubner)	Lepidoptera	Noctuidae
Leaf defoliator	*Monolepta signata* Olivier	Coleoptera	Galerucidae
Large brown cricket	*Brachytrypes portentosus* (Litchtenstein)	Orthoptera	Gryllotalpidae
Sap sucking insects	*Ferrisia virgata* (Cockrell)	Homoptera	Pseudo-coccidae
	Saissetia coffeae (Walker)	Homoptera	Coccidae
	Aspidiotus destructor Signoret, *Lepidosaphes piperis* (Green)	Homoptera	Diaspididae

Chilli thrips : It is a polyphagous pest and is having a wide range of host plants. Both the nymphs and adults lacerate the leaf tissues and suck sap from tender leaves, growing shoots, flower buds and developing fruits. The infested leaves curl crump and prematurely shed whereas flower buds become brittle and drop down. The entire plant may dry and wither away if there are no rains and the weather remains dry. This pest also transmits leaf curl disease. Under severe conditions, 30-50 per cent crop may be damaged.

The eggs are laid under leaf tissues as much as 40-50 eggs. They are minute and dirty white in colour. The eggs hatch in 5 days. Nymphs and adults are also very small, slender, fragile and yellowish in colour. The wings are heavily fringed and are grey in colour. The larval feeding continues for 7-8 days and the pupal period is 2-4 days. Reproduction is both sexual and parthenogenetic. The longevity of the adult thrips is about 30 days. Several overlapping generations have been reported in a year.

Management : Removal of weed hosts is helpful in checking the infestation. Do not grow mixed crop of onion and chilli. Encourage predatory thrips, *Scolothrips indicus* Priesner and *Frankliothrips megalops* Back to feed upon *S. dorsalis*. Spray dimethoate (0.03%), Achook (1.5%) and neem oil (3-5%) in case of severe infestation.

Aphids : Both the species are polyphagous and distributed throughout India. These aphids are found infesting chilli.

The infested leaves curl and dry up. The aphids also secrete honey dew on which the sooty mould grows. This coating of mould interferes with the photosynthesis causing retardation in growth and fruiting capacity.

Small ovate, soft bodied nymphs and adults are found in large number on underside of the tender leaves and shoots, sucking cell sap from the tissues. The female of *A. gossypii* gives birth to 8-22 nymphs a day. The nymphs moult four times to become adults, thus completing the life cycle in 7-10 days. Cloudy weather is very conducive for the rapid multiplication of these pests whereas heavy showers cause a sudden drop in their population.

Management : To check the aphid infestation, clip off and destroy the affected shots and twigs in the initial stage of attack. The crop may be sprayed with acephate (0.05%) or oxy demeton methyl (0.025%) or endosulphan (0.05%). The coccinellid beetles, *Coccinella septempunctata* Linnaeus and *Menochilus sexmaculatus* (Fabricius) are the common predators on aphids.

Termites : Termites or white ants as they are commonly but erroneously called are cosmopolitan pests and the species of agricultural importance have a wide range of host plants. These abound in sandy or sandy loam soils and do not thrive well under conditions of bad aeration and poor drainage. The ravages by this pest can start the moment seed is placed in the soil and may continue till the harvest of the crop. The pest being subterranean gnaws the roots and stems below ground level and tunnels upwards through stems, eating all the inner tissues. The affected plants become pale, wither and ultimately dry away.

Roots of chilli are occasionally damaged by the termite, *Odontotermes obesus* Rambur. Cream coloured tiny insects resembling ants with dark coloured head, cause withering of plants and extensive damage to roots.

Management : Burn crop residue to discourage termite buildup. Avoid using undecomposed or partially decomposed farmyard manure. Apply chlorpyriphos dust @ 20 kg/ha in soil at the time of sowing. Apply 4.0 litres endosulphan 35 EC or chlorpyriphos 20 EC /ha with irrigation water.

II. INSECT-PESTS OF WINTER VEGETABLES

1. Cabbage and Cauliflower : These are cold season crops and thrive best under temperate climates. Cabbage is a rich source of vitamin A, B_1 and C and also contains various minerals like calcium, phosphorus, potassium, sodium and iron. Cauliflower is less hardy than cabbage and is rich in iron, magnesium, phosphorus, potassium and sodium. The regular pests which invade these crops are listed in Table 6.

Table 6 : Insect-pests of cabbage and cauliflower

Common Names	Scientific Names	Order	Family
Cabbage caterpillar	*Pieris brassicae* (Linnaeus)	Lepidoptera	Pieridae
Diamondback moth	*Plutella xylostella* (Linnaeus)	Lepidoptera	Plutellidae
Cabbage head borer	*Hellula undalis* (Fabricius)	Lepidoptera	Pyralidae
Leaf webber	*Crocidolomia pavonana* (F.)	Lepidoptera	Pyralidae
Tobacco caterpillar	*Spodoptera litura* (Fabricius)	Lepidoptera	Noctuidae
Cabbage flea beetle	*Phyllotreta cruciferae* (Goeze), *P. chotanica* Duviv,*P. birmanica* Harold, *P. oncera* Maulik, *P. downesi* Baly	Coleoptera	Chysomelidae

Contd.

Cabbage aphid	*Brevicoryne brassicae* (Linnaeus)	Homoptera	Aphididae
Aphids	*Lipaphis erysimi* Kalt., *Myzus persicae* (Sulzer)	Homoptera	Aphididae
Cabbage semi-looper	*Thysanopulsia orichalcea* (Fabricius), *Autographa nigrisigna* (Walker)	Lepidoptera	Noctuidae
Bihar hairy caterpillar	*Spilosoma oblique* (Walker)	Lepidoptera	Arctiidae
Cutworms	*Agrotis ipsilon* (Hufnagel), *A. segetum* (Dennis & Schiff)	Lepidoptera	Noctuidae
Mustard sawfly	*Athalia lugens proxima* Kulg	Hymenoptera	Tenthredinidae
Painted bugs	*Bagrada cruciferarum* Kirkaldy, *B. hilaris* (Burmester)	Heteroptera	Pentatomidae
Termites	*Microtermes obesi* Homlgreen	Isoptera	Termitidae
Earwig	*Euborellia annulipes* (Lucas)	Dermaptera	Forficulidae
Large brown cricket	*Brachytrypes protentaosus* (Lichtenstein)	Orthoptera	Gryllotalpidae
Grasshoppers	*Atractomorpha crenulata* Fabricius	Orthoptera	Acrididae
Leaf eating weevils	*Tanymecus circumdatus* (Weidemann), *Myllocerus blandus* Faust	Coleoptera	Curculionidae
Leaf eating caterpillar	*Trichoplusia ni* (Hubner)	Lepidoptera	Noctuidae
Pea leaf miner	*Chromatomyia horticola* Goureau	Diptera	Agromyzidae
Thrips	*Caliothrips indicus* (Begnall), *Thrips tabaci* Lindemann, *Frankliniella schultzei* (Trybom)	Thysanoptera	Thripidae

Cabbage caterpillar : It is an oligophagous pest with a limited host range of cruciferous crops like cabbage, cauliflower, radish, turnip, brassica oilseeds etc. It is distributed all along the Himalayan regions.

The caterpillars feed gregariously during early stage and disperse in the fourth instar. The young larvae scrape the leaf surface whereas the old larvae eat up the leaves from the margin inwards leaving the main veins only. They skeletonize leaves and bore into heads of cabbage and cauliflower.

The butterflies are pale white and have a smoky shade on the dorsal side of the body. The females measure 65 mm across the spread wings and have two conspicuous black circular dots on the dorsal side of each forewing. Males are smaller than the females and have black spots on the underside of each forewing. Female lays yellowish conical eggs in clusters of 50-90 on the leaves. The eggs hatch in 11-17 days in November-February and 3-7 days in March. The full grown caterpillars are greenish-yellow in colour and measures about 40-50 mm in length. The larvae pass through five stages and are full-fed in 15-22 days. The larvae pupate at some distance from the host plants. The pupal stage lasts for 7-14 days in March-April.

Management : For the management of the caterpillars, collect and destroy the yellow egg masses of *Pieris brassicae* in early stages of infestation. Hand picking and mechanical destruction of caterpillars during early stages of attack can reduce the infestation. *Cotesia glomeratus* parasitizes the larvae of *Pieris brassicae.* Foliar application of Bt formulation @ 500 g a.i./ha alongwith sticker controls the larvae effectively. Spray endosulfan (0.05%) or malathion (0.05%) or deltamethrin (0.0028%) or cypermethrin (0.0075%) or fenvalerate (0.01%). Repeat after 10-day intervals if required. Observe a waiting period of 7 days after spraying.

Diamondback moth : It is a major pest of all cruciferous vegetable crops and cabbage and cauliflower are major host crops. It is cosmopolitan in distribution and enjoys worldwide distribution.

Larva feeds on foliage and causes serious damage by defoliation. First instar larve mine epidermal surface of leaves producing typical white patches. Second instar onwards feed externally making holes on the leaves and soil them with excreta. Heavy infestations leave little more than the leaf veins. The larvae also enter the head/ curd affecting the production and the quality of produce.

The adult moth is brown or grey with three clear white spots on the forewings which form diamond like pattern when the moth is at rest. Creamish eggs of the size of pinheads are laid singly or in batches of 2-40 on the underside of leaves. The eggs hatch in 3 - 9 days. The newly hatched caterpillars bore into the tissues from the underside of leaves and feed in these tunnels. At first, their appearance is detected only from the blackish excreta that appear at the mouth of each tunnel, but in second instar, mines become more prominent. The larvae of fourth instar feed from the underside of leaves, leaving intact a transparent cuticular layer on the dorsal surface. When full grown, the larvae measure about 8 mm in length and are pale-yellowish green with fine black hair all over the body. The larvae become full grown in about 16-17 days. The pupal stage lasts for 4-5 days under favourable conditions.

The total life cycle is completed in 2-3 weeks and there are 8 to 12 generations in a year.

Management : Remove and destroy all debris and stubbles after harvest of crop and plough the field. Indian mustard as the trap crop is effective in suppressing the incidence of diamondback moth and cabbage aphid. Grow paired rows of mustard after every 25 rows of cabbage and cauliflower. Of the two mustard rows one is sown 15 days prior and another 25 days after cabbage transplanting. This planting ensures adequate mustard foliage throughout the cropping period. Periodically spray mustard crop with dichlorvos (0.04%) at 10 or 15 days interval to avoid dispersal of larvae.

Install pheromone traps @ 12/ha for monitoring. Spray cartap hydrochloride (0.5%) or *Bacillus thuringiensis* (2 g/l) or quinalphos (0.05%) at primordial stage (between 17 and 25 days after planting) or head initiation stage when economic threshold level crosses 2 larvae/ plant.

Release parasitoids, *Diadegma semiclausum* in hills or *Cotesia plutellae* in plains. Release of 1,00,000 adults/ha @ 20,000 adults/release commencing from 20 days after planting reduces the damage.

Cabbage head borer : Cabbage head borer has world wide distribution. It is one of the serious pests of cruciferous crops. The larvae attack cabbage, cauliflower, radish, knol-khol and the weed, *Gynadropsis pentaphylla* (Capparidaceae).The pest breeds throughout the year but becomes visible when the cruciferous crops are sown.

The caterpillars first mine into the leaves. Later on, they feed on the leaf surface, sheltered within the silken passages. As they grow bigger, they bore into the heads of cauliflower and cabbage. When the attack is severe, the plants are riddled with worms and the heads look deformed.

The female moth lays eggs singly but frequently in clusters on the under surface of leaves. The colour of the egg is pinkish and their shape is oval. The eggs hatch in 2-3 days. The caterpillars feed in the heart of the cabbage and become full grown in 7-12 days, after undergoing four moults. The caterpillar is about 12-25 mm long and creamy-yellow with a pinkish tinge and has seven purplish-brown longitudinal stripes. The full grown caterpillar spins a cocoon among the leaves touching the ground. The pupal period is about 6 days and the life cycle is completed in 15-25 days. The adult moth is slender, pale yellowish brown, having grey wavy lines on the forewings. The hind wings are pale dusky.

Management : Collection and mechanical destruction of caterpillars in early stage of attack helps to check the infestation. Because the attack is mostly in the nursery and on young plants in the field, spray endosulphan (0.05%) or carbaryl (0.1%) and repeat if the attack persists. Follow a waiting period of 7 days for picking of heads.

Leaf webber : Leaf webber is a serious pest of cabbage, radish, mustard and other crucifers. It is distributed throughout India, South-east Asia, Australia and Africa.

The larvae web the leaves with silken strands and feed on the lower surface of the leaves completely skeletonising them. In cauliflower, larvae nibble the growing tip of seedlings and bore into the curd resulting in discoloration of the curd. Even a single mature larva per plant is capable of inflicting economic loss to cabbage at pre- and post-heading stages.

The pest is active with the availability of cruciferous crops in the field. The larva is green with red head and longitudinal red stripes on the body. It is about 2 cm in length. The moths are small with light brownish forewings. The moth lays eggs in masses of 40-100 each, on the lower surface of leaves. The eggs hatch in 5-15 days. Young larvae feed gregariously on the leaf tissues. As they grow older, they start webbing the leaves and feeding on them. The larval stage is completed in 24-27 days in summer and in about 50 days in winter. The pupation takes place in soil in earthen cocoon. The adults emerge in 14-40 days. The entire life cycle is completed in 43-82 days.

Management : Remove and destroy the webbed leaves with larvae inside. Use of Indian mustard as trap crop for DBM management also attracts the leaf webber moths for egg laying and reduce the infestation in the main crop. Foliar application of neem seed kernel extract (4%) along with sticker or Bt formulations (500 g a.i./ha) is also effective. Application of cypermethrin (0.0075%) or malathion (0.05%) is effective against the pest.

Cabbage flea beetle : The cabbage flea beetles attack almost all the cruciferous plants in Europe, erstwhile USSR, North and South Amercia, Australia, Japan, and India. The common field crops like mustard, raya, toria, and vegetables like radish, turnip, cabbage, cauliflower and knol-khol are severely damaged by adult beetles. Some ornamental plants and flowers are also attacked.

The adults mostly feed on the leaves by making innumerable round holes in the host plants. The old eaten away leaves dry up, while the young leaves are rendered unfit for consumption. A peculiar kind of decaying odour is emitted by the cabbage plants attacked by this pest.

The dorsum of the adult beetle is metallic blue with a greenish hue. The body is elongate, narrow in front, broad distally and round at the anal end. The female beetle measures 2.0 mm in length. The female lays 50-80 creamy-white eggs singly in soil around the host plants. The incubation period ranges from 5-10 days. The larva is dirty white in colour with pale white head and measures about 5.0 mm in length. The larvae are active and feed on tender roots of the host plants. The larval period is 9-15 days. Pupal stage lasts for 8-14 days and pupation takes place in the vicinity of the infested plants.

Deep ploughings during summer helps to expose and kill the over wintered population of the pest. Spray 0.1% carbaryl or 0.05% endosulphan or dichlorvos (0.1%) for the control of this pest.

Cabbage aphid : Aphids in general have a very high rate of reproduction and a short life-span as a result of which they are serious pests of many economic plants. Cosmopolitan in distribution, this is a pest of cabbage, cauliflower, radish and many other crucifers, appearing in the cold season.

Both nymphs and adults suck cell sap from the plants especially the tender parts resulting in devitalization of the plants. They also produce honeydew, which attract sooty moulds resulting in the hindrance in photosynthesis. In case of severe infestation plants may completely dry up and die away. Feeding damage from large numbers of aphids can kill seedlings and young transplants. On larger plants, damage results in curling and yellowing of leaves, stunted plant growth, and deformed heads. White cast skin will be present at the base of the plant.

The cabbage aphid is a soft bodied, mealy-grey insect covered with waxy filaments. They reproduce parthenogenetically giving birth to young ones and a female lays upto 35 young ones in her life span of 8-10 days. The nymphs become full-fed in 10-16 days. The nymphs measure 1.5 to 2 mm and the adults are 2 to 3 mm long. Cloudy days coupled with high humidity are most favourable for its multiplication. High temperature & low humidity, and overcrowding results in the production of nymphs deemed to become winged adults, which migrate to suitable areas.

Management : Regular surveillance of the crop is very important. Remove the infested plant parts, early in the season and destroy along with aphids. Use yellow sticky traps to attract winged aphids. Encourage activity of predators, Coccinellids and Syrphids; and parasitoid, Diaeretiella rapae.

Whenever 5% or more plants are found infested with this pest, remedial measures like spray malathion 0.05% or endosulfan 0.05% or dimethoate 0.03% or oxy demeton methyl 0.025% or phosphamidon 0.03% or neem oil 3%. Wetting agent in spray fluid should be added for better adherence on aphid body as the aphid skin is covered with waxy filaments.

Cabbage semi-looper : These two species are widely distributed in north western India and are minor pests of cabbage, cauliflower and other winter vegetables. They are polyphagous and attack a number of plants, including groundnut and sunflower.

The caterpillars are plump and pale green. They cause damage by biting round holes into cabbage leaves. On walking they form characteristic half-loops and are often seen mixed with cabbage caterpillars.

The adults of *T. orichalcea* are light pale brown with a large golden patch on each fore-wing. They measure about 42 mm across the spread wings. The adults of *A. nigrisigna* are darker and have dark-brown and dirty-white patches on the fore wings. The insects are active during the winter. Females lay eggs on the leaves of their host plants and larvae feed individually, biting holes of varying size according to the stage of their development. When full grown, they pupate in debris lying on the ground. The moths are active at dusk on flowers in gardens and public parks. The total generation time is 24-35 days.

2. Pea : Pea is relatively a hardy crop and can be grown in colder climates. It is one of the popular vegetables which is grown extensively all over the world. The seeds are excellent source of vitamin A, C and B1 and also rich in vitamin G besides being a good source of phosphorus, iron and carbohydrates. The important insect-pests infesting pea are given in Table 7.

Table 7 : Insect-pests of Pea

Common Names	Scientific Names	Order	Family
Pea leaf miner	*Chromatomyia horticloa* (Goureau)	Diptera	Agromyzidae
Pea stem fly	*Ophiomyia phaseoli* (Tryon)	Diptera	Anthomyiidae
Pod borer	*Etiella zinckenella* Treitschke	Lepidoptera	Pyralidae
Pea blue butterfly	*Lampides boeticus* (Linnaeus)	Lepidoptera	Lycaenidae
Cutworms	*Agrotis ipsilon* (Hufnagel) *A. segetum* (Dennis & Schiff)	Lepidoptera	Noctuidae
Bean aphid	*Aphis craccivora* Genn.	Homoptera	Aphididae
Pea aphid	*Acyrthosiphum pisum* (Harris)	Homoptera	Aphididae

Pea leaf miner : Pea leaf miner is found throughout the temperate region of the world. It is a polyphagous pest feeding on leguminous crops, cucurbits, crucifers, tomato and lettuce.

The larvae make prominent whitish tunnels in the leaves. If the attacked leaves are held against bright light, the minute slender larvae can be seen feeding within the tunnels. The large numbers of tunnels made by the larvae interfere with photosynthesis and proper growth of plants.

The adults are two-winged flies having greyish-black mesonotum and yellowish frons. They are active from December to April or May. The adults emerge at the beginning of December and after mating start egg laying singly in leaf tissues. The eggs hatch in 2-3 days and larvae feed between the lower and upper epidermis by making zig-zag tunnels. The larvae become full grown in about 5 days and pupate within the galleries, which lasts for 6 days. The insect completes one generation in 13-14 days and pass through several generations during the period of its activity.

Management : Spray 0.004% lambda-cyhalothrin or deltamethrin (0.0028%), if 40 per cent leaf infestation is there. Repeat after 7 days if required. Observe a waiting period of 7 days for picking of green pods. Lines viz, 65-88-1, IP-3, JP-169-1, JP-71, JP-83, JP-179, JP-747, JP-854, ML-21 and ML-22 have been reported to be tolerant to pea leaf miner (Brar and Kaur, 2005).

Pea stem fly : It is also a polyphagous pest and feeds on almost all parts of beans, gram and pea. It is widely distributed in India, Srilanka, Philippines and China.

The maggots on emergence feed on leaf tissue at first but later on bore into the terminal stems causing withering and ultimate drying of the affected shoots, thus reducing the bearing capacity of the host plants. The adults also cause damage by puncturing the leaves and the injured parts turn yellow. The damage is more serious on seedlings.

The adult flies are metallic black, 2.0-2.5 mm long with a wing expanse of 5 mm and are active in summer. The female lays 14-64 elongate, oval and white eggs into the leaf tissue. The eggs hatch in 2-4 days. Maggots are initially white in colour late becoming yellowish. These are small in size being less than 1.0 mm in length. The larval development is completed in 6-7 days. The larva pupates within its gallery and the pupal period last for 5-9 days. Pupae are barrel-shaped and brown in colour.

Management : Avoid sowing of the crop earlier than mid October to check the attack of the pest. Apply 7.5 kg of phorate 10 G or 25 kg of carbofuran 3 G per hectare in furrows at the time of sowing. Spray the crop with 0.025 % oxy-demeton methyl just after germination and can be repeated, if required.

Pea pod borer : It has been reported from USA, Mexico, West Indies South America, Hungary, France, Austria, Southern Russia, UAE, parts of Africa, Indian sub-continent, Indonesia, Japan and Australia. It is serious pest of green pea and lentils in northern India and also attacks other pulses in various parts of the country.

The caterpillars bore inside the green pods and feed within, generally one caterpillar is found in one pod. The damaged pod has a large emergence hole made by the pupating larva.

The moths are grey with a wing expanse of 25 mm. The moths emerge in February/March and are nocturnal. A female lays generally 50 to 180 eggs in 5 to 6 batches. The eggs are laid both singly or in clusters usually at the junction of calyx and pod or occasionally on young pods. The eggs are small, 0.5 to 0.7 mm long, elliptical and glistening white in colour. Under laboratory conditions, the eggs hatch in 5 days. The tiny greenish caterpillars bore into the pods and eat away the young grains. The full grown larvae are rosy with a purplish tinge. The larval stage is completed in 10-27 days. Pupation takes place in soil which lasts for 10-15 days.

At flower initiation, spray the crop with 0.05 % endosulfan or 0.1 % carbaryl or 0.0075 % cypermethrin. Repeat the treatment, if necessary.

Bean aphid : These aphids are distributed particularly in the tropics. Leguminaceous crops are the major host plants of this pest. Young colonies of A. craccivora concentrate on growing points of plants and are often tended by ants. This symbiotic association with ants helps in dispersal of aphids from plant to plant. Parthenogenetic reproduction is common and noticed throughout the year.

Both nymphs and adults suck the sap from the ventral surface of tender leaves, growing shoots, flower stalks and pods. The infested leaves turn pale yellow, the shoots wither, flower buds fall off whereas the pods shrivel and become deformed. Yield losses are severe when aphid colonies concentrate on the growing tips of the plants. Indirect damage is caused due to the production of honeydew, which hampers normal photosynthesis. This aphid is also known to transmit several plant viruses, leading to complete crop losses.

The aphid is ovo-viviparous, with females retaining eggs inside their bodies and giving birth to nymphs. Males are winged and females are both winged and wingless. The total life cycle varies from 10-12 days depending on the weather conditions.

Management : Cut and destroy the infested shoot mechanically as soon as aphid infestation appears. Foliar application of oxy demeton methyl (0.025%) or endosulphan (0.05%) when 5 per cent plants are infested. Repeat foliar spraying fortnightly. When predators like *Menochilus sexmaculatus* (Fabricius), *Coccinella septempunctata* Linnaeus and *Ischiodan scutellaris* Fabricius are present in sufficient number in the field, then do not spray in the field.

3. Onion and Garlic : Onion is generally used as a supplement with various other vegetables. It is potentially a biennial crop producing large bulbs and hollow leaves during the first year and it can flower and produce seeds only if kept for the next year. Onion is rich in carbohydrates & minerals and also contains proteins as well as vitamin C. Garlic can be grown year-round in mild climates. Garlic is widely used around the world for its pungent flavor, as a seasoning or condiment. When crushed, garlic yields allicin, a powerful antibiotic and anti-fungal compound (phytoncide). It also contains alliin, ajoene, enzymes, vitamin B, minerals, and flavonoids. There are only few insect-pests of economic importance which are listed in Table 8.

Table 8 : Insect-pests of onion and garlic

Common Names	Scientific Names	Order	Family
Onion thrips	*Thrips tabaci* Lindemann	Thysanoptera	Thripidae
Onion maggot	*Delia antiqua* (Meigen)	Diptera	Anthomyiidae
Leaf eating caterpillars	*Agrotis ipsilon* (Hufnagel), *Spodoptera litura* Fabricius, *S. exigua* (Hubner)	Lepidoptera	Noctuidae
Groundnut earwig	*Euborellia annulipes* Lucas	Dermaptera	Forficulidae

Onion thrips : Onion thrips has now spread to all parts of world and has become cosmopolitan in distribution. It is highly polyphagous with a wide range of host plants in India. Besides onion and garlic, it attacks cole crops, cotton, pea, cucurbits, tobacco, tomato, turnip, pine apple and ornamentals like carnation, lilies and roses, *etc.*

The nymphs and adults feed by lacerating the tissues and imbibing the oozing cell sap. On onion and garlic, they are usually congregated at the base of leaf or in the flowers. Infested onion develops a spotted appearance on the leaves; subsequently turning into pale white blotches. In case of severe attack, leaves dry from tip to downward. Development of onion or garlic bulb is affected to a greater extent. Thrips may also serve as vectors of some viruses and other plant diseases, especially the fungus, purple blotch (*Alternaria porri*).

The pest is active throughout the year and breeds on onion and garlic from November to May. In October, it is found on cabbage and cauliflower. The adults are slender, yellowish-brown and measure about 1 mm in length. The males are wingless, whereas the females have long, narrow strap-like wings, which are furnished with long hair along the hind margins. Eggs are laid only in the tender leaves. A single female lays 40-50 kidney-shaped eggs in slits which are made in leaf tissue with its sharp ovipositor. The eggs hatch in 4 to 9 days. The nymphs pass through four stages and are full-fed in 4-6 days, after which they descend to the ground and pupate in soil. The nymphs resemble the adults but are wingless and slightly smaller. The pupal period lasts for 3-5 days. The life cycle is completed in 11-21 days and many generations are completed in a year.

Management : Maintain plant vigour. Clean cultivation, regular hoeing and flooding of infested field will check the thrips population. Since the insects feed between leaves near the base of the plant, they are hard to reach with insecticides. Insecticides should be applied with sufficient water to ensure thorough coverage.

All varieties can tolerate populations of 25 thrips per plant. In well managed, irrigated onion crops, plants can tolerate high populations of thrips without reducing yields. Sprays need to be based on high populations but before feeding damage is readily apparent. Spray malathion (0.05%) or dimethoate (0.03%) as soon as the pest appears. A waiting period of 7 days should be observed before harvest.

Onion maggot : It is another destructive pest of onion and garlic. It is widely distributed in different countries and northern regions of India.

Small maggots burrow down into the underground portion of stem and often into the onion bulb. Each maggot carves out a small cavity, which results in rotting of the bulbs in storage. Due to burrowing action, the plant withers off. The damage predisposes the plants to soft rot. The damage caused by the pest is generally followed by the attack of *Bacillus carotovorus*, causing soft rot of onions. Larger onions may survive an attack but the injured bulbs will often rot in the field or in storage. The attacked plants become yellowish brown from tip downwards.

The flies are slender, greyish large winged, about 6 mm in body length. The eggs are white, elongate, laid near the base of the plants, in cracks in the soil. The eggs hatch in 2-7 days. Maggots are small, white and about 8 mm in length. The maggots crawl up to the plants and enter inside the leaf sheath and reach the bulbs. They feed there and become full grown in 2-3 weeks. The maggots then crawl out of the bulb and pupate in the soil. After 2-3 weeks, the adults emerge and start new generation. There are three to four generations each year depending on the weather. The first brood is always more injurious to plants.

Management : Never grow onions in the same location repeatedly, rotate with other crops. Use a minimum three year rotation and quickly compost onion debris. Garlic appears to be totally resistant. Remove and destroy infested plants. There is a predatory nematode (Neoplectana) available which is reported to give good biological control.

Spraying at fortnightly interval with 0.05 % malathion is effective. Grow resistant varieties of onion. Apply 10 kg of carbaryl 4 G or lindane 6 G or phorate 10 G per hectare to the soil followed by light irrigation.

Leaf eating caterpillars : Greasy cutworm, *Agrotis ipsilon* (Hufnagel), tobacco caterpillar, *Spodoptera littoralis,* Fabricius and lucerne caterpillar, *S. exigua* (Hubner) are sporadic pests that cause severe damage especially to the seedlings. They are polyphagous pests having a wide range of host plants.

Caterpillars are nocturnal in habit. Those of *Agrotis ipsilon* remain in soil during day time, come out at night and cut the seedlings at ground level. Caterpillars of *Spodoptera* species feed gregariously and move in swarms destroying the young seedlings and later feeding voraciously on leaves. During day time the caterpillars hide in hollow tubular leaves of onion but their presence is indicated by leaves and faecal matter.

To protect the crop from the ravages of these caterpillars, especially in nurseries, adopt clean cultivation followed by regular hoeing and flooding the fields. Soil application of chlorpyriphos is effective against *Agrotis ipsilon*. For *Spodoptera,* collect and destroy the egg masses and gregarious caterpillars; also clip the hollow tubular infested leaves of onion with the caterpillars within and destroy them. In case of severe infestation dust 5% carbaryl or 4% endosulphan or spray 0.05 % dichlorvos or endosulphan.

Another caterpillar found feeding on these crops is gram pod borer, *Helicoverpa armigera* (Hubner). Though a minor pest of onion, it has been reported causing havoc in onion crop raised for seed purpose. The caterpillars attack the umbels and feed on inflorescences, later they move downwards, cut the pedicels of flowers and feed on the stalks. When full fed, the caterpillars bore into the stalks, enter scape and pupate therein. Spray 0.05 % endosulphan if the attack is severe.

4. Potato : Off all the vegetables, potato has the maximum per capita consumption and it is an excellent source of carbohydrates. It provides more proteins per hectare per day as compared to wheat, maize and rice. In India, potato is mostly grown under sub-tropical conditions during winter season. Potato is vulnerable to the attack of pests both in fields as well as in stores. Insect-pests of major importance are given below in Table 9.

Table 9 : Insect-pests of potato

Common Names	Scientific Names	Order	Family
Potato tuber moth	*Phthorimaea operculella* (Zeller)	Lepidoptera	Gelechiidae
Jassids	*Amrasca biguttula biguttula* (Ishida)	Homoptera	Cicadellidae
Aphids	*Aphis gossypii* (Glover), *Myzus persicae* (Sulzer)	Homoptera	Aphididae
Tobacco caterpillar	*Spodoptera litura* Fabricius	Lepidoptera	Noctuidae
Cutworms	*Agrotis ipsilon* (Hufnagel) *A. segetum* (Dennis & Schiff.)	Lepidoptera	Noctuidae
Tomato fruit borer	*Helicoverpa armigera* (Hubner)	Lepidoptera	Noctuidae
Green bug	*Nezara viridula* Linnaeus	Heteroptera	Pentatomidae
Whiteflies	*Trialeurodes vaporariorum* (Westwood), *Bemisa tabaci* (Gennadius)	Homoptera	Aleyrodidae
Mealy bug	*Ferrisia virgata* (Cockerell)	Homoptera	Pseudoco-ccidae
Defoliating beetles	*Brahmina coriacea* (Hope), *Holotrichia longipennis* (Blanchard) *Holotrichia serrata* (Fab.)	Coleoptera	Scarabaeidae
Spotted leaf beetles	*Henosepilachna viginoctopunctata* (Fabricius)	Coleoptera	Coccinellidae
Wireworms	*Drasterius* spp. *Agronichis* spp. *Lacon* spp.	Coleoptera	Elateridae
Termites	*Microtermes obesi*, *Odontotermes obesus* and *Eromotermes* spp.	Isoptera	Termitidae
Red ants	*Dorylus orientalis* Westwood *D. labiatus* Shuck	Hymenoptera	Formicidae
Leaf weevil	*Myllocerus subfasciatus* Guerin-Meneville	Coleoptera	Curculionidae

Potato tuber moth : Potato tuber moth is one of the most widespread and destructive pests of potato in the field and stores in warmer climates of India. Besides potato, its larvae also feed on other solanaceous plants. Damage is particularly severe in rustic non-refrigerated stores during the summer storage periods. Field infested tubers are the primary sources of infestation in the storage. Damage by this pest is known to range from 30-70 per cent in stored potatoes (Chandel *et al.*, 2005).

The damaging stage of the pest is larva which feeds on potato foliage and attack tubers in the field before and shortly after harvesting. The foliage feeding larvae make transparent galleries in infested leaves. The tuber mining larvae bore mostly near the eyes of tubers filling the tunnels with excreta. Fungus grows in burrow and discolours it. Later the skin of potato partially dries and sinks so that the scars become very prominent; this is called sub-epidermal injury (Chandel and Chandla, 2003).

The pest is active throughout the year in plateau region and north-eastern parts of India but is most active from April to early August. In H.P., pest is active from October and over winters as full fed larva in volunteer tubers from December-March. The highest population level occurs during July in stores causing heavy damage to tubers. The adult is very small narrow-winged nocturnal moth measuring about 13 mm across the wings. A female on an average lays 150-200 eggs singly on the underside of leaves or on the exposed tubers. The eggs hatch in 4-7 days. The larvae are full fed in 15-20 days, pupate in greyish, silken dirt covered cocoons in the dead leaves or among trash etc. on the ground and emerge adults in 6-8 days.

Management : To control this pest, healthy seed tubers should be planted at 10cm depth and potato fields should be kept well-cultivated and deeply hilled during their growth. Irrigation may be done at regular intervals to avoid cracking of the soil. Irrigating the field at 60 - 80 per cent field capacity has been observed to reduce tuber infestation. It is observed that tuber infestation is more in furrow irrigation than with sprinkler irrigation.

Infestations in growing potatoes may be controlled after mid-March when worms begin to infest leaves by spraying with acephate (0.05%) or synthetic pyrethroids. The infested wilting vines should be cut and removed from the field a few days before digging and never piled over dug potatoes. Harvesting should be done before three-fourth of the foliage dries up. The tubers should not be left exposed to egg-laying moths during late-afternoon or over-night.

Aphids : Aphids can injure a potato plant directly by sap feeding and are capable of transmitting several important potato viruses. High aphid populations can have substantial direct effects on yield, but such populations are uncommon in commercial potato production. The primary concern with aphids is usually their role as vectors in transmitting viruses. The honey dew excreted by aphids is deposited on the plant where it provides a good medium for sooty moulds which interferes with normal photosynthesis.

As a result of sucking cell sap the crop remains stunted and has blighted appearance. Besides sucking cell sap, it transmits viruses and lowers the quality of tubers of seed crop.

Five aphid species are commonly found on potato crops in the India; the green peach aphid (*Myzus persicae*), cotton aphid (*Aphis gossypii*), bean aphid (*Aphis fabae*), potato root aphid (*Rhopalosiphum rufiabdominalis*), tuber aphid (*Rhopalosiphoninus latysiphon*) (Verma and Chandla, 1999). The cosmopolitan aphid, *M. persicae* is an important pest on potatoes and other crops. These species all feed on the foliage but *Rhopalosiphoninus latysiphon* feeds on the underground portions of the stem and roots.

In *M. persicae* only eggs are produced by sexual reproduction whereas all subsequent reproduction is viviparous and parthenogenetic (Harris, 1992). The aphids have both winged and wingless forms. Wingless forms are predominant on potato during most part of the year. *Myzus persicae* overwinters as eggs on a very restricted number of primary host species, often woody plants (peach etc.). Aphids moult four times, mean number of offspring's per wingless aphid ranges from 60-75 (Chandla and Verma, 2000). Optimum temperature for reproduction is around 70 F. There may be several generations on the primary host, but eventually winged adults (spring migrants) develop which fly away to colonize "secondary" often herbaceous, host plants.

Transmission of viruses by aphids : Both PLRV and PVY are transmitted by several aphid species, but the most important vector throughout India is the *M. persicae*. It is the most efficient vector of PLRV and also the most abundant species on potato. Both winged and wingless green peach aphids transmit viruses efficiently.

Management : Young plants are particularly susceptible to infection with a virus and seed plots must, therefore, be kept free of aphids where possible. The selection and rouging of plants infected with a virus must also be done under aphid free conditions. An action threshold of 20 aphids/100 compound leaves has been strictly used in seed production (Chandla *et al.*, 2004). The most vulnerable period in the aphid life cycle is passed on peach. The application of chemical defoliant to peach trees in fall, to deny foliage to fall migrants and pruning of trees to remove most over wintering eggs is useful.

Many insecticides repel aphids but are not lethal to them and this tends to promote plant to plant movement and increased flight activity. Pyrethroid insecticides are generally not effective against aphids and their use may be avoided even for control of other insects, in potato. Both foliar sprays of dimethoate 0.03% or oxy demeton methyl 0.025% and phorate 10G @ 20 kg/ha applied in-furrows at planting are widely recommended in seed potato for aphid control (Chandla *et al.*, 2004). Imidacloprid and thiamethoxam also provide effective protection for about two weeks.

White grubs : Apart from potato, the grubs are reported to damage crops like paddy, finger millet, French bean, black gram, chilli, tomato, brinjal and okra. In Himachal Pradesh, the extent of tuber damage varies from 40-50 per cent (Chandel *et al.*, 2003).The adult beetles feed on foliage of apple, pear, plum, apricot, cherry, almond, walnut, ber and chest nut. White grubs are most destructive and troublesome soil insects, threatening potato production in hilly states. These white grubs are present in the soil at a depth of 5 -20 cm during the crop season.

Large holes are made in the tubers which ultimately may be entirely transversed by wide deep mines. They live concealed, feed on the roots or decaying vegetative matter and suddenly increase their population in places having enough food and least disturbance of soil. This is a polyphagous pest both in grub and adult stage and inflicts heavy damage on various fruits / forest trees, their nurseries, vegetables, lawns and field crops.

In India, 20 species of white grubs have been reported on potato. Out of these *Brahmina coriacea* (Hope) and *Holotrichia longipennis* (Blanchard) are most destructive in north western hills. *Holotrichia serrata* (Fab.) damages potato in Karnataka (Misra and Chandel, 2003).

White grubs have always existed in nature feeding on roots of both weeds and crops. However, there are large number of factors which collectively led to sudden increase in their population and attainment of a key status in different areas. These are larval stages of beetles which damage the potato tubers by making small to large and shallow to deep circular holes on them in the hilly areas where the crop is grown during summer as rainfed crop. The white grubs feed on the roots of a wide variety of cultivated as well as wild plants. The young grubs after hatching move to the root and start feeding on them. The length of life cycle of these beetles depends upon the climate and species. Most species complete a life cycle in one year while some require 2-3 years.

Management : As a rule, these beetles will not deposit eggs in fields of clovers unless there is a considerable mixture of grasses or weeds in such fields. Thus, grow clovers or alfalfa if possible to prevent the beetles from laying eggs. One of the best ways to clean grubs out of field is to pasture the land with pigs. When pigs are allowed to run on heavily infested land they will usually root out and eat the grubs nearly making the land free of grubs.

Ploughing infested fields especially when most of the grubs are pupating, kills many of the pupae and newly formed adults. The beetles can be collected at night and killed in kerosenized water. The collection should be done on community basis. The FYM used should be well rotten. Soon after the attack is noticed, methyl parathion (0.05%) or carbaryl (0.1%) or acephate (0.05%) can be sprayed. In potato, their damage can be minimized by application of phorate at the time of earthing up. Chlorpyriphos is also equally effective. To obtain best results, insecticide application should be properly timed. Hence, apply insecticides soon after emergence of beetles coinciding with their egg laying or egg hatching.

Cut worms : Cut worms, (*Agrotis* spp.), are polyphagous insects of cosmopolitan distribution. In India, *Agrotis ipsilon* (Hufnagel), *A. segetum* (Schiff.), *A. flammatra* Schiff., *A. interacta* Walker, and *A. spinifera* Hubner occur on potato. But *A. segetum* and *A. ipsilon* are more serious. The former is common in the hills and latter in the plains.

The cut worms eat off the plants just above or at a short distance below the surface of soil. Most of the plant is not consumed, merely being eaten enough to cause it to fall over. Consequently these caterpillars have great capacity for causing damage. Tuber damage is manifested in the form of deep irregular holes in the flesh which reduce tuber quality and may allow secondary pathogens and pests to invade and cause further damage. The holes can look like those caused by slugs, but slugs are usually problem in wet, heavy soils. Cut worm damage is usually significant only in non irrigated crop on the lighter soils in hot and dry summers.

Surface cut worm, *Agrotis spinifera* occurs in Punjab, Bihar, Andhra Pradesh and Karnataka. Greasy cut worm, *Agrotis ipsilon* is generally a cool climate pest. In plains it is active from October onwards and with the onset of summer, it migrates to hilly regions. In India, it is more serious in northern region than in the south. Black cut worm, *Agrotis segetum* is also a pest of cool climate in the hills. Gram cut worm, *Agrotis flammatra* is distributed in Punjab and sub Himalayan Region. The pest in active from October - April in the plains and migrates to mountains in summer. *Agrotis interacta* moths are more or less similar in appearance and size as *A. segetum*. It is exclusively subterranean and feed generally on roots and tubers by chewing inside cavities.

In plains, *Agrotis ipsilon* is active from October onwards and with the onset of summer, it migrates to hilly regions. In India, it is more serious in northern region than in the south. The female starts laying eggs 4 - 6 days after emergence and lays 649 -1711 eggs in 4 - 11 days. Eggs are laid during night at 2100 hrs either singly or in batches of 7 - 42 eggs. Eggs are laid on ventral surface of leaves or moist soil. Freshly ploughed fields are preferred for oviposition (Butani and Jotwani, 1984). Egg hatches in 3 -5 days. The caterpillars are light brown with reddish tinge which turns greenish thereafter. Full fed larvae are, 40 - 50mm long. These are greasy to touch that why are often called as greasy cut worms. Larval period is 22 - 30 days. Occasionally the caterpillars may also nibble tubers. Pupation takes place in soil and lasts for 12 - 15 days during March. Life cycle is completed in 39 - 53 days. Moths are medium sized, stout, dark greenish brown with reddish tinge and have greyish - brown wavy lines and spots on forewings; hind wings are hyaline having dark terminal fringe.

Management : Cut worms are subject to attack by other predators. Thus, fork the soil during period of attack so that the larvae are exposed and readily fed by their avian predators. Efficient control of cut worms can only be achieved by properly applied sprays when the young caterpillars are still on the haulms and therefore vulnerable. Once below ground, cut worms are unlikely to be significantly affected by the insecticides applied to the soil or to foliage. Older caterpillars are generally less susceptible to insecticides than young caterpillars. Young caterpillars can be killed by rain or irrigation and an insecticidal spray would probably be unnecessary in these conditions (Hancock, 1990).

Treatments should be applied when the soil is dry and the weather is warm. Good control of cutworms depend on a thorough coverage of the foliage with a high volume application, preferably using at least 1000L / ha of water. Chemicals like chlorpyriphos 0.04% cypermethrin 0.01% and triazophos 0.04% can be used for the control of cut worms on potatoes.

Leaf hoppers : Leaf hoppers are strong flies and are much more mobile than aphids. Unlike aphids, leaf hoppers are important mainly because of the direct feeding damage they cause. The potato leaf hopper

(*Empoasca devastans*) is the most important species and has long been recognized as a major pest of potato. Several other species are important on potato in certain regions. These include *Amrasca biguttula biguttula, Alebroides nigroscutulatus, Seriana equate, Empoasca solanifolia, E. fabae* and *E. punjabensis*.

Prolonged feeding by the potato leaf hopper causes a condition known as "hopper burn" manifested in the form of brown triangular lesion at the tip of the leaf. Both adults and nymphs are injurious, but late instar nymphs can reduce yields more than twice as much as an equal number of adults. Damage results from disruption of phloem (Trivedi and Rajagopal, 1999). Toxins in the saliva of potato leaf hopper induce swelling of cells, which eventually crushes the phloem. There is depletion of plant reserves due to increase in plant respiration subjected to hopper attack. Infestations are most damaging during early tuber bulking.

The leaf hoppers have broad host range. Leaf hoppers pass through three life stages; egg, nymph and adults. On potato, they usually complete 2-4 generations in a year. The population density is dependent upon the date of arrival on the crop and the temperature. Over wintering takes place in adult stage, often at great distances away from the potato crop. In summer, potato leaf hopper adults live for 30-40 days, and some over 90 days, with the result that generations overlap. The females lay an average of 2-3 eggs/day and can lay as many as 200 eggs in her life time. In summers, eggs hatch in about 10 days. The nymphs are yellowish green when young but closely resemble the adult leaf hopper when older. They require about 12 days developing into adults. New adults begin laying eggs when they are 6 days old.

Transmission of diseases : Empoasca spp. is not known to transmit any potato pathogens. However, some species of leaf hoppers are important to potato health because of their role as vectors (Misra, 1995). Potato yellow dwarf virus (PYDV) and beet curly top virus (BCTV) are transmitted by leaf hoppers. Purple top in potato which is caused by aster yellows mycoplasma like organisms (AY-MLO) is also transmitted by leaf hoppers.

Management : Early maturing cultivars are generally more susceptible, but these cultivars bulk more rapidly and their yield may actually be affected less. It is more important to control leaf hoppers under stress because potato is more susceptible to leaf hopper injury under these conditions. In seed crop, systemic insecticides applied in furrows at planting or side dressed at plant emergence give 6-8 weeks of control. On fresh market potatoes, standard practice is to apply foliar sprays. Dimethoate @ 0.03 % and oxy demeton methyl @ 0.025 %, applied on the appearance of the pest, and phorate 10 G @ 20 kg/ha applied at planting are recommended for the control of leaf hoppers.

5. Sweet Potato : Sweet potato is grown all over tropics and sub-tropics. This plant is a herbaceous perennial vine. It is cultivated as annual. The edible tuberous root is long and tapered, with a smooth skin whose color ranges between red, purple, brown and white. The tubers contain 16 per cent starch and 4 per cent sugar. The crop is attacked by a number of pests both in field and in storage, the improitant ones are listed below in the Table 10.

Table 10 : Insect-pests infesting sweet potato

Common Names	Scientific Names	Order	Family
Sweet potato weevil	*Cylas formicarius* (Fabricius)	Coleoptera	Curculionidae
Leaf-eating caterpillars	*Theretra oldenlandiae* Fabricius	Lepidoptera	Sphingidae
Bihar hairy caterpillar	*Spilosoma obliqua* (Walker)	Lepidoptera	Arctiidae
Leaf feeding beetles	*Aspidomorpha miliaris* (Fabricius), *A. furcata* Thunberg, *A. indica*, *A.sanctaecrucis* (Fabricius), *Cassida indicola* Duvivier, *Chiridia sexnotata* (Fabricius), *Glyphocassis trilineata* (Hope), *Metriona circumdata* (Herbest) and *M. varians* (Herbest)	Coleoptera	Chrysomelidae
Sap sucking bug	*Riptortus linearis* Fabricius	Heteroptera	Coreidae
Sap sucking bug	*Halticus minutus* Reuter	Heteroptera	Miridae
Defoliators	*Brachmia arotraea* (Meyrick) and *B. convolvuli* Walsingham	Lepidoptera	Gelechiidae
	Cretonia vegata (Swinh.)	Lepidoptera	Noctuidae
Root mealy bug	*Geococcus coffeae* (Green)	Homoptera	Pseudo-coccidae
Leaf miner	*Acrocercops prosacta* Meyrick	Lepidoptera	Gracillariidae
Flower and fruit feeder	*Oxycetonia versicolor* F.	Coleoptera	Scarabaeidae
	Mylabris thunbergi Billberg	Coleoptera	Meloidae

Sweet potato weevil : It is a specific pest of sweet potato, extensively pan tropical in distribution. Grubs bore into stems, cause tunneling inside and feed on soft tissues. Grubs as well as adults bore into tubers both in field and in godowns. The affected tubers develop dark patches which may later on start rotting. Occasionally adults feed on stems and leaves. Loss of tubers to the extent of 60-70 per cent has been reported by Nair (1975). The pest is disseminated from field to field through infested vines and is carried over from season to season by breeding in damaged tubers left in the fields after harvest.

Grubs are fattish, about 8-10mm long, legless, pale yellowish-whitish in colour. Adult weevils are ant-like due to which they have been given specific name formicarius and are 6 to 8 mm long, slender bodied having elongated snout like bluish brown head with non-geniculate antennae, bright red thorax and legs and brownish-red abdomen. Adults are active and fly long distances during night in search of food. Females make small cavities on the tubers or stems and lay eggs singly in these cavities. Each female lays 100 to 200 eggs in 50 to 100 days. Incubation, grub and pupal stages last for 5 to 10, 16 to 20 and 4 to 8 days, respectively. Pupation takes place in larval burrows from where the freshly emerged weevils cut their way out. Life cycle is completed in 4 to 5 weeks.

Care should be taken that no infested or rotting tuber is left in the fields. Do not use the seed material from infested areas and plant only healthy cuttings. Spraying 0.2% carbaryl followed by 0.15 % malathion at 10 days interval is quite effective in controlling weevils. In godowns, treat the outside of the bags containing the tubers with 5 % malathion or carbaryl dust.

Leaf-eating caterpillars : Sphinx caterpillars, *Agrius convolvuli* (Linnaeus) and *Theretra oldenlandiae* Fabricius have been reported damaging sweet potato vines. The former is a polyphagous pest that attacks a number of crops including fruit trees, some legumes, arum and sweet potato. These are widely distributed in the old world and have been reported from Europe, Africa, Iran, Indian subcontinent, Southeast Asia, South China, Australia and New Zealand.

The caterpillars feed voraciously on leaves often defoliating the vines completely. The pest is active during monsoon season when moths can be seen hovering around light at night. The female lays conspicuous seed like shiny eggs, singly on the tender parts of plants. Eggs of Agrius convolvuli are spherical in shape and about one mm in diameter. Full grown caterpillars are robust, about 80-100 mm long, dark brown in colour with radish patches on sides and conspicuous curved horn-like process at the anal end. Pupae are reddish-brown in colour. Moths are stout, pale grey in colour having a well developed proboscis and frenulum and pale grey wings with transverse violet bands on abdomen.

Bihar hairy caterpillar, *Spilosoma obliqua* (Walker) (Arctidae) is a serious pest of jute in eastern India and Bangladesh. It is highly polyphagous pest having a wide range of host plants. It has been reported to damage fruit trees, tobacco, pulses and vegetables including cabbage, cauliflower, cowpea, lettuce, potato, soybean and sweet potato. Eggs are laid in clusters on the ventral side of leaf in parallel rows on 3 to 4 consecutive nights. A single female may lay 400 to 1000 eggs in its life time. On hatching the caterpillars feed voraciously on leaves, skeletonising the same. The pupation takes place in soil. Full grown caterpillars are stout 25 to 40 mm long and have 7 broad orange coloured transverse bands with tufts of yellow hair, dark at both ends. Moths have crimson coloured body with black dots; antennae are black in colour, pectinate in males and filiform in females. There may as many as 8 generations in a year.

Management : After harvesting the crop, give deep ploughings and if possible flood the infested fields to kill the pupae present in the soil and thereby prevent the carry over of the pest. Collect and destroy the egg clusters and leaves bearing the caterpillars to prevent the population build up. Dusting with 5 % malathion dust is quite effective against young caterpillars during early stages of crop growth. Spraying 0.05 % dichlorvos or endosulphan can effectively control the wide spread infestation of caterpillars at advanced stage.

Leaf-feeding beetles : These include *Aspidomorpha miliaris* (Fabricius), *A. furcata* Thunberg, *A. indica, A.sanctaecrucis* (Fabricius), *Cassida indicola* Duvivier, *Chiridia sexnotata* (Fabricius), *Glyphocassis trilineata* (Hope), *Metriona circumdata* (Herbest) and *M. varians* (Herbest). All are specific pests of sweet potato; *A. miliaris* and *M. circumdata* are relatively more common. A. miliaris is comparatively more destructive. These beetles are more active during monsoons. Eggs are laid on ventral leaf surface. Grubs are nocturnal in habit and feed on epidermal tissues at leaves, skeletonising them. Later, the grubs as well as the adults gnaw holes in the leaf lamina. Pupation takes place in ventral surface of leaves.

If and when the high infestation of these beetles occurs, spray the crop with 0.2 % carbaryl. In nature, *A. miliaris* is parasitized by *Cassidocida aspidomorphae* Crawford and *Tetrastichus colemani* Crawford.

6. Root crops (radish, carrot and turnip) : Radish, carrot and turnip are the main root crops grown commercially in India. Radish greens are rich in vitamin A, B and C with small amount of iron and proteins as well. Carrot roots are exceptionally rich in iron and are used as vegetable. Carrot juice is a rich source of carotene which is used for colouring butter and other food articles. Turnip greens are excellent source of vitamin A, B and C and also have calcium in appreciable quantities. The insect-pests reported to cause economic losses have been listed in Table 11.

Table 11: Insect-pests infesting radish and turnip

Common Names	Scientific Names	Order	Family
Mustard Aphid	*Lipaphis erysimi* Kalt.	Homoptera	Aphididae
Whitefly	*Bemisia tabaci* (Gennadius)	Homoptera	Aleyrodidae
Painted bug	*Bagrada cruciferarum* Kirkaldy	Heteroptera	Pentatomidae
Thrips	*Thrips tabaci* Lindemann	Thysanoptera	Thripidae
Leaf miner	*Chromatomyia horticola* Goureau	Diptera	Agromyzidae
Mustard saw fly	*Athalia lugens proxima* (Klug.)	Hymenoptera	Tenthredinidae
Diamondback moth	*Plutella xylostella* (Linnaeus)	Lepidoptera	Plutellidae
Cabbage looper	*Hellula undalis* (Fabricius)	Lepidoptera	Pyralidae
Cutworms	*Agrotis* spp.	Lepidoptera	Noctuidae
Tobacco caterpillar	*Spodoptera litura* (Fabricius)	Lepidoptera	Noctuidae
Bihar hairy caterpillar	*Spilosoma oblique* Walker	Lepidoptera	Noctuidae
Greater moth	*Crocidolomia binotallis* Zeller	Lepidoptera	Pyralidae
Flea beetles	*Phyllotreta cruciferae* Goeze, *P. chotanica* Duvier	Coleoptera	Chrysomelidae

Aphids : Aphids feed on radish foliage include cabbage aphid, *Brevicoryne brassicae* mustard aphid, *Lipaphis erysimi*, peach green aphid, *Myzus persicae* and *Toxoptera aurantii* (Boyer de Fonsco.) The first one prefers cabbage and cauliflower; second one is a serious pest of crucifers. *M. persicae* and *T. aurantii* are highly polyphagous pests having a wide range of host plants.

The colonies of aphids consisting of various stages of nymphs and adults suck the cell sap from tender stems and underside of leaves. The affected plant part fades, curl and dry up. The damage caused by sucking the sap from pods adversely affects the seed quality. In addition, the sooty mold which develops on the honeydew secreted by the aphids interferes with the normal photosynthesis of the plants. Remove and destroy the affected plant parts with aphids thereon.

If more than 10 per cent of the plants have aphid incidence on the central shoot in case of seed crop, spray the crop with oxy demeton methyl (0.025%) or dimethoate (0.03%). If the pods of radish are used as vegetable purpose, then spray the crop with malathion (0.05%).

Mustard sawfly : It is an oligophagous pest attacking various winter cruciferous vegetables. The pest appears on radish leaves by the end of July and the activity keeps on increasing and maximum in during September to December. The larva is greenish-black and feeds on leaves. The damage is more pronounced on seedlings as compared to grown up crop. This insect has a high degree of gustatory preference for turnip crop. Collect the larvae at dusk and dawn, when these are not active and destroy them. Apply malathion (0.05%) or endosulphan (0.05%) or carbaryl (0.1%).

Flea beetles : The flea beetles are regular pests of crucifers. The adult beetles feed on the foliage by making holes. The damage is more pronounced at seedling stage. Cover the germinating seedlings with nylon nets. Apply malathion (0.05%) or endosulphan (0.05%) or carbaryl (0.1%).

7. Spinach : Spinach is grown for production of edible leaves which provides necessary roughages in the diet. The leaves are chief source of vitamin A, B1 and C, and excellent source of iron and also contain proteins and carbohydrates. Spinach is comparatively attacked by blue beetle and different species of aphids which are described below:

Blue beetle, *Altica caerulescens* (Baly) (Chrysomelidae:Coleoptera): Blue beetle has been reported as a pest of cabbage and spinach. Besides it it has been recorded as a minor pest on strawberry and plums. The grubs feed on tender cotyledon leaves as well as the fleshy older ones. Adults nibble the leaf margins causing very little damage.

On hatching, the freshly emerged grubs scrap and feed on chlorophyll containing tissues, later they mine inside the leaves, feed on mesophyll tissues and pupate therein. Eggs are laid in soil. Grubs are 8-10 mm long, dark brown in colour. Pupae are 12-15 mm long and brown in colour when freshly formed turning blackish-brown. Adults are medium-sized, 5-7 mm long, steel-blue in colour with hind femora strongly thickened which enable the beetles in jumping movements.

To control these beetles, dust with 4 % carbaryl or spray with 0.1 % carbaryl or 0.05% endosulfan or malathion. Observe a waiting period of 7 days for consumption of leaves.

Aphids : *Lipaphis erysimi* (Kaltenbach), *Myzus persicae* (Sulzer) and *Hyadaphis indobrassicae* (Das) have been found infesting leaves and causing damage to the crop; the last one being more common. These aphids are polyphagous in habit. The damage caused by these aphids by sucking the plant sap results in yellowing of leaves and the infested leaves become unfit for consumption.

The adults of *H. indobrassicae* are dull green in colour with two median rows of shining spots on dorsal side; antennae are dark coloured except at the base. The cornicles are longer than cauda and are dark tipped. The pest appears in October and the activity continues till end of March.

To check the aphid infestation, cut and destroy all the infested leaves alongwith the colonies of aphids during the early stage of attack. Any one of the insecticides suggested for the control of blue beetle can be used for these aphids also.

8. Drumsticks (*Moringa oleifera* Lamarck) : Drumsticks grows wild along entire sub-Himalayan range but it is cultivated in the plains of north, central and south India. Generally tender pods are used in culinary preparations especially curries. All the plant parts have some medicinal value. The destrucitve insect-pests of drumsticks are enlisted in Table 12.

Table 12 : Insect-pests of drumstick

Common Names	Scientific Names	Order	Family
Leaf eating caterpillars	*Eupterote mollifera* Walker, *E. geminata* Walker	Lepidoptera	Bombycidae
	Pericallia ricini (Fabricius)	Lepidoptera	Arctiidae
	Noorda blitealis Walker, *Actias selene* Hubner, *Metanastria hyrtaca* (Cramer)	Lepidoptera	Pyralidae
	Streblote siva (Lefevre)	Lepidoptera	Lasiocampidae
Bark caterpillar	*Indarbela tetraonis* (Moore)	Lepidoptera	Metarbelidae
Pod fly	*Giotna* sp.	Diptera	Drosophilidae
Stem borer	*Batocera rubus*	Coleoptera	Cerambycidae
Scales	*Ceroplastodes cajani*	Homoptera	Diaspidae
Aphids	*Aphis gossypii*	Homoptera	Aphididae
Bud midge	*Stictodiplosis moringae*	Diptera	Ceciodmyiidae

Leaf eating caterpillars : *Eupterote mollifera* Walker is the most destructive and specific pest of drumstick trees. The caterpillars feed gregariously by scrapping the bark and gnawing the foliage. A severe infestation may result in complete defoliation of the tree.

Eggs are laid in clusters on leaves and tender stems. The caterpillars when full grown measure 40 to 50 mm in length, these are brownish in colour and densely hairy. Adults are large-sized moths uniformly light yellowish-brown in colour. Wing expanse is 60-70 mm and 75-85 mm in case of males and females, respectively. Moths appear with the onset of monsoon and lay eggs which hatch in 6 days. The larval period lasts for 12-14 days and pupal period for 8-10 weeks. There is only one generation in a year.

Another species *E. geminata* Walker has also been recorded damaging drumstick trees in Tamil Nadu. The damage symptoms are same as those of *E.mollifera.*

Pericallia ricini (Fabricius) popularly known as black hairy caterpillar is a foliage pest, which occasionally appears in large numbers. Besides drumstick trees, the other host plants include banana, cotton, cucurbits, castor, cowpea, black gram, soybean, tea, yam etc. The caterpillars feed on leaf lamina, initially by scrapping epidermal layers and later by cutting the blades.

Noorda blitealis Walker is a sporadically serious pest of drumstick trees especially in South India.Eggs are laid in batches usually on ventral surface of leaves. Caterpillars feed on leaf lamina and pupate in soil. Peak period of infestation occurs during March April and December-January when even a medium infestation may defoliate the entire tree.

Actias selene Hubner, popularly known as moon moth, is widely distributed all over India and is found in almost all the areas growing temperate fruit trees. Caterpillars of this pest are voracious feeders and defoliate the trees in no time. Full grown caterpillars are smooth, apple-green in colour and 100 to 150 mm long. Pupation takes place in dense, firm silken cocoons. Adults are big, pinkish moths.

The other caterpillars include *Metanastria hyrtaca* (Cramer) and *Streblote siva* (Lefevre) which are also polyphagous in nature. If and when a serious infestation is observed, check the damage of these lepidopterous pests by spraying the trees with carbaryl 0.2% or malathion 0.1% or endosulfan 0.05%.

Trunk borers : *Indarbela tetraonis* (Moore) commonly known as bark eating caterpillar has a number of host plants including drumstick trees of which it is a minor pest, except for South India where it often causes sever damage.

The caterpillars feed superficially below the bark making zigzag galleries, later they bore inside the burrows during day but often come out at night and feed on bark. The conspicuous attack symptom is the presence of huge silken webbed masses comprising of chewed wooden particles and excreta of the larvae hanging loosely on the tree trunks covering the entry holes.

To kill these trunk and stem boring caterpillars and beetles, clean the affected portion of the tree by removing all the webbed material, excreta etc. and insert in each hole cotton-wool soaked in any fumigant like carbon tetrachloride, chloroform or even petrol and seal the treated hole with mud.

9. Elephant's Foot (*Amorphophallus companulatus* Blume) or Yams: It is found through out the plains of Indiann sub continent both growing wild as well as cultivated.The corms are either cooked and taken as vegetable or used for making pickles. Corms are generally are not attacked by insects and the foliage feeders are listed in Table 13.

Table 13 : Insect-pests infesting Yams

Common Names	Scientific Names	Order	Family
Spotted beetles	*Galerucida bicolor* Hope, *Crioceris impressa* Fb., *Lema lacordairei* Baly.	Coleoptera	Chrysomelidae
Leaf eating caterpillars	*Pericallia ricini* Fabricius, *Spodoptera litura* Fabricius, *Rhycholoba acteus* (Cramer), *Hippotion celerio* (Linnaeus) and *Theretra gnoma* (Fabricius)	Lepidoptera	Noctuidae
Aphid	*Aphis gossypii, Pentalonia nigronervosa* Coq., *P. galadii D., A. durranti* Das. *A. punicae* Passerini, *Hydronaphid colocassiae* Ray	Homoptera	Aphididae
Scale insect	*Aspidiella hartii* (Ckll.)		
Mealy bugs	*Ferrisia virgata* (Ckll.)	Homoptera	Coccidae
Thrips	*Caliothrips indicus* Bagnall., *Helinothrips kadaliphilus* (Ram. and Mar.)	Thysanoptera	Thripidae

Contd.

White grubs	*Leucopholis coneaophora* Burm	Coleoptera	Scarabaeidae
Sawfly	*Senocolidia dioscreae* R.	Hymenoptera	Tenthredinidae
Whitefly	*Bemisia tabaci* Genn.	Homoptera	Aleyrodidae
Tingids	*Stephantis typicus* D.	Hemiptera	Tingidae
Jassids	*Empoasca* sp, *Hecalus* sp.	Hemiptera	Cicadellidae
Termites	*Odontotermes obesus* (Rambur)	Isoptera	Termitidae

Leaf eating beetle : It is a pest of regular occurrence in South India. The grubs feed gregariously on leaves and after skeletonising the leaves they gnaw into the petioles; adults feed by nibbling the leaves and cause much less damage than the grubs. Adult is stout 7-9 mm, fulvous beetle. Female lays 250-300 eggs in groups of 3 to 5 on leaves or stem. The grubs emerge out in 3 to 4 days and pass through three instars. It pupates inside the soil and the adults come out after 10-13 days. Spraying with 0.05 % endosulfan or 0.2% carbaryl can effectively check this beetle infestation.

Leaf eating caterpillars : *Pericallia ricini* Fabricius- a hairy caterpillar-is a polyphagous pest which in addition to elephant's foot has also been reported damaging brinjal, colocasia, drumsticks, sweet potato and yam. Tobacco caterpillar, *Spodoptera litura* Fabricius is yet another polyphagous pest of this crop that has been recorded on a large number of other crops of economic importance. Both these caterpillars occur sporadically and can cause considerably damage to leaves. Besides these two pests, caterpillars of several species of sphingid moths including *Rhycholoba acteus* (Cramer), *Hippotion celerio* (Linnaeus) and *Theretra gnoma* (Fabricius) have been found feeding on leaves of elephant's foot but the damage caused is of minor importance.

Eggs of *Rhycholoba acteus* are smooth, shining, broadly ovoid in shape, 1.5 to 2.0 mm long and bright green in colour. Full grown caterpillars are stout, 68 to 76 mm long and have two different colour patterns-green and brown. Horn is prominent and 6 to 8 mm long stout at the base, orange in colour and covered with minute tubercles. Pupae are 46 to 54 mm long having head and thoracic region brown dorsally and pinkish ventrally. Adults have purplish grey head; thorax and abdomen with a dorso-lateral green strip from vertex of head to anal end. Fore wings are purplish-grey with a green oblique central area from just below the apex to inner margin. Hind wings are fuscous in colour with an anal patch and submarginal band ochraceous. Wing expanse is 64 to 76 and 70 to 80 mm in case of males and females, respectively.

Hippotion celero resembles *R. acteus* but is slightly smaller in size. Full grown caterpillars are 56 to 64 mm long with a small, 5 mm long, straight dark brown or blackish horn. Adults have head and thorax brown with white lateral stripe, abdomen is brown with white dorso-lateral spot on each segment.

The eggs of *Theretra gnoma* are broadly-ovoid in shape, about 2 mm long and shining pale green in colour. Full grown caterpillars are 80 to 85 mm long, having glaucous-green head and yellowish-green body speckled with dark green and narrow broken dark green dorsal stripe. Horn is smooth, stout and about 8 mm long. Pupae are 58 to 62 mm long and soiled-brown in colour. Adults have greenish brown head and thorax with a white lateral stripe. Abdomen is brown with a black dorsal patch. Wing expanse is 80 to 96 and 84 to 109 mm in case of males and females, respectively.

Generally no chemical control measures are adopted against these caterpillars. Hand picking and mechanical destruction of the larvae in the initial stage of attack can effectively check the pest population. If necessary, spraying with 0.2 % carbaryl or 0.05 % endosulfan can successfully control the population of these caterpillars.

10. Cassava : Cassava is a shrubby, tropical, perennial plant that is not well known in the temperate zone, most commonly associated with tapioca. The edible parts are the tuberous root and leaves. Cassava's starchy roots produce more food energy per unit of land than any other staple crop. Its leaves, commonly eaten as a vegetable in parts of Asia and Africa, provide vitamins and protein. Nutritionally, the cassava is comparable to potatoes, except that it has twice the fiber content and a higher level of potassium. Almost all parts of cassava plants during various stages of growth are infested by insect-pests which are given in Table 14.

Table 14 : Insect-pests of cassava

Common Names	Scientific Names	Order	Family
Soft scale	*Aonidomytilus albus* Ckll.	Homoptera	Diaspidae
Mealy bug	*Pseudococcus filamentosus* Ckll.	Homoptera	Coccidae
Cockchafer grub	*Leucopholis coneophora* Burm.	Coleoptera	Melolonthidae
Helmet scale	*Saissetia coffeae* Wlk.	Homoptera	Coccidae
Stem borer	*Pterolophia melanura* Pasc. Sybra praeusta Pasc.	Coleoptera	Cerambycidae
Termites	*Odontotermes* spp.	Isoptera	Termitidae
Aphid	*Rhopalosiphum esculentum* sp.n.	Homoptera	Aphididae
Whitefly	*Bemisia tabaci* (Genn.)	Homoptera	Aleyrodidae
Grey weevil	*Myllocerus curvicornis* F.	Coleaoptera	Curculionidae
Hairy caterpillars	*Euproctis fraterna* (Moore)	Lepidoptera	Lymantridae
Hairy caterpillar	*Euchromia polymena* Linn.	Lepidoptera	Amatidae
Ground beetle	*Gonocephalum* spp.	Coleaoptera	Tenebrionidae

Soft scale : The soft scale is prevalent throughout southern India. Though the incidence is sporadic and localized under field condition, it is very severe in storage throughout Kerala when the stems are stored after harvest as planting material.

The infestation spreads in the field through the contaminated planting material and the infested setts do not sprout properly. When sprouted, the main twig breaks and side shoots appear in the form of a bunch. The attacked plants become weak, the stems get dried up and tubers become unpalatable. The infestation occurs in dry situation and aggravates during prolonged moisture stress situations.

Life cycle is short and completed in 20-25 days. The female is mussel-shaped and covered with a white waxy secretion. The eggs are deposited between the upper scale covering and lower cottony secretion. After oviposition, the female shrinks and shrivels up. Eggs hatch in four days, the first instar nymphs crawl and disperse and later become sedentary. The spread occurs by wind or by crawling of immature stages and through infested planting materials.

For effective control, only scale free stems should be selected for storing and planting. Dipping the setts in dimethoate or oxy demeton methyl (0.05%) solution for 10 minutes has been found effective.

Mealy bug : It is also sporadic pest and sometimes becomes serious. The nymphs and adults colonize in large numbers on the leaves, petioles and stems. The honey dew secretion of the mealy bugs causes sooty mould. The infested stems dry up soon. As the infested stems can be easily seen, it is easy to detect them and destroy them. Spraying the stems with dimethoate or oxy demeton methyl (0.05%) is effective.

Cockchafer grub : It is a specific pest of coconut, but has been found attacking cassava and other tuber crops that are inter-cultivated with coconut. The pest infestation is severe in light sandy soils receiving relatively high rainfall. The female lays eggs during June-July in soil to a depth of 8-15 cm very close to the roots of grasses. The eggs hatch in about 20 days. The grubs that emerge out at first feed on the roots of grasses and organic matter, and later from second instar onwards start feeding on the roots of cassava, coconut and other tuber crops. The roots of cassava seedlings and the bark are eaten voraciously by the grubs. The attacked plants become pale, wilt and finally dry up. By October-November, the larval period is completed and the larvae go deep into the soil to pupate in earthen cocoons. The final emergence takes place during June in monsoon season. The control measures suggested in potato will also be applicable in this case.

11. Beans (French bean, Indian bean, Winged bean, Cluster bean) : French bean is an important crop whose tender pods are used as fresh vegetable while dry leaves are used as pulses. There are several species of cultivated beans but those commonly grown for vegetable purposes are French bean, Indian bean, Goa bean and Cluster bean. The pests recorded on beans are listed in Table 15.

Table 15 : Insect-pests infesting beans

Common Names	Scientific Names	Order	Family
Black bean bug	*Chauliops nigrescens* Distant, *C. fallax* Scott	Heteroptera	Lygaeidae
Blister beetle	*Mylabris pustulata* Thunberg, *M. phalerata* Pallas, *M. mecilenta* Marshal	Coleoptera	Meloidae
Whitefly	*Bemisia tabaci* Genn., *Trialeurodes vaporariorum* (Westwood)	Homoptera	Aleyrodidae
Bean stem fly	*Ophiomyia phaseoli* (Tryon)	Diptera	Agromyzidae

Contd.

Bean aphid	*Aphis craccivora* Koch.	Homoptera	Aphididae
Aphids	*Aphis fabae* Scopoli, *A. gossypii* Glover, *A. rumicis* Linnaeus, *A. adusta* Zehntner, *Acyrthosiphon pisum* (Harris), *Macrosiphum euphorbiae* (Thomas)	Homoptera	Aphididae
Pod borers	*Helicoverpa armigera* (Hubner), *Maruca testulalis* (Geyer)	Lepdioptera	Noctuidae
Leaf eating caterpillars	*Amsacta* spp., *Acherontia styx* (Westwood)	Lepidoptera	
Thrips	*Caliothrips indicus* (Bagnall), *Frankliniella schultzei* (Trybom)	Thysanoptera	Thripidae
Surface grasshopper	*Chrogonus trachyperous* Blanch.	Orthoptera	Acrididae
Bean bug	*Riptortis pedestris* (F.), *R. linearis* (F.)	Heteroptera	Coreidae

The black bean bug : It is widely distributed in Srilanka, Japan, and South-eastern Australia. In India, the pest is distributed in the northern states on various beans. This species has since been reported to feed on soybean, moth bean, cowpea, French bean, horse gram and the green gram.

Both nymphs and adults suck sap from the plant usually from the lower surface of the leaves. As a result of sap sucking, the chlorophyll content gets reduced which ultimately affects the quality and yield of crop. Severely affected leaves show several small, black pustules of excreta. The damaged leaves gradually wither and fall off. Adults appear in June (with the onset of monsoon) and remain on the crop until final harvest.

The eggs are cylindrical, dark brown and glued to the leaf surface. The chorion exhibits fine reticulations. The eggs are generally laid singly but sometimes in pairs near the mid ribs or veinlets. Before hatching the eggs become dark. The incubation period averages 8.7 days. The bug has five nymphal instars. Just after hatching, the nymph is light orange in colour but after feeding, the thoracic parts of the body become dark red while the abdominal parts turn red having light greenish tinge. The total nymphal development takes 18-23 days. Females are larger than the males. The average length and breadth of male is 2.77 and 1.0

mm while that of females 2.86 and 1.25 mm, respectively. Pre-oviposition period averages 8.7 days while the oviposition period averages 15.5 days. Adult females live for an average duration of 25.4 days and males for 20.9 days. The average fecundity is 45 eggs. The bug has three overlapping generations in a year. It overwinters in the adult stage (Sharma et al., 1994).

Application of phosphamidon, dichlorvos, endosulphan, dimethoate or oxy demeton methyl at 0.05 % and malathion at 0.1 % at fortnightly interval will be effective in controlling the pest.

Aphids : Aphids, which are generally polyphagous, are perhaps the most common pest found all over the world on a vast majority of crops including beans. Both nymphs and adults suck the sap from ventral surface of tender leaves, growing shoots, flower stalks and pods. The infested leaves turn pale yellow, the shoots wither, flower buds fall off whereas the pods shrivel and become malformed. The growth of the vines is retarded and ultimately the yield is adversely affected. A number of aphid species have been reported damaging beans. The most destructive one is Aphis craccivora Koch. popularly known as bean aphid. This is cosmopolitan in distribution and has been reported as a severe pest of Indian bean and cluster bean.

INSECTICIDE RESISTANCE IN MAJOR VEGETABLE PESTS

A study conducted by (Ramakrishnan *et al.* (1984) in Andhra Pradesh showed that the *Spodoptera litura* larvae were 5.73-fold resistant to malathion, 14.73-fold to pyrethrum, 16.25-fold to lindane and 85.91-fold to endosulfan. *Armes et al.* (1997) reported resistance levels ranging from 0.2 to 197-fold to cypermethrin, 8 to 121-fold to fenvalerate, 1 to 13-fold to endosulfan, 1 to 29-fold to quinalphos, 2 to 362-fold to monocrotophos and 0.7 to 19-fold to methomyl.

Diamondback moth has developed resistance to nearly all classes of insecticides used against it all over the world (Sun, 1990). A high degree of resistance to pyrethroids, cypermethrin, fenvalerate, deltamthrin, and quinalphos has been reported in this pest from various parts of the country (Saxena *et al.*, 1989). Mustard aphid (*Lipaphis erysimi*) has developed resistance to methyl parathion, dimethoate, malathion and lindane (Dhingra, 1991). Appearance of pesticide resistance to *Helicoverpa armigera* to pyrethroids is now well established in various parts of the country (Reed and Pawar, 1982).

Survey on insecticide resistance monitoring for a period of one year (June 2002-April 2003) indicated that the *Helicoverpa armigera* population of Tamil Nadu developed very high level of resistance to synthetic pyrethroids, medium level of resistance to chlorpyrifos and quinalphos, low level of resistance to endosulfan, thiodicarb and profenofos and 100% susceptibility to the new chemical spinosad. Synergistic studies with piperonyl butoxide, pungam [*Pongamia* sp.] oil and profenofos clearly indicated that the common MFO mechanism was responsible for the high level resistance observed to synthetic pyrethroids (Ramasubramanian and Regupathy, 2004).

Vastrad *et al.* (2004) studies insecticide resistance in diamondback moth (*Plutella xylostella*) which was monitored throughout the year using diagnostic doses at different locations (Dharwad, Belgaum, Haveri, Bidar, Gadad and Davanagere) in Karnataka, India. The resistance level at different periods in all locations fluctuated within a narrow range. The resistance levels for endosulfan, monocrotophos, methomyl, fenvalerate, cartap hydrochloride and Bacillus thuringiensis subsp. kurstaki was 69.40, 78.42, 71.24, 71.19, 11.89 and 3.80% in the Dharwad population; 74.38, 81.75, 76.58, 81.64, 10.87 and 3.35% in the Belgaum population; 79.96, 83.90, 82.43, 87.36, 11.89 and 5.74% in the Haveri population; and 61.23, 81.43, 79.18, 81.55, 20.63 and 7.17% in the Bidar population, respectively. All the *P. xylostella* populations monitored continuously at 3 locations and randomly at 12 locations indicate low to moderate levels of resistance to cartap hydrochloride and high levels of resistance to monocrotophos, fenvalerate methomyl and endosulfan. Resistance to conventional insecticides was relatively high during the winter whereas seasonal variations in the resistance to B. thuringiensis and cartap hydrochloride were not observed.

Resistance to six insecticides (endosulfan, monocrotophos, malathion, acephate, carbaryl and cartap hydrochloride) was tested using topical and leaf residue bioassay methods against diamondback moth, *Plutella xylostella* in and around Hyderabad, Andhra Pradesh. High resistance was recorded to monocrotophos (Nirmal *et al.*, 2004).

Sood *et al.* (2003) reported the development of resistance in the field population of the greenhouse whitefly to dimethoate from Himachal Pradesh. For the management of this problem, Gill (2005) studies the cross resistance spectrum of laboratory selected dimethoate and imidacloprid resistant strains of *Trialeurodes vaporariorum* to various insecticides by comparing the LC_{50} values of different insecticides for the resistant and susceptible strains. The dimethoate resistant strain revealed low level of cross resistance which did not differ significantly. Similarly, in case of imidacloprid resistant strain, the cross reistance pattern of *T. vaporariorum* also exhibited low level of cross reistance.

INSECTICIDE CAUSING PEST RESURGENCE

Insect-pests	Insecticides causing resurgence
Aphis gossypii	Carbaryl, cypermethrin, deltamethrin, endosulfan, fenpropathrin, fenvalerate, flucythrinate, fluvalinate, monocrotophos, permethrin, phorate
Aphis malvae	Malthion, cypermethrin, deltamethirn, permethrin
Amrasca devastans	Deltamethrin, dimethoate, methyl parathion, phorate
Amrasca biguttula biguttula	Chlorpyriphos, quinalphos
Myzus persicae	Cypermethrin, deltamethrin, fenvalerate
Bemisia tabaci	Cypermethrin, deltamethrin, dimethoate, endosulfan, fenvalerate, monocrotophos, phosalone
Polyphagotarsonemus latus	Acephate, chlorpyriphos, oxy demeton methyl, monocrotophos, neem cake extract, phosalone, phosphamidon
Ferrisia virgata	Cypermethrin, deltamethrin, fenvalerate, permethrin
Leucinodes orbonalis	Acephate
Lipaphis erysimi	Phorate, dimethoate
Plutella xylostella	Methomyl

References

Armes, N.J., Wightman, J.A., Yadhav, D.R. and Rao, G.V.R. 1997. Status of insecticide resistance in Spodoptera litura in Andhra Pradesh, India. *Pesticide Science* **50**:240-48.

Atwal, A.S. and Dhaliwal, G.S. 2005. *Agricultural Pests of South Asia and their Management.* Kalyani Publishers, Ludhiana.

Brar, K.S. and Kaur, Ramandeep 2005. Advances in integrated pest management of vegetable crops (cucurbits, pea, onion and garlic). pp: 86-92. In: *Advances in the Integrated Pest Management of Horticultural, Spices and Plantation Crops* (eds. Chhillar, B.S., Kalra, V.K., Sharma, S.S. and Ram Singh)

Butani, D.K. and Jotwani, M.G. 1984. *Insects in Vegetables.* Colour Publications, Mumbai: 356p.

Chadha, K.L. and Nayar, G.C. 1994. *Advances in Horticulture.* Malhotra Publishing house, New Delhi.

Chandel, R.S. and Chandla, V.K. 2003. Managing tuber damaging pests of potato. *Indian Hort.*, **48** (2): 15-17.

Chandel, R.S., Chandla, V.K. and Sharma, A. 2003. Population dynamics of potato white grubs in Shimla hills. *J. Indian Potato Assoc.*, **30** (12): 151-152.

Chandel, R.S., Chandla, V.K. and Singh, B.P. 2005. Potato tuber moth, *Phthorimaea operculella* (Zeller). *Tech. Bull.* No. 65, CPRI, Shimla, 42 p.

Chandel, R.S., Rajnish Kumar and Kashyap, N.P. 2001. Bioecology of potato tuber moth, *Phthorimaea operculella* (Zeller) in mid hills of Himachal Pradesh. *J. Ent. Res.*, **25** (3):195-203.

Chandla, V.K. and Verma, K.D. 2000. Potato aphids pp. 48-52. In: *Diseases and Pests of Potato - a manual.* (Ed. Khurana, S.M. Paul) CPRI, Shimla:

Chandla, V.K., Khurana, S.M. Paul and Garg, I.D. 2004. Aphids, their importance, monitoring and management in seed potato crop. *Tech. Bull.* No. 61, CPRI, Shimla, 12 p.

Dhillon, M.K., Singh, Ram, Naresh, J.S. and Sharma, H.C. 2005. The melon fruit fly, *Bactrocera cucurbitae:* A review of its biology and Management. *J. Insect Sci.* **5**: 40-46.

Dhingra, S. 1991. Periodic checking to detect development of resistance in mustard aphid *Lipaphis erysimi* Kalt. To recommend insecticides used for more than two decades. *J. Entomol. Res.*, **15:**88-92.

Gill, D. 2005. Development of insecticide resistance in greenhouse whitefly, *Trialeurodes vaporariorum* (Westwood). M.Sc. Thesis, CSK Himachal Pradesh Krishi Vishvavidyalaya, Palampur.

Gopalakrishnan, T.R. 2007. *Vegetable Crops.* New India Publishing Agency, New Delhi.

Hancock, M. 1990. Pests of the potato crop. pp. 79-89. In: *Crop protection hand book-Potatoes* (ed. John S Gunn), Lavenham Press Ltd., Lavenham, Great Britain:

Harris, P.M. 1992. *The Potato Crop.* Second Edition. Chapman & Hall, London, 909 p.

Lal, O.P., Sinha, S.R. and Srivastava, Y.N. 2004. Perspectives of host plant resistance in vegetable crops. pp: 399-438. In: *Host Plant Resistance to Insects: Concepts and Applications* (eds. Dhaliwal, G.S. and Singh, R.)

Malik, Kamlesh, Chandel, R.S., Singh, B.P. and Chandla, V.K. 2005. Studies on potato apical leaf curl virus disease and its white fly vector *Bemisia tabaci* (Gennadius). pp. 17. In: *Proc. Ann. Meet of Indian Soc. of Plant Pathologists & Centenary Sym. on Plant Path.*, 7-8 April 2005.CPRI, Shimla: 17.

Misra, S.S. 1995. Potato pests and their management. *Tech. Bull.* No. 45, CPRI, Shimla, 57 p.

Misra, S.S. and Chandel, R.S. 2003. Potato white grubs in India. *Tech. Bull.* No. 60, CPRI, Shimla.

Nair, M.R.G.K. 1975. *Insects and Mites of Crops in India,* Indian Council of Agricultural Research, New Delhi, 404p.

Nirmal, B., Singh, T. V. K. 2004. Insecticide resistance in diamondback moth, *Plutella xylostella* Linn. *Indian Pl. Protec.* **32** (2): 17-21.

Ramakrishnan, N, Saxena, V.S. and Dhingra, S. 1984. Insecticide resistance in the populations of *Spodoptera* litura (F.) in Andhra Pradesh. *Pesti.* **18:**22-27.

Ramasubramanian, T. and Regupathy, A. 2004. Magnitude and mechanism of insecticide resistance in *Helicoverpa armigera* Hub. population of Tamil Nadu, India. *Asian J. Plant Sci.* **3**(1): 94-100.

Rani, Mamta and Kaur, Sandeep 2005. Advances in integrated pest management of tomato, brinjal and okra crops. pp: 93-103. In: *Advances in the Integrated Pest Management of Horticultural, Spices and Plantation Crops* (eds. Chhillar, B.S., Kalra, V.K., Sharma, S.S. and Ram Singh), CCSHAU, Hisar

Reed, W. and Pawar, C.S. 1982. Heliothis: a global problem. In *Proc. Intl. Workshop on Heliothis Management.* ICRISAT, Patancheru, India, p 9.

Regupathy, A., Palanisamy, S., Chandramohan, N and Gunathilagiraj, K.1997. *A guide on crop pests* Sooriya Desktop Publishers, Coimbatore, 290 p.

Saxena, J.D., Rai, S., Srivastava, K.M. and Sinha, S.R. 1989. Resistance in the filed population of the diamondback moth to some commonly used synthetic pyrethroids. *India J. Ento.*, **51**:265-68.

Shivalingaswamy, T.M. and Satpathy, S. 2007. Integrated pest management in vegetable crops. pp : 353-375. In: *Entomology: Novel Approaches* (eds Jain, P.C. and Bhargava, M.C.), New India Publishing Agency, New Delhi.

Singh, R.B., Khurana, S.M. Paul; Pandey, S.K. and Srivastava, K.K. 2000. Tuber treatment with imidacloprid is effective for control of potato stem necrosis disease. *Indian Phytopath.*, **53:** 142-145.

Singh, S.P. 1987. Studies on some aspects of biology-ecology of potato cut worms in India. *J. Soil. Biol. Ecol.*, **7:** 135-143.

Singh, S.V., Mishra, A., Bisan, R.S. and Malik, Y.P. 2000. Host preference of red pumpkin beetle, *Aulacophora foveicollis* and melon fruit fly, *Dacus cucurbitae. Indian J. Ento.*, **62:**242-246.

Sood, S. 2002. *Bionomics and management of the whitefly on summer vegetables.* M.Sc. Thesis, CSK Himachal Pradesh Krishi Vishvavidyalaya, Palampur.

Sood, A.K., Sood, S., Mehta, P.K. and Verma, K.S. 2003. Detection of resistance to dimethoate in the greenhouse whitefly. *In National Sym. Frontier Areas Entomol. Res.*, November 5-7, Abstract, p 529. 2003. IARI, New Delhi.

Srivastava, K.P. 2002. *A Text Book of Applied Entomology*. Kalyani Publishers, New Delhi.

Srivastava, K.P. and Butani, D.K. 1998. *Pest Management in Vegetables*. Research Periodicals and Book Publishing House, Texas, USA.

Sun, C.N. 1990. Insecticide resistance in diamondback moth what can we do with existing formulation? In: *2nd Intl. Workshop on Management of diamondback moth and other crucifer pests*, Abstract Volume, AVRDC, Shanhua, Taiwan.

Trivedi, T.P. and Rajagopal, D. 1999. Integrated pest management in potato. In : 299 - 313 *IPM System in Agriculture, Cash Crops*, Vol.6. (eds. Rajeev K. Upadhyay, K.G. Mukerji and O.P. Dubey), Aditya Books Pvt. Ltd., New Delhi.

Vastrad, A. S., Lingappa, S. and Basavanagoud, K. 2004. Monitoring insecticide resistance in diamondback moth, *Plutella xylostella* (L.) in Karnataka, India. *Resistant Pest Mgmt. Newsletter,* **13** (2): 22-24.

Verma, K.D. and Chandla, V.K. 1999. Potato aphids and their management. *Tech. Bull.* No. 26(Revised) CPRI, Shimla, 34 p.

Chapter-3

Mites of Fruit Crops, Mushroom and their Management

Divender Gupta and P.R. Gupta
Department of Entomology and Apiculture
Dr. Y.S. Parmar University of Horticulture and Forestry, Nauni (Solan) HP

Mites are small arthropods which attack number of crops damaging different parts of the plants (leaves, buds, flowers and fruits) and due to short duration of life cycle and high fecundity, their population increases at a fast rate and when feeding in large numbers they become serious pest of many crops. Since last few years the menace of mites is on the increase in Indian sub-continent causing enormous losses. One of the probable reasons for it is attributed to excessive use of broad-spectrum pesticides, which have led to the development of resistance in mites. Further, owing to their deleterious effect on non target organisms, especially natural enemies of the mites in the ecosystem there is often a spurt in the population of mites which otherwise may be of minor pest category. Because of their small size, the infestation goes unnoticed in the beginning and by the time prominent symptoms appear a lot of damage has already been done.

FRUIT CROPS

In fruit crops, being perennial in nature, the infestation of mites not only affects the trees at a particular time but their overall vigour gets reduced which is reflected in reduction in production as well as poor quality of the fruit in the years to come. The identification characters, nature of damage and the symptoms must be well understood in order to manage the mite pests effectively. A number of mites are associated with various fruit crops and those infesting major fruit crops are discussed herewith.

Apple

Mites were considered as a minor pest due to its low population count till 1989 but thereafter the outbreak appeared in the entire apple growing regions of Himachal Pradesh (Kumar and Bhalla, 1993a; Khajuria and Sharma, 1996; Singh *et al.*, 2000). Their incidence has also been reported in apple growing regions of Jammu and Kashmir and Uttarakhand. The European red mite, *Panonychus ulmi* (Koch) and two spotted mite, *Tetranychus urticae* (Koch) are the major mite species associated with apple crop, with the former often predominant one and their relative population varies from locality to locality.

European Red Mite (ERM), *Panonychus ulmi* (Koch) (Tetranychidae : Prostigmata)

In India it was reported for the first time in 1974 from north western Himalayan region including Jammu and Kashmir, Himachal Pradesh and Uttarakhand. Besides apple, it also infests pear, peach, plum, apricot, etc. (Prasad, 1974). It is chocolate brown in colour and bears white tubercles on its body. A bristle arises out of each tubercle, which distinguish it from other species. The adult measures 0.4 mm in length and can be seen with the help of magnifying glass. It lay eggs on the lower surface of the leaves preferably near the mid rib and veins, which are spherical (0.14mm dia) and pinkish to red in colour with a hair arising from its center. These hatch in 5-10 days to produce six legged larvae, which by moulting twice becomes full fed in 5-15 days depending upon the season. The total developmental period from egg to adult is completed in 10-25 days. Adult survives for 8-20 days and a female lays 20-50 eggs. The pest completes 5-8 generations in a year (Rather *et al.*, 1994; Khajuria and Sharma, 1996). The development is highly temperature dependent. Temperatures below 15°C and above 35°C have not been found suitable for the development of *P. ulmi*. The eggs hatched in 8, 6 and 5 days, respectively, at temperature of 20, 25 and 30°C, whereas, the immature stages took 9, 6 and 4 days, at the respective temperature (Thakur, 2002). The mite passes winter in the egg stage, which are laid during August to November. Overwintering pink to red bead like eggs can be seen on twigs, spurs, branches, cracks and crevices and sometimes even on the fruits. Low temperature and short day length induce females to lay winter eggs. Dry and hot weather with moderate rainfall is necessary for the build up of the mite population but heavy rainfall washes out the mites and a decline in the population is observed (Bhardwaj *et al.*, 2006). The majority of overwintering eggs (ca 70 %) hatch by mid April coinciding with the pink bud stage of the crop growth and rest of the eggs hatch in a month (Khajuria and Sharma, 1996; Bhardwaj, 2004; Rather *et al.*, 2007).

Two Spotted Mite (TSM), *Tetranychus urticae* (Koch) (Tetranychidae: Prostigmata)

It is a polyphagous pest, which feeds on more than 150 species of economic importance. In green houses it is a major problem. The mite is variable in colour (yellowish green, amber or rusty green and ferruginous red), oval to pyriform (0.5mm x 0.3 mm) and bears two red spots on the anterior part of the body. The female lays 60-80 spherical eggs (pale green) on lower surface of leaves, which are without any bristle and hatch in 5-14 days (Bhardwaj, 2005) depending upon the temperature. The life cycle is completed in 10-20 days during the active period. These mites overwinter as adults in debris, fallen leaves and under loose bark just below soil surface or in the cracks and crevices. In summer, mites are lighter in colour whereas the hibernating mites produced during autumn are entirely pink. During the season the mite completes 5-7 overlapping generations. After overwintering the pest resumes activity in spring first feeds on the water sprouts or on the weed touching lower branches and then climbs up the tree and start sucking sap from the newly formed leaves. It makes silken webs on the lower surface of leaves.

Damage : Both the above mentioned mite species suck cell sap from leaves which results in stippling of the leaves, followed by turning into light green, yellow and ultimately bronze. In case of heavy infestation mites are recorded on both the surfaces. In severe conditions the leaves may fall from the tree. The photosynthetic activity is hampered, the tree looses vigour resulting in poor fruit set and poor quality fruit. Its infestation adversely affects the fruit bud formation and the crop the following year. White exuviae of mites can be seen sticking on the leaf surface, which gives it a powdery look. Economic losses caused by the mites have been estimated at 30 per cent (Khajuria and Sharma, 1996; Bhardwaj, 2005).

Khajuria and Sharma (1996) reported *Stethorus punctum* (Le Conte) (Coccinellidae), *Chrysoperla carnea* (Stephens) (Chrysopidae), *Orius* sp.

(Anthocoridae), and predatory mites *Amblyseius fallacis* (Garmar) (Phytoseiidae) and *Zetzellia mali* (Ewing) (Stigmaeidae) as predators. Amongst these, predatory mites constitute 47.8 per cent of the total natural enemy population (Khajuria and Sharma, 2001). Dokras *et al.* (2002) reported five groups of natural enemies of mites belonging to seven genera namely predatory mites, *Typhlodromus*, *Amblyseius* and *Zetzellia* which make active movements than the pest, green lace wing *Chrysoperla carnea* (Stephens), anthocorid *Orius* sp., coccinellid, *Stethourus* sp. and predatory thrips *Leptothrips* sp. feeding on *P. ulmi* in apple orchards of Himachal Pradesh. Among predacious mites, *Typhlodromus* was most commonly observed. They further reported that the relative abundance of chrysopids was the maximum (16.3), followed by predatory thrips (12.4), predacious mites (10.8), anthocorid bugs (9.2) and coccinellids (6.9) in a sample of 125 apple leaves. Overall, significantly higher number of predators (18.7) was recorded in unsprayed orchards, which was about 3 times more in comparison to heavily sprayed orchards (5.8). In the moderately sprayed orchards, the predator count (8.8) was significantly higher than that of heavily sprayed orchards.

Management : For the effective management, various types of methods need to be implemented. In the case of *T. urticae*, which overwinters in the fallen debris or soil, weed-free tree basins reduce the chance of recolonisation in ensuing season. The branches touching the ground or adjacent branches of other trees should be cut in away so that the movement of mite is restricted. During the peak period, predator activity is also good and hence indiscriminate use of pesticides should be avoided which results in killing of predators.

Dormant spray oil applied at 2% concentration at tight cluster stage kills overwintering eggs of the European red mite (ERM). Besides some adults of two spotted mite, which hide in cracks and crevices, also get killed. After hatching of the eggs, the mite population should be closely monitored with the help of field lens in the orchard. Bhardwaj and Bhardwaj (2000) reported that when 10 per cent spurs are found with the winter eggs, acaricide treatment should be given. They devised an action threshold of 4-5 mites/leaf on the basis of 100 randomly drawn leaves from the apple orchard. If the population is low (1-2 mites/ leaf) at petal fall, Horticultural mineral oils (Summer oils) like DC Tron Plus, Orchex-796, Shelter-909, etc. should be sprayed at a concentration of 1%, which can be repeated after 20 days. Negi and Gupta (2007) reported effectiveness of summer spray oils namely DC-Tron Plus, Orchex-796, IPOL orchard spray oil and Servo orchard spray oil at a concentration of 1% in managing the mixed infestation (both *P. ulmi* and *T. urticae*) of phytophagous apple mites. These oils are safe to natural enemies and environment but the care must be taken not to use spray in excessive humid (foggy) or drought conditions, otherwise it may cause phytotoxicity/ fruit injury. Similarly dormant oils should not be used in summer on foliage otherwise it will also result in phytotoxicity. If the population is high, then persistent acaricide like fenazaquin (0.0025%) or propargite (0.057%) should be sprayed, which keeps the mite population under check for 20-30 days. In addition, fenpyroximate (0.0025%), bifenthrin (0.008%) and profenofos (0.05%) were also found quite effective in checking the mixed mite populations (Gupta *et al.*, 2005; Khajuria *et al.*, 2006). Fenazaquin and propargite were found relatively safe to the phytoseiid predator, *Amblyseius fallacis* (Garman) (Khajuria *et al.*, 2006).

Citrus

As many as 54 species of mites have been reported infesting citrus in India (Garg, 1978) out of which *Eutetranychus orientalis* is the major one.

Oriental Red Mite, *Eutetranychus orientalis* (Klein) (Prostigmata: Tetranychidae)

This mite has been reported damaging 60 plant species in India and is commonly known as citrus mite because of its common occurrence in large numbers on *Citrus* (Singh, 1986). The adult mite is pale and bears bristles situated on white tubercles. The female lays around 50 eggs along the mid rib and lateral veins on upper surface of leaves. The eggs are sub spherical and pale brown and measure about 0.14mm and are visible if seen critically with unaided eye. Eggs hatch to produce light brown six legged larvae and after few days they moult into eight legged orange brown nymphs and later to adults. The total life cycle is completed in a fortnight in summer and may take about three weeks in the post monsoon period, depending upon the temperature. All the active stages feed and moult on upper surface of leaves and the shed skin (exuviae) can be seen attached to the leaf surface appearing as small specks. The developmental threshold reported for the mite is 11°C (Jeppson *et al.*, 1975).

Different motile stages feed on the chlorophyll, which result in stippling of leaves, and later the leaves turn yellow. In nursery and when the plants are young, the affect is more pronounced and the leaves may even fall. The overall vigour of the tree is affected causing loss in production and quality. Premature fruit dropping also occurs.

The other mite species namely *Eutetranychus banksi* (Mc Gregor), *Panonychus citri* (Mc Gregor), *Tetranychus cinnabarinus* (Bois du val), *Tetranychus sexmaculatus* Riley and *T. fijiensis* have been reported feeding on citrus with their peak activity during January - May and August-September in different parts of the country (Tandon, 1993). Infestation by *T. fijiensis* results in curling of leaves in the downward direction. All these mites cause yellowing of leaves thereby devitalizing the plants.

Management : Coccinellid (*Scymnus gracilus* Motsch), thrips (*Scolothrips indicus* Priesner) and four species of mites *Pronematus* sp., *Amblyseius cucumeris, A. hibisci* and *Agistemus* sp. have been reported predating on *E. orientalis* (Sadana and Kanta, 1971). Spraying of dicofol (0.04%), dimethoate (0.05 %), oxydemeton methyl (0.025%) and wettable sulphur (0.2%) has been found very effective against these mites.

Eriophyid Mites

Two eriophyid mites namely citrus rust mite *Phyllocptruta oleivora* (Ashmead) and citrus bud mite *Aceria sheldoni* Ewing appear as sporadic pests and sometimes cause problems. Citrus rust mite is deep yellow and occurs worldwide. The adult measures 0.12mm in length. The female lays about 20-30 minute whitish spherical eggs in depressions on the fruits and leaves. These eggs hatch in 3-7 days and the worm like nymphs with only two anterior pairs of short legs start feeding on the fruits or leaves. The life cycle is completed in 1-2 weeks in summer but development slows down in winter. Most of the rust mite damage is noticed from late spring to late summer. As a result of feeding by the mite, the fruits become silvery in colour in lemon and rust brown in oranges. The skin of the fruit also becomes thicker and affected fruits remain small in size. When the fruits have attained a diameter of 20-30 mm it is more prone to the attack of rust mite. The other eriophyid mite, i.e. citrus bud mite also looks similar to rust mite but is light yellow or pinkish in colour and measures 0.17mm in length. They inhabit the bud, the developing blossom and the fruit and lay about 50 pearly white spherical eggs. The period from egg to adult is about 10 days and in winter it goes to 20-30 days. The twigs, leaves and fruits become malformed as a result of damage by the bud mite. The twigs may become bunchy and assume rosette type of growth.

Management : Monitoring of the mites in the field is required from early spring to summer. The observations on these minute creatures in the field can be made by using field lens (10X or 15X). The rust mite generally feeds in protected places like pedicel end of the fruit and later when the population becomes high it cover the entire fruit. Spraying of the trees with acaricides suggested for the other mites will take care of these mites. No effective natural enemies are known.

Litchi

Litchi Mite, *Aceria litchii* (Keifer) (Eriophyidae: Prostigmata)

Red rust of litchi is a major constraint in litchi production. This mite is not only present in India (Mathur and Tandon, 1974, Lall and Rehman, 1975, Prasad and Bagle, 1981, Sood *et al.*, 1987, Sharma *et al.*, 1996, Gupta *et al.*, 1997 and Babu *et al.*, 2004) but also reported from Florida and Hawaii (Nishida and Holdaway, 1945), Bangla Desh (Alam and Wadud, 1963), Taiwan (Huang, 1985) and Australia (Pinese, 1981; Waite and McAlpine, 1992). In India it has been reported from West Bengal, Bihar, Jharkhand, Uttar Pradesh, Uttarakhand, Haryana, Himachal Pradesh and Punjab. Red rust was reported to be caused by an eriophyid mite *Aceria litchii* (Keifer) (=*Eriophyes litchii*) way back in 1912 by Misra but Sharma *et al.* (1972) reported an alga, *Cephaleuros virescens* Kunze as the causal organism. Thereafter Somachaudhary *et al.* (1988), Kumar and Bhalla (1993b) and Gupta *et al.* (1997) reported mite *A. litchii* as well as an alga *C. virescens,* associated with this malady. The mite is minute, vermiform and pinkish white creature measuring 0.15-0.20 mm in length and bears two anterior pairs of legs. The female lays spherical and translucent eggs singly on the ventral surface of the leaflet, out of which the larvae hatches in 3- 4 days that is followed by further two nymphal stages, which occupy 8-12 days. The life cycle is completed in about a fortnight and the pest completes many generations in a year.

In the beginning when the new flush starts appearing, small pits are observed on the ventral side of the leaflet on which light green velvety layer called erineum develops. Later these pits increase in size and coalesce together causing blistering of leaves when observed from the dorsal side. In extreme cases the outer edges of the leaves curl. The colour of the erineum later changes to light brown, brown and finally chocolate brown. Such leaves become leathery and brittle. The maximum population is observed on brown and chocolate brown erinea. When the erinea is observed under microscope, interwoven hyphal strands of alga, *C. virescens* are visible (Sharma, 1991; Gupta *et al.*, 1997). In case of severity, petiole of the leaves, tender stems, panicles and even immature fruits are not spared. The fresh leaves are more prone to attack and the infestation decreases as the leaves start hardening. The severely infested orchard appears brick red and the infestation can be spotted from a distance. The maximum infestation is observed on the spring flush (February-March) and the population of mites ranged from 15-60/leaflet (Prasad and Singh, 1981). A survey study carried out in the Kangra valley of Himachal Pradesh revealed 50-90 per cent infestation in the orchards situated at an elevation range of 750-950 metres amsl (Gupta *et al.*, 1997). As a result of this problem, the tree vigour gets reduced, resulting in lesser fruit set, premature fruit drop and poor fruit quality.

The pest is carried by taking goottee plants from the affected orchards and by the movement of mites mainly through touch from one leaf to another and from one tree to the adjacent tree. Honey bees (*Apis mellifera* Linnaeus) have been reported to carry litchi erinose mite from flowers of affected trees to the healthy ones (Waite and McAlpine, 1992). Five species of predatory mites belonging to family Phytoseiidae namely *Amblyseius largoensis* (Muma), *A. paraaerialis* Muma, *Phytoseius intermedius* Evans and Macfarlane, *Typhlodromus homalii* Gupta and *T. flescherni* Chant have been reported predating upon *A. litchii*, the population build up starts in March and increases gradually till June followed by decline (Sharma and Thakur, 1992).

Management : A number of practices for the management have been suggested. While planting new orchards, rust free nursery plants should be purchased from the registered nurseries. The architect of trees should be such that they do not touch one another and the ground. Cutting and burning of affected parts has been advocated by many workers (Prasad and Singh, 1981; Gupta *et al.*, 1997). Pruning and burning of affected parts alone during January resulted in 40 per cent reduction in the infestation over control. Further 3-4 sprays of dimethoate (0.03%) and dicofol (0.05%) starting from the initiation of problem have also been found very effective. An additional spray can be given on post monsoon flush (Gupta *et al.*, 1997). Babu *et al.* (2004) reported 86 per cent reduction in infestation with three sprays of wettable sulphur (0.15%). The predatory mites take care of the pest to some extent and sulphur and dicofol have been reported to be comparatively safe to the predatory mites (Sharma and Thakur, 1992).

Mango

Mango Bud Mite, *Aceria mangiferae* Keifer (Eriophyidae: Prostigmata)

The bud mite is a major problem in Northern India mainly in Punjab, Delhi, Haryana and Uttar Pradesh (Singh and Mukherjee, 1989). The mite is worm like in appearance and bears only two pairs of legs situated anteriorly. The population starts building up with the warming up of the season and reaches at the peak in February. Thereafter the population declines and reaches a low by May. A further spurt in the population is observed during June-July and then the population comes down and reaches quite low in October-December (Atwal and Dhaliwal, 2007).

The mites suck the sap from the buds and when the population is high, the bud may be killed. Most of the varieties are affected by the mite.

Management : Removal and destruction of affected inflorescence followed by spray of miticides like dicofol (0.05%), dimethoate (0.03%) has been found quite effective (Atwal and Dhaliwal, 2007). These miticides can be repeated during June-July when an increase in the population is recorded. Control should be anticipated through monitoring and knowledge of phenology and flushing behaviour of trees so that management measures can be taken accordingly.

Grape

Mites are important pest of grapevines throughout the country. Five species of mites belonging to Tetranychidae have been reported to be associated with the grapevine crop. Tandon and Verghese (1994) reported *Oligonychus mangiferus* Rahman and Saptra, *O. punicae* Hirsto, *Tetranychus cinnabarinus* Andre and *T. urticae* infesting grapevines. Oriental red mite, *Eutetranychus orientalis* Klein has been reported from North India (Khangura *et al.*, 1991). The life cycle of these mites gets completed in 17-20 days and number of generations are completed in a year. These mites suck the sap from the upper surface of the leaves. The affected leaves turn yellow, brown and ultimately bronze. The activity of the mites is observed from April to October under north Indian conditions whereas in the southern part the main population is observed during July-October and January-March (Chadha and Shikhamany, 1999). Dhooria and Sandhu (1975) studied varietal preference of grapevines to different species of the mites. They reported that the varieties more prone to attack of *O. mangiferus* were attacked less by the *Eutetranychus truncatus* and vice-versa. Important varieties like Anab-e-Shahi, Delight and Gulabi were highly susceptible to *O. mangiferus*. Khabli, Beauty seedless and Arkavati were moderately susceptible whereas Black Champa, Perlette, Thompson seedless and Tas-A-Ganesh were less susceptible (Batra *et al.*, 1992).

Management : Spray of acaricides like dicofol (0.05%), wettable sulphur (0.2%) on the initiation of attack manages the mite population. If required it can be repeated after 15 days.

Papaya

Two Spotted Spider Mite, *Tetranychus urticae* (Koch) (Tetranychidae: Prostigmata)

It is a polyphagous pest, which feeds on number of fruit, vegetable and ornamental crops and is distributed throughout the country. The mite is oval to pyriform (0.5mm x 0.3mm), bears two black spots on the back and is generally yellowish green but the colour may vary from ferruginous red, amber or rusty green. The eggs are spherical and with warming up of the weather the female lays eggs singly on the lower surface of the leaf. After an incubation period of 2-6 days, the larva bearing three pairs of legs hatches out of the egg and starts sucking sap from the leaves. Within 3-4 days they change into nymphs with four pairs of legs. In two stages it grows to the adult stage. The total life cycle during active period is completed in 10-20 days. The pest completes number of generations in a year. These mites feed by making webs. As a result of sucking of cell sap, the infested leaves show symptoms of stippling in the beginning and later these become yellow and ultimately dry up. It results in poor fruit setting and premature fruit drop. Further the quality of the fruit is also impaired.

Management : Spray of miticides like dicofol (0.05%) and dimethoate (0.03%) have been found effective. The spray should be directed towards the lower surface of the leaves, as mites have a tendency to settle and multiply abaxially.

Mushrooms

A variety of mite species are reported to be associated with cultivation of mushroom right from spawn preparation, initial stage of composting to harvest of the crop. These mites belong to Acaridae (*Aeroglyphus, Blattisocius, Caloglyphus, Eberherdia, Oppia, Rhizoglyphus, Tyroglyphus, Tyrophagus*), Anoetdidae (*Glyphanoetus, Histiostoma*), Ascidae (*Arctoseius*), Digamasellidae (*Digamasellus*), Eupodidae (*Linopodes*), Glycyphagidae (*Glycyphagus*), Laelapidae (*Hypoaspis*), Macrochelidae (*Macrochelus*), Pyemotidae (*Macrodispodides, Pygmephorus, Pseudopygmephorus*), Scutacardiae (*Scutacarus*), Tarsonemidae (*Tarsonemus*) and Tydeidae (*Pronematus*) families. The initial stage of composting is most vulnerable and sometimes huge population of mites can be encountered. However during phase I of composting and pasteurization, most of them die. Subsequent appearance of mite infestation in the crop is an indicator of poor quality of composting material and unhygienic conditions prevailing in its cultivation. Although more than 60 species of mites are reported from various parts of the world (Kumar *et al.*, 2004), yet 16 species are of economic importance (Suman and Sharma, 2005). Mites associated with mushroom-cultivation can be grouped into three categories, pestiferous mycophagous mites, saprophagous mites and predatory mites. Kumar *et al.* (2004) have given a comprehensive review of the work done on mushroom mites and their management.

(a) Mycophagous mite species : Feeding behaviour and nature of damage varies from mite species to species and injury inflicted on and damage done to the crop depends upon the species of the mite, their absolute population, and susceptibility of the mushroom crop. Pestiferous mites initially feed on mycelial mat of the mushroom by sucking juices from hyphae as a consequence of which fruiting body formation may be hampered. When tiny button formation occurs, sporophores and caps may be affected, as these mites may make small irregular pits, riddle them and hollow out tiny buttons and stipes. Mite infested mushroom may be speckled with brown lesions, become leathery and have sickly appearance. Tarsonemid mites are pale brown and very minute and difficult to see with unaided eye. They feed on mycelium and one comes to know of its infestation as appearance of reddish brown discoloration at base of the stem of the fruiting bodies. Tyroglyphids, on the other hand, are slow moving, translucent and with long hair on their body. When in abundance, their attack results in small pits on the cap and stalk. They also feed on mycelium. Many species of mites spread under overcrowding conditions by developing migratory stage, the hypopus, which is a special deutonymph stage becoming flattened and are phoretic clinging with other insects especially flies in the mushroom houses.

Commonly found pestiferous species are *Tarsonemus tarsalis, Tyroglyphus dimidiatus, T. putrescentiae, Tyrophagus longior, Histiostoma, Caloglyphus berlesi, Oppia,* etc, which have been noticed to feed on cultivated mushrooms. Their intensity of attack may vary from season to season and there may be succession of species in a region, as reported by Das (1986) in West Bengal where *Rhizoglyphus echinopus* is first to appear in January to March because of its preference to mild climatic conditions, followed by *Tyrophagus dimidiatus* during April to June, *Histiostoma heinemanni* in June to October and *H. miles* from October to November. Mites *Rhizoglyphus echinopus, Tyroglyphus dimidiatus, Histiostoma heinemanni* and *H. miles* attack *Pleurotus sajorcaju, P. ostreatus* and *Volvariella volvacea*, while *Brennandania lambi* attacks mycelium of *Auricularia, Hericium erinaccus, Tremella fuciformis* and *Agaricus bisporus*.

Some pyemotid mites, like red pepper mite, *Pygmephorus* species develop only on competing weed fungi (*Trichoderma, Chaetomium, Penicillium*, etc.) and create nuisance to mushroom growers due to contamination of the crop. Some of these mites are also known to cause allergic reactions. They are very minute (ca 0.25mm) but their rate of reproduction is fast (1 female may lay up to 160 eggs in 5 days). Okabe (1999) demonstrated that spores of the weed fungus *Hypocrea nigricans* passing through the alimentary canals of the three acarid mite species *Tyrophagus putrescentiae, Histiogaster* sp. and *Schwiebea* sp. into their excreta, were viable and these fungivorous mites disseminated spores thus acting important vectors of weed fungal spores in indoor mushroom production units. These mites are conducive to the spread of weed fungi in the infected rooms as well as in the whole farm. Okabe *et al.* (2001) observed that even mites *Tyrophagus putrescentiae, Pygmephorus mesembrinae* and *Tarsonemus* sp., which are common mushroom pest mites in Japan, could not increase in numbers when transferred to a mushroom cultivation medium of sawdust and rice bran that has been pre-colonized with the mushroom *Hypsozygus marmoreus*. However, *T. putrescentiae* and *P. mesembrinae* established their populations in both the media not colonized with *H. marmoreus* and those with only a weed fungus, *Trichoderma harzianum*. The initial numbers of mites inoculated to the medium was crucial for certain species. Only 10 female *T. putrescentiae* and 100 females of *P. mesembrinae* were required for population establishment. Initial conditions such as invasion by numbers of females and/or fungal contamination before mite migration accelerated population increase. No contamination occurred in media without mites.

Another pygmephorid, *Luciaphorus auriculariae*, a destructive mite to commercial production of jew's ear mushroom *Auricularia polytricha*, causing 10-50 per cent crop loss in Fujian province of China, can also feed on mycelia and sporophores of *A. auricula* and *Flammulina velutipes*, but not on the mycelia of *Agaricus bisporus, Pleurotus ostreatus, Tremella fuciformis*, or *Lentinus edodes* or the sporophores of *L. edodes* (Zou *et al.*, 1993)

(b) Saprophagous mites : Many colourless and translucent saprophytic mites of *Tyrophagus, Caloglyphus* and *Histiostoma* species feed on decaying vegetative matter and sometimes their infestation can be seen on mushrooms. These mites are slightly bigger in size and slow moving. Their infestation in mushroom house indicates use of poor quality compost that is not well suited for mycelium growth of the mushroom.

(c) Predatory mites : Beside myceliophagous and saprophagous mites, a variety of predatory mites are also noticed in the mushroom growing rooms. These mites are bigger in size and faster in movement as compared with other mites. The predatory mites namely *Geolaeles aculeifer, Hypoapsis miles, Parasitus fimetorum, Arctoseius cetratus* and *Macrocheles merdarius* feeds on eggs and larvae of flies, *Digamasellus fallax, Arctoseius cetratus, Parasitus bituberosis* on nematodes and *Linopodes antennaepes, Arctoseius cetratus, Macrocheles merdarius* on spring tails and smaller mites, infesting mushroom. Although useful in reducing pestiferous species, yet their presence is an indicator of pest infestation, defective compost preparation and poor sanitation encouraging mycophagous mite species.

Biology of Mycophagous Mites

Mycophagous mites of concern mainly belong to families Acaridae, Tarsomenidae and Pyemotidae. Most of mites complete life cycle after passing through egg, larva, protonymph, deutonymph and tritonymph stage to become adult, often each post-embryonic stage with a feeding and resting stage. Duration of development, fecundity and number of generations completed depend upon temperature and humidity and quality of food.

Among acarids *Tyrophagus* and *Tyroglyphus* take longer duration, while *Caloglyphus* and *Eberhardia* complete life cycle in short period. *Tyrophagus dimidiata* lays 40-60 eggs and duration of life cycle is 17-24 days at 22°C. *Tyroglyphus putrescentiae* multiplies at faster rate at 22-23°C and 90% relative humidity (RH) and its duration of development decreases at the cost of fecundity as the temperature rises. At 30°C and 80% RH, it completes its life cycle in 17 days. *Caloglyphus and Eberhardia* sp. need 8-15 days at 20-21 °C for completion of life cycle. Tarsonemid *Tarsonemus myceliophagus* develops at a faster rate and completes life cycle in 8 days at 24 °C and 12 days at 10 °C.

Pigmy mites *Pygmephorus mesembrinae* has high fecundity laying 20-25 eggs per clutch and more than 200 eggs, which hatch within a day at 22 °C. Generation time is 4, 5 and 7.5 days at 25, 20 and 15 °C, respectively. Their biology is unique and different from other mycophagous mites. Zou *et al.* (1993) studied biology of pygmephorid *Luciaphorus auriculariae* in detail. The newly emerged and fertilized female after inserting its gnathosoma into the host and commence feeding starts swelling in hysterosoma region which becomes pin head like globular and eggs develop directly into adults in the physogastric hysterosoma. Mating may take place inside or outside hysterosoma after it bursts to release the adults. Females are monogamous, while males can mate with as many as 20 females. On an average, embryonic development gets completed in 13.6, 6.6, 4.5 and 3 days, while adult survives for 26.6, 12.6, 8.8 and 7 days at 20, 25, 30 and 35°C, respectively, with a progeny production of 125, 128, 163 and 77 per female. Under unfavourable conditions, hysterosoma becomes hardened delaying egg development and protecting the newly formed adults.

Management of Mushroom Mites

Although many insecticides/acaricides have been found effective for suppression of mites infesting mushroom (Kumar *et al.*, 2004; Kumar and Sharma, 2005), yet their use alone will never give effective control of mushroom pests until a holistic management strategy is adopted. Prophylactic or preventive measures if followed in right spirit will give remedy from most of the pest problems. Practising good agricultural practices, their incidence can be minimized. For management of mites following strategy be implemented.

- Maintenance of proper hygiene and sanitation should be the basic essential step. Before cropping, the mushroom house/ rooms/tunnels should be properly cleaned by removing debris. Attention should be given for sterilization of empty trays, utensils and other containers.
- If in mushroom house mite infestation has been observed in earlier cropping season, it would be advisable to treat it by spraying dicofol @0.1% (Sandhu, 1995) or burning sulphur (35-50g/m^3 space).
- All air inlets, ventilators, etc. should be covered with mesh (aperture size <0.6mm) to prevent entry of flying insects, especially flies, which are source of spread of phoretic mites. Doors should be opened for minimum possible time and should tightly close without leaving any space for entry of such creatures.
- Continuous monitoring of the crop for pest infestation is very essential as at the onset of infestation, it can effectively be contained and its spread prevented. One should avoid use of pesticide when the pest load is low, especially during winter and early spring.

- Compost should be prepared by scientific and efficient method at a place remote from raising the crop. Composting material be free from mite-infestation. Surface steaming of the compost (at 66-71°C for several hours) eliminates many mite species. If mite infestation appears in compost, spray of diazinon (0.003-0.004% *i.e.* 1.5-2ml diazinon 20EC in 10 litre water) can be made.
- If insecticides are to be used, select an appropriate one, apply at recommended dose at right time. Higher doses are sometimes detrimental for growth of the mushroom mycelium and may also result in insecticidal residues in the crop. For suppression of mite infestation after spawning, propargite (Omite 57%) can be sprayed at 0.66-0.88g a.i./m^2 (1.5-2ml of formulated product/m^2). Spraying with chlorfenvinphos, fenitrothion, fenthion, ethion, oxy-demeton methyl, etc. (1g a.i./m^2) also gives satisfactory control (Kumar *et al.*, 2004).
- Recently, Bussaman *et al.* (2006) have shown feasibility of use of some strains of entomopathogenic bacteria *Photorhabdus* and *Xenorhabdus* against the mushroom mite, *Luciaphorus* sp., which is a serious pest of tropical mushrooms *Lentinus squarrosulus*. They found supernatants of two strains, GPS12 and GPS11, of *Photorhabdus luminescens* ssp. *laumondii* significantly more toxic than the other species and strains to the female mite, resulting in 90-95% mite mortality within 48 hours. Potential of such products needs to be worked out for using them in pest management.

References

Alam, M. Zahurul and Wadud, M.A. 1963. On the biology of litchi mite, *Aceria litchii* Keifer (Eriophyidae: Acarina) in East Pakistan. *Pakistan J. Sci.*, **18**(5): 232-240.

Atwal, A.S. and Dhaliwal, G.S. 2007. *Agricultural pests of South Asia and their management.* Kalyani Publishers, Ludhiana. 505pp.

Babu, Naresh, Venkateswaralu, P, Verma, R.C. and Bujarbaruah, K.M. 2004. Managing mites on litchi. *Indian Horti.*, **49** (2): 26-27.

Batra, R.C., Cheema, S.S. Brar, S.S., Kaur, H. and Khangura, S. 1992. Incidence of grapevine thrips and grapevine mite on different cultivars of grape. p. 354-357. In : *Proc. of Inter. Sym. Recent advances in viticulture and oenology.* AP Grape Growers Association, Hyderabad.

Bhardwaj, S.P. 2004. Red spider mite: A major threat to apple production. p.293-299. In: *Recent Trends in horticulture in the Himalayas.* (eds K.K. Jindal and R.C. Sharma). Indus Publishing Company, New Delhi.

Bhardwaj, S.P. 2005. Insect and mite pests. p.346-396. In: *The Apple: Improvement, production and post-harvest management* (eds. K.L. Chadha and R.P. Awasthi). Malhotra Publishing House, New Delhi, India.

Bhardwaj, S.P. and Bhardwaj, Sushma. 2000. European red mite-an emerging problem of apple. In: *Horticulture Technology Vision 2000 and Beyond (Vol.2) Protection and Preservation* (eds. Sharma, V.K. and Azad, K.C). Deep and Deep Publications, Pvt. Ltd., New Delhi.

Bhardwaj, Sushma, Sharma, Sangita and Bhardwaj, S.P. 2006. Environmental factors affecting the population build up of *Panonychus ulmi* (Koch) in apple. *Indian J. Plant Protec.*, **34:** 36-39.

Bussaman, P., Sermswan, R. W. and Grewal, P. S. 2006. Toxicity of the entomopathogenic bacteria *Photorhabdus* and *Xenorhabdus* to the mushroom mite (*Luciaphorus* sp.; Acari: *Pygmephoridae*). *Biocontrol Sci. and Tech.*, **16**(3): 245-256.

Chadha, K.L. and Shikhamany, S.D. 1999. *The Grape-Improvement, production and post harvest management.* Malhotra Publishing House, New Delhi. 579pp.

Das, P. 1986. *Economics and control of mushroom mites.* Ph.D thesis. BCKV, Kalyani, West Bengal.196pp.

Dhooria, M.S. and Sandhu, G.S. 1975. Varietal susceptibility of the grapevine to the mites. *Sci. Cul.*, **41:**209-211.

Dokras, Ashish L., Gupta, P.R. and Kumar, Ram. 2002. Natural enemy complex of European red mite *Panonychus ulmi* (Koch) in apple orchards of Himachal Pradesh. *Pest Manag. in Horti. Ecosy.*, **8**(1): 5-11.

Garg, D.P. 1978. *Insect pests of citrus fruits.* Agri. Hort. Publishing House, Nagpur, 176pp.

Gupta, Divender, Bhatia, Ranjeet, Sharma, N.K., Chandel, J.S. and Sharma, Ravinder. 1997. Incidence and Management of red rust of litchi in lower hills of Himachal Pradesh. *Pest Manag. in Horti. Ecosy.*, **3**(2): 70-74.

Gupta, Divender, Dubey, J.K. and Dinabandhoo, C.L. 2005. Bioefficacy of some new acaricides against phytophagous apple mites. p. 387-389. In: *Proceedings of Short Rotation Forestry for Industrial and Rural Development* (eds. K.S. Verma, D.K. Khurana and Lars Christenson). Indian Society of Tree Scientists, Nauni (Solan) India.

Huang, B.K. 1985. Lychee gall mite-*Aceria litchii* Keifer. In: *Control and collection of illustrative plates of diseases and pests of fruit crops.* Fujian Science and Technology Publishing House, Fuzhou.

Jeppson, Lee R., Keifer, Hartford, H. and Baker, Edward W. 1975. *Mites injurious to economic plants.* University of California Press, Berkeley. 614pp.

Khajuria, D.R. and Sharma, J.P. 1996. Outbreak of phytophagous mites on apple in Kullu Valley of Himachal Pradesh. *Indian Plant Protec.*, **24:** 134-138.

Khajuria, D.R. and Sharma, J.P. 2001. Natural enemies of phytophagous mites and their role in integrated mite management in apple ecosystem. In *Proce. of Sym. on Biological Control, Ludhiana.* 182 pp.

Khajuria, D.R., Dubey, J.K. and Gupta, Divender. 2006. Management of phytophagous mites with selective acaricides on apple and sensitivity of phytoseiid mites to acaricides. *Pest Manag. and Econ. Zool.*, **14**(1): 1-9.

Khangura, J.S., Cheema, S.S. and Minhas, P.S.S. 1991. Insect pest problems of grapevines. *Draksha Vritta* **11:** 61-62.

Kumar, R. and Bhalla, O.P. 1993a. An epidemic outbreak of *Panonychus ulmi* (Koch) (Acari: *Tetranychidae*) in apple orchards of Himachal Pradesh, India. *Curr. Sci.*, **64:** 709.

Kumar, R. and Bhalla, O.P. 1993b. Studies on the litchi leaf curl mite disease in Himachal Pradesh. *J. Tree Sci.*, **12**(1): 47-50.

Kumar, S., Gautam, Y. and Sharma, S. R. 2004. Mushroom mites and their management. *Mushroom Res.*, **13**(2): 46-52.

Kumar, S. and Sharma, S. R. 2005. Insect, mite and nematode pests of mushrooms and their management. Technical Bulletin, National Research Centre for Mushroom (ICAR), Solan, 61 pp.

Lall, B.S. and Rehman, M.F. 1975. Studies on the bionomics and control of the erinose mite *Eriophyes litchii* (Acarina: *Eriophyidae*). *Pesticides*, **9**(11): 49-54.

Mathur, A.C. and Tandon, P.L. 1974. Litchi mite can be controlled. *Indian Horticulture*, **19**(2): 11-12.

Misra, C.S.1912. Litchi leaf curl. *Agricultural J. India*, **7**:285-289.

Negi, M.L. and Gupta, P.R. 2007. Management of phytophagous mites, *Panonychus ulmi* and *Tetranychus urticae* in apple. *Indian J. of Plant Protec.*, **35**(2): 268-273.

Nishida, T. and Holdaway, F.G. 1945. The erinose mite of litchi. *Hawaii Agri. Exp. Station Circular* No.48.

Okabe, K. 1999. Vectoring of *Hypocrea nigricans* (Hypocreales: Hypocreaceae) by three fungivorous mite species (Acari: Acaridae). *Exp. and App. Acarol.*, **23**(8): 653-658.

Okabe, K., Miyazaki, K. and Yamamoto, H. 2001. Population increase in mushroom pest mites on cultivated *Hypsozygus marmoreus* and their vectoring of weed fungi between mushroom cultivation media. *Japanese J. of App. Entomol.*, **45**(2): 75-81.

Pinese, B. 1981. Erinose mite- a serious litchi pest. *Queensland Agricultural J.*, **97**:79-81.

Prasad, V. 1974. *A catalogue of mites of India.* Indira Acarology Publishing House, Ludhiana, Punjab. 320pp.

Prasad, V.G. and Bagle, B.G. 1981.Evaluation of acaricides for the control of litchi mite, *Aceria litchii* Keifer (Acarina: Eriophyidae). *Pesticides*, **15**(1): 22-23,27.

Prasad, V.G. and Singh, R.K. 1981. Prevalence and control of litchi mite, *Aceria litchii* Keifer (Acarina: Eriophyidae) in Bihar. *Indian J. Entomol.*, **43**(1): 67-75.

Rather, A.Q., Masoodi, M.A. and Sofi, A.A. 1994. Studies on ecobiology, distribution and host range of *Panonychus ulmi* (Acari: Tetranychidae) in Kashmir. *J. Ecobiology*, **6**: 287-291.

Rather, A.Q., Parrey, M.A., Zaki, F.A. and Wani, N.A. 2007. Management of European red spider mite, *Panonychus ulmi* at the overwintering egg stage on apple. *Indian J. Plant Protec.*, **35**: 36-38.

Sadana, G.K. and Kanta, V. 1971. Predator of the citrus mite, *Eutetranychus orientalis* in India. *Sci. Cul.*, **37**:530

Sandhu, G. S. 1995. Management of mushroom insect pests. p. 239-260 In: *Advances in Horticulture*, Volume13, Mushrooms, (eds. K.L. Chadha and S.R. Sharma), Malhotra Publishing House, New Delhi.

Sharma, V.K., Srivastava, A.K. and Chohan, J.S.1972. A new host record for parasitic alga *Cephaleuros virescens* Kunze. *Sci. and Cul.*, **38**(1): 39-40.

Sharma, D.D. and Thakur, A.P. 1992. Bioefficaccy of eight pesticides against litchi erineum mite (*Aceria litchii*) and its predators. *Indian J. Agricultural Sci.*, **62**(3): 240-242.

Sharma, N.K., Chandel, J.S., Gupta, Divender, Sharma, Ravinder and Badiyala, S.D. 1996. In Himachal Pradesh: Problems and prospects of litchi cultivation. *Indian Horticulture,* **41**(3); 59-61.

Singh, Harcharan. 1986. *Household and kitchen garden pests. Principles and Practices.* Kalyani Publishers, Ludhiana.420pp.

Singh, J. and Mukherjee, I.N. 1989. Pest status of phytophagous mites in some northern states of India. p.192-203. In: *Proc. First Asia Pacific Con. Entomol., Chiang Mai, Thailand.*

Singh, Mohinder, Gupta, P.R. and Rehalia, A.S. 2000. Occurrence of phytophagous mites on apple in Kinnaur district of Himachal Pradesh. *Insect Environment,* **5:** 170-171.

Somachaudhary, A.K., Singh, P. and Mukherjee, A.B.1988. Interrelationship between *Aceria litchii* (Acari: Eriophyidae) and *Cephaleuros virescens,* a parasitic alga in the formation of erineum like structure of litchi leaf. p.147-152 In: *Progress in Acarology, Vol. 2,* (eds. Channabasavanna, G.P. and Viraktamath, C.A.). Oxford and IBH Publishing Co., Pvt. Ltd., New Delhi.

Sood, A.K., Sharma, R.D. and Singh, B.M. 1987. Red rust: An emerging threat to litchi cultivation In Kangra Valley. *Indian Horticulture*: 19-20.

Suman, B.C. and Sharma, V.P. 2005. *Mushroom: Cultivation, processing and uses.* AGROBIOS (India), Jodhpur. 348pp.

Tandon, P.L. 1993. Insect and mite pests of tropical fruits. p.1527-1555. In *Advances in Horticulture Vol.3 –Fruit Crops* (eds. K.L. Chadha and O.P. Pareek). Malhotra Publishing House, New Delhi.

Tandon, P.L. and Verghese, A. 1994. Present status of insect and mite pests in India. *Draksha Vritta,* **14:** 149-157.

Thakur, Yogeeta. 2002. *Studies on the biology and damage of the European red mite, Panonychus ulmi (Koch) on apple and evaluation of chemicals for its control.* Ph.D Thesis. Dr. Y.S Parmar University of Horticulture and Forestry, Nauni (Solan), HP.133pp.

Waite, G.K and McAlpine, J. D.1992. Honey bees as carriers of lychee erinose mite *Eriophyes litchii* (Acari: Eriophyiidae). *Exp. App. Acarol.,* **15:** 299-302.

Zou, Ping, Gao, Jian-Rong and Ma, En-Psi. 1993. Preliminary studies on the biology of the pest mite *Luciaphorus auriculariae* (Acari: Pygmephoridae) infesting Jew's ear mushroom, *Auricularia polytricha* in China. *Exp. Applied Acarol.,* **17**(3): 225-232.

Chapter-4

Insect Pests of Fruit Trees and their Management

TarunVerma and Ram Singh
CCS Haryana Agricultural University, Hisar - 125 004, Haryana

Fruits are very important in human diet, as they are chief sources of vitamins and micronutrients. Variable agro-climatic condition in India provides tremendous scope for cultivation of fruit crops. Among fruits, apple, pear, cherry, plum, peach, apricot, walnut and almond (temperate fruits); mango, grapes, banana, citrus, pineapple (tropical and subtropical fruits); pomegranate (arid zone fruit) are important fruit grown in India. Insect-pests are amongst one of the major constraint for the low productivity of fruit crops in India. Therefore, it is important to have some information of the important insect-pests of fruit tree and their management. Over 1000 species of insects found damaging fruit trees all over the world, of these as many as 800 insect species have been intercepted from India (Butani, 1979). The information on insect pests of apple and other temperate fruits in India have been reviewed by several workers (Pruthi and Batra, 1960; Butani, 1979; Bhalla and Gupta, 1993; Sharma and Thakur, 1996). In India, more than 200 insect species on apple trees (Bhalla and Pawar, 1977), 70 species on pear, 80 species on peach, 60 species on plum, 30 species on apricot (Butani, 1979), 85 species on grapevine (Tandon and Verghese, 1994), 175 species on mango, 182 species on banana (Dhaliwal *et al.*, 2006) and 250 species of insect and mites on citrus (Pruthi and Mani, 1945; Wadhi and Batra, 1964; Rajput and Hari Babu, 1985; Tandon, 1993) have been reported. The important insect-pests of fruits are as follows:

1. San Jose Scale, *Quadraspidiotus perniciosus* (Comstock) (Hemiptera : Diaspididae)

San Jose Scale is a most serious cosmopolitan species having broad host range attacking at least 34 host families and over 700 plant species in temperate region. The preferred fruit plants are apple, plum, pear and peach. It is an armoured scale and except for first nymphal stage and the adult males all life stage live under hard, protective scale covering. The scales of adult female are round, about 1.6-2.0 mm across, grey and flattened with a raised projection in the centre. Beneath her scale covering, the adult female is yellow, flattened, somewhat pear shaped and immobile (Horton *et al.,* 2008). Damage is caused by sucking cell sap by nymph and adults from aerial parts (Sharma and Thakur, 2004). The infested fruits show scaly appearance and spots surrounded by a scarlet/red area. Initially the growth of the plant is checked and later on the infested plants may die due to increase in scale number. In USA, even a single infested fruit on a tree is taken as an economic threshold for initiation of control measures (Whalon and Croft, 1984).

Life cycle : Pest is active from March to December and passes the winter in nymphal stage, again resume activity in spring season. They become full grown by April-May and start to reproduce by mid May. Fertilized female may give birth to 200-400 nymphs which lead a free life, settles down within 12-24 hours (Dhaliwal *et al.,* 2006) and travels 3 m distance in random direction with a travel rate of 9m/hour (Chadha and Awasthi, 2005). Shortly after setting crawlers insert their mouth parts into plant tissue and suck cell sap. Female remain immobile throughout their lives (Horton *et al.,* 2008). They become full grown in 30-40 days and female again start giving birth to young ones within the next 10-14 days. Life span of gravid mother is about 50-53 days. Male nymph has an elliptical or oval scale develops into a winged adult in 25-31 days. There are 4 overlapping generations in a year and fifth-generation nymphs overwinter (Dhaliwal *et al.,* 2006).

Management : Orchard sanitation (removal of alternate hosts, weed hosts, volunteer plants and crops residues) (Reddy, 2009) should be given priority. Collect infested pruned material and burnt them (Dhaliwal *et al.,* 2006).

Soil treatment with carbofuran at 6g/ meter square before raising the nursery and treating the plants with chlorpyriphos 0.05% before planting found effective to protect nursery from scale infestation (Reddy, 2009).

Predators like Coccinellids, *Chilocorus bijugus* Mulsant @ 10-20 beetle/tree (Rawat *et al.*, 1988) and *Pharoscymnus flexibilis* (Sharma and Thakur, 2004) shown the promising results. Repeated releases of two exotic parasitoids, *Encarsia perniciosi* (Tower) and *Aphytis* sp. (proclia group) have given promising results for suppression of the pest on apple (Masoodi and Trali, 1987; Masoodi *et al.*, 1989). The Russian; American and Chinesestrain of *E. periniciosi* gave 89 and 95% parasitism in Himachal Pradesh and Kumaon hills of Uttar Pradesh, respectively (Sahai and Joshi, 1965).

Dormant spray with diesel oil emulsion + Bordeaux mixture (diesel oil 68 litre + copper sulphate 15 kg + unslacked lime 3.75 kg) to be emulsified and diluted 5-6 times before spraying or spray 7.5 litres of ESSO tree spray oil emulsion in 250 litres water/ha during the winter season when trees are completely defoliated. Additional summer sprays with 1.25 litres of diazinon 20EC in 500 litres water/ha is required in case of severe scale infestation in neglected orchard management (Dhaliwal *et al.*, 2006). However, Khajuria and Sharma (1997) recommended 2 per cent miscible oil at half inch stage of apple bud development followed by 0.04% chlorpyrifos, in first week of May, whereas, Awasthi (2007) suggested spraying of trees with diazinon or methyl parathion 0.05% or methyl demeton 0.03% during summer to suppress the population of scale.

2. Woolly Apple Aphid, *Erisoma lanigerum* (Hausmann) (Hemiptera : Aphididae)

One of the most destructive pest of apple, pear , quince and crab apple in the world. The pest was first recorded at Conoor (Tamil Nadu) in 1989 and Kumaon and Shimla hills in 1909 (Lal and Singh, 1947), now widespread in all apple growing areas of India including North-eastern states. Adults are purplish colour, medium sized, 1.5-3.0 mm long covered by a white flocculent waxy secretion given off as woolly mass through the pores of their skin (Chadha and Awasthi, 2005). Damage is caused by sucking the cell sap from both the aerial and subterranean parts of plant as a result, phloem cells, cambium and xylem around feeding site undergo excessive cell division and elongation resulting into gall/knot formation (Sharma and Thakur, 2004). The infested plant show whitish cottony patches on the stem and branches. Young nursery plants affected the worst may die quickly (Dhaliwal *et al.*, 2006). Other symptoms are stunted size, loss of vigour, reduced fruiting capacity. Due to severe infestation and disintegration of the roots tree get easily uprooted by wind (Awasthi, 2007).

Life cycle : The peak activity period is between March to September and multiply at reduced rate during October to December. The pest reproduces by parthenogenetic viviparity. Developmental period is longer (32-51days) during winter with poor progeny per female (20-34 nymphs) while shorter (11-13 days) during summer months with higher fecundity (30-116 nymphs) (Sharma and Thakur, 2004) at the rate of 1-4 nymphs per day in March-April, 1-5 in May-July, 1-6 in August, and only 1-2 per week in winter months. There are 4 nymphal instar and duration of each varies according to the season. Nymphal period lasts for 35-42 days, 29.5 days and 10.5 to 19.5 days in February, August-November and April-July, respectively. About 13 generations are completed in a year. Partial aphid migration is there during December from aerial parts to root roots of infested plant and the case is reverse during the month of May (Dhaliwal *et al.*, 2006).

Management : The pest can be effectively controlled by growing tolerant/resistant root stocks like Golden Delicious, Northern spy, Morton stocks 778, 779,789 and 793 (Dhaliwal *et al.*, 2006) and rootstocks of M 21, M25 and MM series, *Malus baccata* L. var. *himaliaca* and grafted plants of variety Fanny (Thakur and Dogra, 1980; Rai and Tripathi, 1984).

Introducing an exotic parasite, *Aphelinus mali* (Hald.) @ 1000 adults or mummies/infested tree once, as early as infestation is started. About 63.3% parasitization by *A. mali* was noticed from mummies collected (Rawat *et al.*, 1989) at lower altitude whereas at higher altitude release of predators like *Metasyrphus confractor* and *Chrysopa* spp. (Thakur *et al.*, 1988) were found effective.

Injury caused during pruning should be healed or covered with Bordeaux mixture. Remove water shoots and weeds from orchard. To protect re-orientation of aphid from infested nursery always use healthy stocks and treat them with chlorpyriphos or fenitrothion 0.05% (Reddy, 2009).

During leaf fall stage (winter months) spray the tree with 1.25 litre of diazinon 20EC for control of aerial forms and spray with 1.0 litre of oxydemeton methyl 25 EC in 500 litres of water to protect from root forms of aphid (Dhaliwal *et al.*, 2006). Spraying with nicotine sulphate 0.04% or malathion 0.08% or 0.03% dimethoate found effective during summer. Fumigation against the root forms with paradichlorobenzene in a 15 cm deep trench dug around the infested apple tree about 2 metres from it, is very useful. Insecticides application should be avoided when predator and parasites activities started in the field (Awasthi, 2007).

3. Tent Caterpillar *Malacosoma indicum* Walker (Lepidoptera : Lasiocampidae)

Tent caterpillar is an important pest of apple, pear, apricot and walnut. Apple is its preferred host. Full grown caterpillar is 40-45 mm long, blackish-brown in colour. The male moth is light reddish while female moth is light brown in colour. The caterpillars are voracious defoliator. After 2-3 days of hatching caterpillar spin triangular or quadrangular tents of silk at forking of twigs and rest in the tent (hibernaculum) during day and feed on leaves during night, leaving behind only midribs and hard veins. The tent is enlarged as the caterpillar grows in size (Joshi and Agarwal, 1979). Under severe infestation, 40-50 per cent of apple plants may be defoliated producing a poor yield (Awasthi, 2007).

Life cycle : Activity of the pest is between mid March to May and remaining part of the year is passed in egg stage. Female lay about one mm long, shining and elongated thimble shaped eggs around the branches in broad bands or rings of 200-400 eggs. Incubation period, larval period and pupal period is 9 to 10 months, 39 to 68 days and 11 to 12 days, respectively. Adult emergence is during May and June (Sharma and Thakur, 2004).

Management : Collect and destroy egg bands in December-January at pruning. Destruction of silken tents with the help of a pole and rags dipped in kerosene between 12 noon to 3 p.m. and putting kerosenized water in an open vessel under tree to kill falling larvae. Spray the crop with 700 ml endosulfan 35 EC or 2.0 kg carbaryl 50WP in 500 lt. of water per ha (Dhaliwal *et al.*, 2006) or diazinon (0.03%) or carbaryl (0.05%) or endosulfan (0.05%) (Sharma and Thakur, 2004) or malathion 0.5% (Awasthi, 2007).

4. Peach Stem Borer, *Sphenoptera lafertei* Thompson (Coleoptera: Buprestidae)

This pest is widely distributed in northwest India and in southwest Afghanistan, UP , Haryana, Punjab and Pakistan. Its grubs feed by making tunnel under the bark of peach, almond, apricot, cherry, loquat, pear and plum trees. The beetles do not cause serious damage (feed on the foliage) and are blackish bronze and are 10-13 mm long. The grubs (tunnel under bark) and are smoky dark or black, club shaped and attain 18-24 mm body length. Gum globules ooze out of the entrance holes. Leaves of the attacked plants turn pale and their growth is arrested. Lakra *et al.* (1980) reported 45 % damage due to this pest on peach tree. The attacked branches shows small oval holes (about 1 cm long and

6mm broad) (Reddy, 2009), dry up and do not bear fruits. With continued damage the trees dies ultimately.

Life cycle : The beetle lay small, spherical, white eggs singly in crevices all over the tree trunk and the main branches. The eggs hatch in 20 days. The grubs are white with a small yellow head and greatly enlarged prothoracic segment, feed on the bark and as they grow they bore inside. From the entry hole a lump of resinous sum oozes out. The larval stage is completed in two months in summer, but those which overwinter takes 6 months. The larva makes a small chamber in the woody tissue about 10 mm deep from the surface. The pre-pupal and pupal periods lasts 1-2 and 8-12 days, respectively in summer. The winter is passed in the grub stage. Depending on the altitude 2-4 overlapping generation are completed in a year (Reddy, 2009).

Management : Proper training and pruning of the plant helps to keep plant straight, provides good canopy, prevent stem from direct sun exposure, thus tree vigour can be maintained. Covering the south facing parts of stem with muslin cloth or craft paper or gunny bags during winter acts as barrier for oviposition and provides some protection against pest. The dead or heavily gummed branches should be cut and destroyed. Chemical control methods should be synchronized with laying of broods, swab the entire trunk and main branches with 0.25 per cent of chlorpyriphos (Dhaliwal *et al.*, 2006). However, Sharma and Chander (1992) reported completely control of pest when sprayed with methyl parathion 0.05% or chlorpyriphos 0.05% synchronized with egg laying.

5. Peach Leaf Curl Aphid, *Brachycaudus helichrysi* (Kaltenbact) (Hemiptera : Aphididae)

The pest is highly polyphagous, worldwide, infesting about 175 species belonging to 115 genera into 49 plants families in India (Ghosh and Verma, 1988; Thakur *et al.*, 1995) and also vector of *Plum Pox* virus which imparts a sickly appearance to the host plant (Maison and Massonie, 1982). It is a very destructive pest of peach, plum, almond and other temperate fruits prevalent both in the plains and in the mountainous area of India. This plant louse is very small and is generally yellow, with dark stripes on the head. Its colour varies according to the host plant i.e. light green on peach, greenish yellow on golden rod. Nymphs and females damage floral and vegetative buds by sucking sap resulting into curling and distortion of leaves. Badly affected leaf whorl dry and shed. The severely affected plants have only crumpled or twisted leaves and bear very little fruit.

Life cycle : The females produce eggs without fertilization and these hatch inside the body of the mother. Thus, the females give birth to young ones instead of laying eggs. Each viviparous female produces about 50 young ones in her short lifetime of about 13 days. Three to four generations are completed on the fruits plants. With the warming up of the season, winged males and females are also produced. They migrate to other alternatives host such as golden rod and again start reproducing. Four or five generations are completed on golden rod from june to October. Early in November, the winged females are produced again. They migrate back to peach, plum and other fruit trees. In the plains another species, *Myzus persicae* (Sulzer) has been recorded as a serious pest of peach trees.

Management : Removal of weeds (secondary hosts) is essential of aphid management. Gupta *et al.* (1988) recommended regular pruning of current vegetative growth (to reduce egg population) during December. Avoid chemical sprays during March-April to enhance activity of biotic agents. Various predator and parasitoids *viz., Coccinella septempunctata* L., *Hippodamia* sp. (*Adonia variegata*) and *Anthocoris minki* (Reddy, 2009); *Leis dimidiata* F., *Coelopjora sauzeti* Muls., *Ischiodon scutellaris* (F.), *Paragus (Paragus) serratus* (Fabr.) and *Diaeretiella rapae* have been recorded on peach by Arora *et al.* (2009). Spraying of systemic insecticides namely, methyl demeton (0.025% or dimethoate (0.03%) at pink bud stage (Reddy, 2009); or monocrotophos 500 ml or dichlorvos 76EC 190 ml in 625 litre water/ha, one at pre bloom and other as post bloom at 10 day interval found effective (Dhaliwal *et al.*, 2006), whereas, Gupta *et al.* (1986) obtained best control on mature peach trees by application of aldicarb 15 g a.i./tree to the soil below tree canopy before bud burst stage.

6. Walnut Weevil, *Alcidodes porrectirostris* Marshall (Coleoptera: Curculionidae)

This is the most destructive pest of walnuts and is reported from Himalyan and sub mountain region (Gaffar and Bhat, 1990). The adult weevil is 9-12 mm long, jet black when young turning dark brown with age having prominent snout in bending downwards. The grub is legless, creamy white and measure 12-16 mm long with a pale-brown head. Both grubs as well as adults are destructive (Sharma and Khajuria, 2004). The grubs cause damage by feeding on walnut kernel and reduce it to a useless black mass. While the adult weevil feed on petioles, flowers buds, tender shoots and on young fruits and young fruits also (Hussain and Khan, 1949). The grubs are far more destructive than the adults as their feeding cause more than 50-70 per cent fruits drop and affecting the quality of fruit remaining on the trees (Dhaliwal *et al.*, 2006).

Life cycle : Female start laying eggs in April coinciding with the appearance of flower buds. As the fruits are formed, female weevil excavate pits for egg laying. One to 4 eggs (1.0-1.5 mm long and translucent) are laid per fruit (Sharma and Khajuria, 2004) but as many as 15 eggs in one fruit have also been noticed. The incubation period is one week and the tiny grubs bore deeper, feeding on the kernels (Dhaliwal *et al.*, 2006). The grub and pupae are creamy white and measure 12-16 mm and 9-12 mm in length, respectively. Larval and pupal period lasts for 13-22 days and 8-17 days, respectively. The adults emerge from the fruits by biting roundish holes. There are two generations in a year.

Management : Collect and destroy the fallen fruits every 10 days to check the population build up and their destruction simultaneously. Spray affected trees with quinalphos (0.04%) or dichlorvos (0.05%) (Sharma and Khajuria, 2004) or 2.0 kg of carbaryl 50 WP in 500 litres water/ha at fortnightly intervals when the young fruits are being formed (Dhaliwal *et al.*, 2006).

7. Long Horned Walnut Beetle, *Bactocera horsfieldi* Hope (Coleoptera : Cerambycidae)

This is a serious pest of walnut in Darjeeling, Kumaon hills, Kulu valley, Uttar Pradesh and Shimla hills. Beetles are 45-65 mm long, black in with fine ashy yellow gray pubescence and pronotum showing two elongated yellowish white spots. Their elytra have numerous shining black tubercles at the base and several rounded white marks extending upto the apex. The damage is caused by grub (pale yellow colour) by making zig zag tunnels on the inner side of the bark. One or two tunnels are found near base of tree but as many as 16 tunnels have also been observed upto 6 metre length of tree trunk (Beeson, 1941). Later on, the grubs bore down to the surface of sap wood and go even upto centre of the wood. The adult beetle feeds on the bark of young twigs and damage is negligible. The timber is damaged severely and attacked trees become useless except as fire wood, which hardly fetches the cost of cutting and marketing the wood.

Life cycle : The beetles emerge in June and July, and live for about 4 months. Gravid female lays 55-60 eggs singly (Dhaliwal *et al.*, 2006) on the bark of lower one metre portion of the trunk (Sharma and Khajuria, 2004). Incubation period is 8-15 days. The eggs are oval shaped and brown in colour. Initially the grub feed on the bark, then bore inside the wood and complete their development in 20-25 months. Pre-pupal stage and pupal stage lasts for 50-182 days and 40-90 days, respectively. The adult life span is 4-5 months during June-November and life cycle is completed in 23-32 months which is spread over 2-3 winters.

Management : Collection and destruction of grubs and adults check the population build up of the beetle. Plugging the live holes with cotton soaked in kerosene, petroleum or EDCT mixture and then plaster them from outside with mud found effective (Dhaliwal *et al.*, 2006).

8. Mango Hopper, *Amritodus atkinsoni* (Lethierry) (Hemiptera: Cicadellidae)

It is the most serious and destructive pest of mango at the flowering stage which directly affects the crop yield. The pest is prevalent all over India, Malaysia, Taiwan and Indonesia. It is 4.2-5.0 mm long, wedge shaped, broad head and tapering body, light greenish brown colour with yellow and black markings (Dhaliwal *et al.*, 2006). Both adults and nymphs suck sap from inflorescence during spring, infested flowers shrivel, turn brown and ultimately fall off. Heavy egg laying within florets and stalklets cause physical injury, resulting in withering of affected parts. These hoppers also excrete honeydew which encourages growth of fungi, *Capnodium mangiferae* and *Meliola mangiferae*, giving rise to sooty mould (Reddy, 2009) which affects photosynthetic activity of leaves thus adversely affects plant growth and yield (Godase *et al.*, 2004). Fruit setting does not occur on the affected inflorescences (Penat *et al.*, 2009). About 20 to 100 per cent loss of inflorescence (Ramachandra Rao, 1930) and more than 60 per cent (Dhaliwal *et al.*, 2006) crop damage have been observed due to this pest. Approximately, 18 species of mango hoppers have been reported as pests of mango (Pena *et al.*, 2009). Of these *Amritodus atkinsoni* (Leth.), *A. brevistylus, A. splendens, Idioscopus clypealis* Lethierry and *I. nitidulus* (Walker) are important. High humidity and moderately high temperature (February to April and July) for congenial for rapid development and reproduction of hopper (Reddy, 2009). The appearance of black foliage is a clear cut indication of its attack (Singh and Palaniswami, 2010).

Life cycle : Biology of *A. atkinsoni* have been studied by Sohi and Sohi (1990). It is multivoltine. Lay eggs singly embedding into plant tissue. The egg, nymphal and adult stage lasts for 7-9, 15-17 and 3-4 days, respectively. Complete development from egg to adult is completed in 25-30 days and 1-6 generation in India, 4-5 generation in Pakistan are completed in a year (Pena *et al.*, 2009). During flowering season 2 or more broods of the pest may occur (Dhaliwal *et al.*, 2006). Adults hibernate in winter (Awasthi, 2007).

Management : The parasitoid *Pipunculus annulifemer* Brun (November and December) and the egg parasites *viz., Aprostocetus* sp., *Gonatocerus* sp. *Polynema* sp. and *Tetrastichus* sp. were found effective against *I. nitidulus* and *A. atkinsoni* (Fasih and Srivastava, 1990). Natural occurrence of fungus *Isaria tax* on *A. atkinsoni* was also reported from Tripura by Kumar *et al.* (1985). Avoid high density planting, waterlogged or damp conditions as it provides favorable habitat for hopper multiplication, pruning of old dense orchard for better light interception (Dhaliwal *et al.*, 2006). Chemical control has been primarily employed to check the population of the mango leaf hopper. Fogging with malathion and diesel mixture in ratio of 1.5 : 8.5 was found cheaper and less time consuming than conventional spraying as it provided 0 per cent population of hopper after 3-7 days of fogging (Srivastava and Verghese, 1985). Spraying twice with 1.25 ml endosulfan 35EC or 2g carbaryl 50WP or 1.75 ml of malathion 50 EC per litre of water (February end and March) (Sharma, 2005; Dhaliwal *et al.*, 2006) or malathion LVC @ 1.4 litres per ha with aerial/ground pesticide application equipment (Dhaliwal *et al.*, 2006), imidacloprid (0.005%), methyl parathion (0.025%), monocrotophos (0.025%), fenitrothion (0.025%) before the opening of flowers and after fruit set (Singh and Palaniswami, 2010) were also found effective in controlling the pest.

9. Mango Fruit Fly, *Bactrocera dorsalis* (Hendel) (Diptera : Tephritidae)

The mango fruit fly is the most serious pest of all the fruit trees (guava, peach, apricot, cherry, pear, chiku, ber, citrus etc.) causing great problem in the export of fresh fruits. It is widely distributed in India, South-east Asia, Malaysia, Indonesia, Taiwan, the Mauritius, Pakistan, the Philippines, Australia and Hawaii Islands and other countries. High temperature and high humidity during May-July provides congenial condition for rapid multiplication and development (Reddy, 2009). In South India the pest activity remains throughout the year, whereas in northern India it hibernate during winter in pupal stage (Kapoor, 1993). The adult fly is stout larger than housefly, brown colour having transparent wing, yellow legs with dark rust red and black patterns on the thorax. Adult fly puncture the fruit with its sharp ovipositor (puncture attracts fermenting microorganisms) to oviposit below fruit epidermis (1-4 mm deep) (Reddy, 2009) and feed on exudation of ripe fruits (Dhaliwal *et al.*, 2006). The grubs are apodous, 8-9 mm long and 1.5 mm across the posterior end, yellow and opaque colour. These cause damage by feeding on the pulp, contaminating it with frass and providing entry for fungi and bacteria, thus rendering fruit unfit for human consumption.

Life cycle : The flies lays on an average 50 eggs in clusters of 2-15 at one time, deep into the soft skin of the mango fruit. Under favourable conditions 150-200 eggs are laid in one month. The incubation period is 2-3 days, 1-2 days and 10 days during March–April, summer winter, respectively. The maggots become full grown in about 6-29 days after passing through 3 instars in the ripening pulp, after that they leave the fruit and pupate inside the soil below 8-14 cm. The flies emerge in about 7-44 days which again starts multiplication on ripening fruits. The whole life cycle is completed in 2-13 weeks depending on the climatic conditions and there are several generations in a year.

Management : Collection and destruction of infested fallen fruits twice in a week, plough round the tree during winter (helps in reducing the pest carryover), followed by insecticide cover spray reduced fruit fly infestation between 95 to 99% (Verghese *et al.*, 2000). Early harvesting of fruit helps in reducing the damage of fly. Use at least 10 methyl eugenol bottle traps (0.1%) per hectare 45 days prior to harvest to monitor (Reddy, 2009; Mohyuddin and Mahmood, 1993; Dhaliwal *et al.*, 2006) and spray with 1.25 litres malathion 50EC + 12.5 kg gur/sugar in 1250 litres of water per ha (repeat sprays after 7-10 days, if required) (Dhaliwal *et al.*, 2006) or decamethrin 2.8EC at 0.5 ml/l + azadirachtin (0.3%) @ 2ml/l three weeks prior to harvest. Hot water treatment of fruits (at 48°C for 1 hour) maintained by thermostat and 5% salt solution at 55°C (30 minutes) without thermostat gave 100 % fruitfly infestation free fruits in *cvs.* Banganapalli and Totapuri (Reddy, 2009). Lindergren and Vail (1986) have investigated the use of nematode, *Steinernema feltiae* Filipjev against *Bactrocera*.

10. Mango Mealy Bug, *Drosicha mangiferae* Green (Homoptera : Coccidae)

This pest is widely distributed in all the mango growing part of India from Punjab, UP, Bihar, Delhi and also reported from Bangladesh, Pakistan and China. It has also been reported damaging 62 other plants including jack fruit, guava, citrus, jamun and the banyan (Dhaliwal *et al.*, 2006). These have large fleshy flat bodies with a length about 1.5 cm and breadth of about 1 cm covered with ashy-white mealy powders. Females (wingless) are oval shaped, flattened and body covered with white powdery mass while males are winged (1 pair) and crimson red colour. Damage is caused by the nymphs and adult female by sucking sap from shoots and flowers resulting into reduced plant vigour, flowers dry up, premature fruit fall. In severe damage tree retains no fruit at all (Dhaliwal *et al.*, 2006). The young fruits when infested become juiceless and fall off (Sharma, 2005a).

Life cycle : The adult gravid females after fertilization (April-May) start migrating below soil from 3rd week of April till May beginning and lay eggs (oval shaped, initially pink later changed to yellow) in soil upto 5-15 cm depth in silken pouches, each clusters having 300-400 eggs. Egg remain in diapauses in soil from May to December (Srivastava, 2004). Hatching takes place at December end or in January coinciding with fresh growth of flowers on mango trees. The young nymphs soon after hatching crawl about for some time to find suitable site and then congregates on panicle and begin to suck the sap. They moult thrice during their nymphal period which lasts about 3 months or more depending on the environmental conditions. Thereafter, the male forming nymphs undergo some sort of pupation and transform themselves into winged adult males and the female producing nymphs do not undergo any change except in size. Thus there is only one generation during the year.

Management : Digging or scrapping the soil 15 cm deep around trunk during April-May to destroy the eggs (Reddy, 2009; Dhaliwal *et al.*, 2006), application of sticky band incorporated with insecticides round the tree trunk, making wide insecticidal barrier round the trunk (Awasthi, 2007), removal of weeds, applying 8 cm sticky with alkathene (20 cm wide, 400 gauge) or plastic sheets around the trunk about one meter above ground level lead to effective control of the nymph. The nymphs congregating below the bands can be controlled by killing mechanically or by burning or shaking them off into a bucket of kerosinized water or spraying methyl parathion 0.1% (Awasthi, 2007) or 0.05% (Singh and Palaniswami, 2010) , diazinon or monocrotophos 0.04% (Awasthi, 2007) or 0.05% (Singh and Palaniswami, 2010) or soil treatment with methyl parathion 2% dust or aldrin 10% dust @ 250 g/ tree around trunk and spraying with 0.2% carbaryl (Singh and Palaniswami, 2010) found effective in controlling early instars of mealy bug. However, bio-dynamics preparation of karanj and neem provided 100% mortality of mealy bug within 7 and 9 days, respectively. However, garlic oil 0.1% was found at par with monocrotophos 0.05% and carbaryl 0.1% against crawlers (Reddy, 2009).

11. Mango Stone Weevil, *Sternochaetus mangiferae* Fabr. (Coleoptera : Curculionidae)

The pest is considered very important in view of major constraint in restricting the export to foreign countries. It is monophagous pest attacking mainly sweet varieties like Neelum, Alphonso and Bangalora and important pest in South India (late varieties) as it is sensitive to low temperature and low humidity. It is prevalent throughout the tropics. Verghese (1998) reported losses varying from 5 to 80 per cent due to this pest. The adult weevil is about 8 and 4 mm in length and breadth, respectively, grayish brown in colour resembling the general background of the bark of the mango tree. The damage is caused by grub (white colour) which bore through the pulp and feed on seed coat and then cotyledons. During adult emergence the fecal matter from seed dragged out into pulp, which results into spoilage of pulp (Reddy, 2009). The beetle remains hiding under the bark and other niches during non fruiting season (Awasthi, 2007)

Life cycle : The beetle lays eggs singly on surface of fruit (when marble size), wound caused by the ovipositor heals soon after and thus exhibiting no outward sign of infestation. One weevil lays about 300 eggs during a period of 3 months (15 eggs in a day) and as many as 12-36 eggs within a singly fruit. Shukla *et al.* (1988) studied intra-tree distribution of eggs on 'Baganpalli' mango and reported the highest number of eggs per fruit occurred on fruit in the lower region of the tree (upto 0-2 m height). About 87% weevils were found at a height of 0-2 m in the trunk compared to 7% at 2-4 m and only 4% above 4m. Incubation, larval and pupal period lasts for about a week, 5 weeks and 7 days, respectively, and finally transforms into the adult weevil. The life cycle is completed in about 40-50 days, but emerging adults resume breeding in the next season only and thus, one generation is completed in a year (Dhaliwal *et al.*, 2006).

Management : The pest can be suppressed by adopting following tactics : Regular sampling of fallen fruits (for forecasting of pest incidence), field sanitation (including collection and destruction of fallen fruits at weekly interval), ploughing (to kill hibernating weevil), raking of soil in October or November and March (provided partial weevil control), spraying main trunk and branches in Nov/December (prior to flowering) with diazinon (0.05%) or fenthion (0.08%) (for weevil hiding in bark) and fenthion @ 2ml/litre of water or deltamethrin (0.0025-0.003%) in March to mid April (for adult weevil), or monocrotophos (0.05%) followed by decamethrin (0.028%) or carbaryl (0.2%) (provided

> 90% control) (Sharma, 2005a). Among biological control agents, adult weevils may be susceptible to predation by ants, rodents, lizards, birds (Hansen, 1993) while larvae found attacking by a baculovirus (Shukla *et al.*, 1984). Besides this, a non-destructive x-ray inspection method has also been developed to detect weevil-infested fruit, which has a good potential for application in the processing industry and the export trade as a quality control (Thomas *et al.*, 1995). X-ray radiograph of infested mangoes show dark areas in the seed corresponding to disintegrated kernel tissue as a consequence of feeding by developing grubs.

12. Mango Stem Borer, *Bactocera rufomaculata* DeGeer (Coleoptera : Cerabycidae)

It is a quite serious pest of mango, fig and a very large number of other trees particularly when attacks main stem and branches of the trees. Beetle is well built, large, about 5 cm in length, greyish brown colour with white triangular patch at the junction of elytra, long antennae with 11 segments (Srivastava, 2004). While full grown larva is yellowish white, stout and measure about 6 cm in length (Dhaliwal *et al.*, 2006). Damage may be predicted oftenly by falling of leaves and sudden collapse of attacked branches. Borer cause damage by feeding on wood, making irregular cavities and tunnels in the peripheral region or deep down the base of the tree. While feeding, a harmonios sound is produced by its mandibles, thus also known as "violin beetle" (Reddy, 2009). The infested site of the pest can be located from sap or frass comes out of entrance hole (Sharma, 2005a).

Life cycle : The active season of pest is March-April. The pre-oviposition and oviposition period lasts for 24 hours and 20-24 days, respectively. About 200 eggs (brownish white cylindrical 6 x 2 mm, long with narrow rounded ends) are laid by females, one by one in incisions cut in the bark (Awasthi, 2007) at a depth of 2.5 to 4 mm. Eggs are generally deposited under loose bark of dead trees, or diseased trunks and branches of living trees and on the roots of trees exposed to erosion. The larval stage probably lasts for more than a year (as winter is passed in grub stage in tunnel) and pupal period lasts about 3-4 weeks (Dhaliwal *et al.*, 2006). The beetles emerge (50% and 30% in May and June, respectively) from March to August by cutting a circular exit hole (Awathi, 2007).

Management : The pest can be effectively controlled by adopting following measures : The infested branches carrying grubs and puape should be cut and destroyed (Sharma, 2005), killing grubs directly by inserting hard wire into tunnel and sealed with wet clay after applying petrol or kerosene oil (10ml) (Reddy, 2009). Remove frass and then inject 4 ml emulsion of methyl parathion (0.2%) or monocrotophos (0.05%) or dichlorvos (0.05%) and plug with mud (Sharma, 2005a), preventing stem by covering with stout paper coated with coal-tar or with wire gauge 1.5 mm mesh or spraying with Bordeaux mixture during oviposition period (Awasthi, 2007).

13. Bark Eating Caterpillar, *Indarbela* Species (Lepidoptera : Metarbelidae)

It is an important polyphagous pest widely distributed throughout India, Sri Lanka, Myanmar and Bangladesh. It has been found feeding on aonla, pear, jamun, pomegranate, ber, guava, orange, plum, litchi, mango, litchi and peach (Verma and Khurana, 1974), drumstick, mulberry, rose and number of ornamental and forest trees (Dhaliwal *et al.*, 2006). The pest has two species *viz., I. tetraonis* Moore and *I. quadrinotata* Walk. The adult is pale brown, large sized moth with a wing span of about 4 cm (female) and 3 cm (male). The forewings are pale rufous having several dark rufous bands while hind wings are fuscous. The larvae is dirty brown in colour which turns to pale brown when full grown (4 cm), nocturnal in habit. Generally the neglected and old trees are damaged more than vigorously growing trees. The caterpillar causes damage by feeding on the bark, making galleries with loose webbings and holes. As many as 16 holes may be present on a tree, each hole occupying one larvae or pupa. As a result of injury to the plant vessels, movement of sap is checked and both tree vigour as well as fruit bearing capacity adversely affected. The presence of silken galleries full of frass and excreta on bark surface extending from holes are the indication of its presence (Rohilla, 2005). In severe attack the whole tree may die (Awasthi, 2007).

Life cycle : Female start to lay in May-June and as many as 2000 eggs (in group of 15-20 each) are laid under loose bark of tree. The incubation, larval and pupal period lasts for about 8-10 days, 9-11 month and 3-4 weeks, respectively. Pupation takes place within the larval gallery. The adults emerges in summer and their life is quite short. The pest has only one generation in a year (Dhaliwal *et al.*, 2006).

Management : The pest can be controlled effectively by adopting following tactics : Orchard sanitation/cleaniness, locate hole after removing frass from gallery and kill caterpillar mechanically by inserting iron spike (Rohilla, 2005), plugging holes with cotton soaked in carbon bisulphide, chloroform, benzene, petro etc. (Thakur and Gupta, 2004) or in Feb –March, insert insecticide (40 g carbaryl 50WP or 10 ml fenitrothion 50EC or 5 ml methyl parathion 50EC or 30 ml endosulfan 35EC or 10 ml monocrotophos 36SL diluted in 10 lt. of water) soaked cotton plugs with help of metallic spike after removing the webbings and plaster on outer side with mud, treating all alternate host in vicinity of orchard (Dhaliwal *et al.*, 2006).

14. Anar Butterfly, *Virachola isocrates* Fabr. (Lepidoptera : Lycaenidae)

The pest is distributed all over India and adjoining countries wherever, pomegranate is grown. Pomegranate being the preferred host, aonla, mulbery, guava, taramind, loquat, sapota, litchi, citrus peach, pear, plum and apple are also attacked (Lakra, 2004). The adult males are glossy bluish violet while females are brownish violet in colour having an orange patch on forewing (Dhaliwal *et al.*, 2006). The larvae is stoutly built, dirty brown colour, 17-20 mm long with short hair and whitish patches all over the body. The damage is caused by the caterpillars which feed inside the fruit (on pulp and seed) and can destroy upto 50% fruits (Reddy, 2009). Offensive smell and excreta coming out or stuck around entry hole are the characteristics symptoms of its damage. Further such holes provide entry for fungi, bacteria, scavenging insects and birds, ultimately causing fruit rot. The affected fruits rot, show conspicuous hole on surface, finally drop-off prematurely and become unfit for human consumption.

Life cycle : The breeding of the pest is round the year on various host plants. The female butterfly lays eggs (shiny white, oval) singly mostly on calyx of flowers and on button size fruits. The eggs hatches out within 7-10 days and newly hatched larva bore inside fruit (where the rind is thin and soft). The larval and pupal period lasts for 18-47 days and 7-34 days, respectively. Pupation takes place inside the fruit and occasionally on branches/stem. Four generation of the pest are completed in a year (Lakra, 2004).

Management : The pest can be controlled by adopting following control measures : Bagging of fruits before maturity (Dhaliwal *et al.*, 2006), collection and destruction of infested fruits in the beginning of the fruiting season, lure away the adult butterflies (Awasthi, 2007), removal of flowering weeds of compositae family (Reddy, 2009), spray with decamethrin 2.8EC @ 1ml/l (>50% fruit set) followed by carbaryl 50WP @ 4g/l or fenvalerate 20EC @ 0.25ml/l or quinalphos 25EC @

2ml/l (after 2 weeks in non rainy season) (Reddy, 2009) or 0.1% phosalone (in December after pruning) followed by 2 sprays with 0.12% monocrotophos (February and April) (Nanjan and Kumar, 1983) or cypermethrin or fenvalerate @ 50 g a.i./ha or decamethrin @ 7.5 g a.i./ha (Kakar *et al.*, 1987) or weekly spraying of *Bacillus thuringiensis* @ 1 g/l (Reddy, 2009) or 0.05% endosulfan or 0.1% carbaryl (Awasthi, 2007). Besides this, Mani and Krishnamoorthy (1996; 1999) reported 60% parasitism of this pest by egg parasitoids *Telenomus* sp., *Trichogramma chilotraeae* and *Ooencyrtis papilionis*.

15. Lemon Butterfly, *Papalio demoleus* Linnaeus (Lepidoptera : Papilionidae)

This insect is widely distributed throughout India and several other countries including Africa, Taiwan and Japan. It attacks all cultivated and wild varieties of citrus (*Citrus sinensis* and *C. grandis* being preferred more, Des Raj and Nirmala Devi, 2004) and plants belonging to family Rutaceae (Awasthi, 2007). The pest occurred in epidemic form in 1940, 1969 and 1982-83 (Radke *et al.*, 1984) and recent outbreak (July-August, 1996) in Maharashtra (in Vidarbha region) on Nagpur mandarin (Shivankar and Singh, 1998b). The butterfly is fairly large (28 mm in length and 94 mm from wing expanse) with black colour head and thorax, while the underside of the abdomen is creamy yellow. The wings are dull black and ornamented with shiny yellow markings (Awasthi, 2007). The young larvae are brownish black which turn to green when full grown, measures about 40 mm long and 6.5 mm wide and has bifid 'Y' shape structure called osmaterium. The peak activity of the pest coincides with emergence of new flush (April, August and September). The larvae feed voraciously on vegetative growth, biting them from edges towards midrib (Reddy, 2009). In severe infestation complete defoliation is there and such trees bear no fruit (Awasthi, 2007).

Life cycle : Pests peak remain active throughout the year except winter months. A female lays about 75-120 eggs singly or in groups of 2-3 on young leaves and tender shoots within 2-5 days. The freshly laid eggs are pale green which turns to dark grey before hatching. The incubation and larval period lasts for 3-4 days and 8-16 days, respectively under favourable conditions and 5-8 days and 4 weeks, respectively under unfavourable conditions. Pupation takes place on the twigs or sticks and pupal period lasts about 8 days and 9-12 days under favourable and unfavourable conditions, respectively (Awasthi, 2007). Life cycle is completed in 13-15 weeks during winter while in 3-6 weeks during summer. Winter is passed in pupal stage. Four to six overlapping generation are completed in a year depending upon different locations (Des Raj and Nirmala Devi, 2004). Under laboratory conditions 10 (Atwal, 1964) and 9 generation (Bhan, 1994) of this pest has been observed.

Management : The pest can be controlled effectively by adopting following control measures : Hand picking of different stages of pest, release *Trichogramma chilonis* @ 500 adults/tree, spraying with *B. thuringiensis* var. *Kurstaki* @ 1 ml/l along with spreader or or dimethoate 30 EC @ 1.5 ml/l or fenitrothion 50 EC @ 1 ml/l at pest initiation (Reddy, 2009) or nematode DD-136strain or 3% neem seed extract (Dhaliwal *et al.*, 2006) or 0.05% malathion or 0.05% endosulfan or 0.1% carbaryl (Awasthi, 2007).

16. Citrus Red Scale, *Aonidiella aurantii* (Maskell) (Homoptera : Diaspididae)

It is amongst one of the most destructive pest (polyphagous) of citrus and thrives best in semi-arid climate. The females are red, circular, legless and flat centre whereas males are winged and elongated in shape. All plant parts (leaves, branches and fruits) above ground are affected by this pest. The damage is caused by both nymphs and adults by sucking cell sap from aerial plant parts resulting into yellow spots. The chlorophyll content of attacked branches is reduced which turn scurfy and start die.

Life cycle : The peak activity of pest is during March-April and in August-October. The reproduction is ovoviviparous (Sharma, 2005b). The first stage crawlers have well developed legs and antennae. Crawlers cover themselves with white waxy secretion. The female nymphs undergo two moult at an interval of 10-20 days, lose their legs and antennae. Casted skins incorporated into the waxy covering, becomes circular and depressed scales. In about 10-15 weeks, they become sexually mature, without wings and live for several months. Male scales develop into winged adults in about 1-2 months and then fly around to fertilize the sedentary females. Several generations of the pest are completed in a year.

Management : The scale can be controlled effectively by adopting the following control measures : Releasing *Aphytis melinus* and *C. nigritus* @ 15 adults/tree, spraying the trees with methyl parathion @ 0.05% or dichlorvos @ 0.04% or 4 ml chlorpyriphos 0.08% or monocrotophos 0.04% (Sharma, 2005b) or phosphamidon 0.03% or dimethoate or malathion 0.05% (Awasthi, 2007) as and when required.

17. Citrus Leaf Miner, *Phyllocnistis citrella*, Stainton (Lepidoptera : Phyllocnistidae)

This is a major pest of citrus nurseries in Assam, Haryana, Tamil Nadu, Madhya Pradesh, Uttar Pradesh, Punjab and in Pakistan. It is also reported from China, Malaya, Philippines, Formosa and East Indies (Rajput and Hari Babu, 1985). The adult moth is tiny silvery white with heavily fringed wings (wingspan 8-10 mm) while larvae are dull greenish yellow, cylindrical, measuring 5 mm in length and apodous (Sharma, 2005b). As high as 87.41% incidence of this pest has been reported in *ambia* flush in Maharashtra (Des Raj and Nirmala Devi, 2004). The larvae cause damage by making zigzag silvery serpentine mines in the young leaves and epidermis of tender twigs. The leaves of severely affected plants become distorted, crumpled, dries and finally fall off. Leaves of fresh growth, young seedlings and varieties having soft succulent leaves are suffering greater damage (Awasthi, 2007). Die back also occur when attacks is confined to twigs in young plants. Besides this, pest also helps in spreading mealybug infestation. About 30% damage and yield reduction upto 50% is accounted only by this insect out of total pest complex of citrus (Reddy, 2009)

Life cycle : The peak period of activity is April-May and the September-October. It prefers dry and hot weather. Female moth lays minute, flattened, yellowish-green eggs singly on the underside of the young leaves. The incubation, larval and pupal period lasts for about 2-7 days, 5-25 days and 5-21 days, respectively. Pupation occurs within the mines. The life cycle is completed in about 14-58 days and there are several generations in a year.

Management : The pest can be controlled by adopting following control measures : Pruning of infested branches from the inner canopy during winter months, avoiding frequent irrigation and application of nitrogen in split dosages (Reddy, 2009), growing resistant varieties like Carrizo, Sacaton, Savge, Troyer, Yama citrange, citrumelo, Campbell, Valencia, Pomary and Rubidoux (Batra *et al.*, 1992). It can also be effectively controlled by spraying with fenpropathrin and bifenthrin @ 0.01% (provided 100% reduction upto 7 days) or monocrotophos, quinalphos and acephate each @ 0.05% (give 80% reduction) (Shivankar and Singh, 1998a) or 250 ml cypermethrin 10EC or 875 ml decamethrin or 370 ml monocrotophos 36 SL in 250 litre water per ha (fortnightly) or 2% NSKE (Dhaliwal *et al.*, 2006) or fenvalerate 0.01% or permethrin 0.01% (Radke and Kandalker, 1990 Singh and Sharma, 1990) or dimethoate 0.05% (Tandon, 1993) or Lufenuron (14g a.i./ha) (Patel *et al.*,

1998). Among biocontrol agents, Shivnakar *et al.* (2008) reported 8.5-10.2% and 7.9-9.9% parasitization due to *Citrostichus phyllocnistoides* and *Cirrospilus quadristriatus*, respectively during October and November. Whereas among predators *Mallada boninsis* and *Menochilus sexmaculatus* were found efficient consuming 140.3-151.4 larvae and 184.7-201.8 pupae, respectively during month of January and February. Singh, 1995 recommended release of *Mallada boninensis* @ 20-50/plant depending on crop condition. While Reddy (2009) recommended combined release of *M. boninensis* and *Tamarixia radiate* @ 30 larvae and 40 adults /tree, respectively which gives 23-26% reduction of leaf miner.

18. Citrus Psylla, *Diaphorina citri* Kuwayana (Hemiptera : Aphalaridae)

Psylla is most destructive pest of citrus and several plants belonging to family Rutaceae and has been reported from India, Taiwan, China, Myanmar, Japan, Sri Lanka, the East Indies and New Guinea. The adults are brownish, 3 mm in length, with raised hind ends, rests on the leaf surface with closed wings. The male are distinguishable by upturned tip of its abdomen. The nymphs are circular, flat, orange-yellow in colour. Both adults and nymphs cause damage by sucking the cell sap in millions from floral buds, tender shoots and newly emerged leaves (Sharma, 2005b). About 83-95% loss has been reported due to this pest alone (Reddy, 2009). The affected plant loses its vitality and the growth is also checked. The infested branch start die off from tip backwards and branches adjacent to these dries up due to toxins injected by the insect. Besides this, nymphs also secrete some fluid on which black fungus develops which ultimately affects photosynthesis.

Life cycle : The pest breeds throughout the year (Bindra, 1957; Lakra *et al.*, 1983; Khan *et al.*, 1984). Female lays about 800-900 almond-shaped, orange colour, stalked eggs on half open leaves, floral buds and tender shoots of citrus trees either singly or in cluster of 2-3. (DesRaj and Nirmala Devi, 2004). As many as 50 eggs (arranged in straight line) may be laid in one place. The incubation period is 4-6 days and 10-20 days in summer and winter, respectively. Development is completed in 10-15 days (summer) and 34-36 days (winter) after passing through 5 nymphal stages. About 8-9 overlapping generations are completed in a year.

Management : Removal of collateral host like curry leaf, conservation of natural enemies like *Tamarixia radiate* Waterston (parasitizing 95% and 30-90% in Punjab and Maharashtra) (Reddy, 2009),

Tetrastichus phyllocnitoids (Narayanan), *Diaphorencyrtus* sp., lady bird beetles, *viz., Coccinella septempunctata* Linnaeus, *Menochilus sexmaculatus* (Fabricius) and *Brumoides suturalis* (Fabricius) and *Chrysoperla carnea* (Stephens) (Sharma, 2005b) helps in reduction of pest. Pest can be effectively controlled by spray 1.70 litres of dimethoate 30EC or 1.25 litres of malathion 50 EC or 500 ml of fenitrothion 50 EC in 250 litres water/ha during February-March (spring flush), May -June (before rains) and July-August (after rainy season) (Dhaliwal *et al.*, 2006) or phosphamidon, quinalphos and thiometon @ 0.025% (100% mortality) at bud burst or as and when required (Reddy, 2009).

19. Grapevine Thrips, *Rhipiphorothrips cruentatus* Hood (Thysanoptera : Heliothripidae)

It is an important pest of grapevine and highly polyphagous, become active in March -May and September-October. The adults are bright yellow (male) to dark brown (females) in colour, 1.4 mm long, while nymphs are reddish in colour. Rasping and sucking insects cause damage (nymphs and adults) by feeding on ventral side of leaves, flower stalks and mustard sized berries and sucking the oozing cell sap (Reddy, 2009). The characteristics symptoms shown are silvery white leaves and scortchy patches with curly tips. The feeding on berries reduces market and export value due to scab formation. The leaves drop off from infested plant and fruits formation on these vines either not occur or fruit formed drop off prematurely.

Life cycle : The pest continues to breeds throughout the year except in winter. About 50 eggs (dirty white and bean shaped) are laid by female by making small slits in the plant tissue on undersides of leaves (placing one egg in one slit). The incubation period, nymphal period and pupal period lasts for 3-8 days, 9-20 days and 2-5 days, respectively. Both sexual (giving rise to females) and parthenogenetic reproduction (giving rise to only males) is there. From December -March it hibernate as "pupa" in the soil upto depth of 15 cm. several generations of the pest are completed in a year (Awasthi, 2007).

Management : The pest can be controlled by adopting the following control measures : Removal of grasses and pruning of infested leaves (Dhaliwal *et al.*, 2006), raking the soil periodically and spraying the vines with acephate 75SP @ 1g/l (before flowering) followed by endosulfan 35EC @ 0.07% or carbaryl 50 WP @ 0.2% (Post berry formation) (Reddy, 2009) or nicotine sulphate 0.04% or malathion 0.05% (at fruit setting) (Awasthi, 2007) or 1.5 Kg carbaryl 50 WP in 500 litres. of water/100 vines (preferred in Perlett var. due to thinning effect on berries) (Dhaliwal *et al.*, 2006) or spray spinosad 45 EC @ 0.25ml/litre (in exportable grapes as it has shorter preharvest interval and safety to natural enemies, *Crytolaemus montrouzieri*) (Reddy, 2009).

20. Grapevine Girdler, *Sthenias grisator* Fabricious (Coleoptera : Cerambycidae)

Grapevine girdler occurs throughout the grape growing areas India and Sri Lanka. It has also been found damaging almond, cashew, jack fruit, mango, mulberry and several garden plants. The adult is a medium sized, 24 mm long, grayish brown with irregular white and brown markings and eye patch on elytra (Verghese, 2004) resembling a mottled bark. Full grown grub measures 10-12 mm have dark brown head, globular thorax and a pair of prominent mandibles. The damage is caused by the adult making girdle (1.25-2.5 cm thickness) on the young green branches with its strong mandible before laying eggs, resulting into vines dry up above the level of girdling.

Life cycle : Peak period of damage is between August -October. Female deposits eggs (oval shape, 3-4 mm long and 1 mm wide) in cluster of 2-5 at night underneath the bark of girdled vines. Hatching takes place in about 7-8 days. Soon after hatching tiny (2-4 mm) grub tunnels into wood and start feeding on wood. The pupation occurs in the tunnel. The complete life cycle takes more than a year (Awasthi, 2007). Adults hibernate during winter (Dhaliwal *et al.*, 2006).

Management : Cutting and burning of attacked branches below the girdling point, hand picking of adult beetles (Dhaliwal *et al.*, 2006) and spraying the vine at ground level with 0.05% malathion (Awasthi, 2007) or monocrotophos or chlorpyriphos (Sharma, 2005b) were found effective.

21. Grapevine Leaf Roller, *Sylepta lunalis* Guenee (Lepidoptera : Pyralididae)

This is a serious of grapevine in Karnataka, Tamil Nadu, Andhra Pradesh, Maharashtra, Uttar Pradesh and Punjab. The moths are dirty brown in colour and are provided with white spots on wings. Wing span of males and females moths are 21-25 mm and 24-28mm, respectively. The full grown larvae measures about 26 mm, brownish black head and body covered with grey. The first to three instar caterpillar feed on the lower epidermis of the leaves, skeletonize them while 4th and 5th instars roll up the leaf margin and each roll contains only one caterpillar .

Life cycle : The pest is active during August-October (Sharma, 2005b). The female moth lays about 100-120 oval, creamy white, about 1.5 mm size eggs on the under surface of the leaves. Incubation period, larval period and pupal period lasts for 2-3 days, 14-20 days and 5-7 days, respectively. The pupation occur within rolled leaf. Sometime pupa drops down onto the fallen dry leaves and debris (Verghese, 2004). One generation is completed in a year.

Management : The pest can be effectively controlled by removing and destructing the rolled leaf having larvae or pupae within or spraying the vines with malathion 0.05%, dimethoate 0.04%, phosphamidon 0.04% (Awasthi, 2007).

22. Banana Weevil, *Cosmopolites sordidus* (Germar) (Coleoptera : Curculionidae)

It is an oligophagous and most destructive pest of banana. Pest originated within the Indo-Malayan regions (Waterhouse, 1993), currently distributed throughout India, South East Asia, parts of Australia, tropical America, tropical and South Africa and the Hawaii Islands. In India it has been reported from Assam, Tamil Nadu, Kerala, Delhi, Karnataka and Maharashtra (Reddy, 2009). The adult weevils are sluggish, shiny black, head more or less prolonged into a snout and body is covered with scales (Dhaliwal *et al.*, 2006). The weevil causes damage by destructing the corm tissue. The vigour and yield of plant is affected when attacks during late stages of plant growth, while cessation of growth and gradual death occur during early infestation stage. Seasonal varieties are attacked less as comparison to perennial varieties of plantain (Faleiro, 2005).

Life cycle : The female weevil bites a small hole, lays 10-50 eggs singly either on collar region (above ground or rhizome under ground) and continuous this activity on other plants throughout the year. The incubation period is about a week. After hatching the larvae bore into the corm, and feed by making a tunnel. The larval and pupal period lasts for 2-6 weeks and about a week, respectively. Pupation occurs in the soil. After emergence the adults for sometimes feed on the underground parts of the plants, then approaches the growing point for oviposition (Dhaliwal *et al.*, 2006). The adults feed during night and remain hidden in rotting pseudostem during day time.

Management : The pest can be controlled effectively by adopting following control measures : Planting healthy, uninfested suckers or rhizome; clean cultivation; planting less susceptible varities like Poovan,

Kadali, Kunnan, Chetti, Basrai, Sawaii, Pisang Seribu, and Poomkalli, dipping suckers in chlorpyriphos 20EC @ 2.5ml/litre (Reddy, 2009) or monocrotophos (0.05%) before planting (Faleiro, 2005) or 0.1 per cent quinalphos emulsion; destroying the sheltering and feeding places of the adult weevils; chopping of pseudostem after cutting bunches and their scattering in plantation; cutting pseudostem close to ground level or spraying 625 ml dimethoate 30 EC or fenitrothion 50EC/ha around the base of the plants or clumps (Dhaliwal *et al.*, 2006) or installing pheromone lure (cosmoloure) @ 5 traps/ha (Faleiro, 2005).

Apart from the insect-pests discussed above a large number of the other insect-pests are also found attacking fruit trees during part of the year, which are given as under :

Common Name	Scientific Name	Order & Family	Fruit Trees
Gypsy moths	*Lymantria dispar* (Linnaeus) *L. obfuscata* Walker	Lepidoptera : Lymantriidae Lepidoptera :	Temperate fruits
Codling moth	*Cydia pomonella* (Linnaeus)	Tortricidae	
Apple stem borer Cherry stem borer Apple root borer Almond weevil Flat-headed apple tree borer	*Apriona cinerea* Cheverlot *Aeolesthes holosericea* Fabricius *Darysthenes hugelii* *Myllocerus lactivirens* Redtenbacker *Chrysobothris* Marshal *femorata* (Olivier)	Coleoptera : Cerambycidae Coleoptera : Cerambycidae Coleoptera : Cerambycidae Coleoptera : Curculionidae Coleoptera : Coleoptera : Buprestidae	Temperate fruits
Green aphid Melon aphid Cherry aphid Peach green aphid	*Aphis poni* DeGeer *Aphis gossypii* Glover *Myzus cerasi* (Fabricius) *Myzus persicae* (Sulzer)	Hemiptera: Aphididae	Temperate fruits
Peach mealy aphid Peach black aphid	*Hyalopterus pruni* (Geoffroy) *Pterochlorus persicae* Cholodkovisky	Hemiptera: Aphididae	Temperate fruits
Apricot chalcid	*Eurytoma samsonovi* Vasiljev	Hymenoptera : Eurytomidae	
Citrus aphid	*Toxoptera citricidus*	Hemiptera : Aphididae	Citrus
Soft scale	*Coccus hesperidum* L., *C. formicarii*	Hemiptera : Coccidae	Citrus Mango

Contd.

Mango coccid	*Rastrococcus iceryoides* Green		Mango
Hemispherical scale	*Saissetia coffeae*	Hemiptera : Coccidae	Citrus
Citrus blackfly	*Aleurocanthus woglumi* Ashley	Hemiptera : Aleyrodidae	Citrus
Cotton cushion scale	*Icerya purchase* Maskell	Hemiptera : Margarodidae	Citrus
Mormon butterfly	*Papilio memnon* Linnaeus *P. polytes* Linnaeus	Lepidoptera : Papilionidae	Citrus
Fruit sucking moth	*Ophideres* spp.	Lepidoptera : Noctuidae	Citrus
Armyworm	*Spodoptera litura* (Fabricius)	Lepidoptera : Noctuidae	Citrus
Blossm midge	*Dasineura citri* Grover	Diptera : Cecidomyiidae	Citrus
Melon fly	*Bactrocera cucurbitae* (Coquillett)	Diptera : Tephritidae	Citrus
Fire ant	*Selenopsis germinate* (Fabricius)	Hymenoptera : Formicidae	Citrus
Grapevine leafhopper	*Erythroneura* sp., *Arboridia viniferata* Sohi & Sandhu	Homoptera : Cicadellidae	Grape
Mealy bug Stripe mealy bug	*Pseudococcus* spp. *Ferrisia virgata* (Cockerell)	Hemiptera : Pseudococcidae	Grape, Banana Pomegranate
Horn worm	*Theretra alecto* Linnaeus	Lepidoptera : Sphingidae	Grape
Lac insect	*Kerria communis*	Hemiptera : Lacciferidae	Grape
Grapevine beetle	*Sinoxylon anale* Lesne	Coleoptera : Bostrychidae	Grape
Defoliating beetle	*Anomala dimidiate* (Hope), *Adoretus* spp.	Coleoptera : Scarabaeidae	Grape
Wasps	*Polistes hebraeus* (Fabricius), *Vespa orientlis* Linnaeus	Hymenoptera : Vespidae	Grape
The Hard scales	*Aspidiotus destructor* Signoret, *Chionaspis vitis* Green	Hemiptera : Diaspididae	Mango
Red tree ant	*Oecophylla maragdina* (Fabricius)	Hymenoptera : Formicidae	Mango
Leaf galls	*Amaraemyia* spp.	Diptera : Cynipidae	Mango

Contd.

Mango leaf twisting beetle	*Apoderus tranquebaricus* Fabricius	Coleoptera : Anthribidae	Mango
Mango leaf mining weevil	*Rhynchaenus mangiferae* Marshall	Coleoptera : Curculionidae	Mango
Leaf webber	*Orthaga equdrusalis* Walker	Lepidoptera : Pyralidae	Mango
Mango shoot gall	*Apsylla cistella* (Buckton)	Hemiptera : Psyllidae	Mango
Whitefly	*Siphoninus phillyreae*	Hemiptera : Aleyrodidae	Pomegranate
Aphid	*Aphis punicae* Passerini	Hemiptera : Aphididae	Pomegranate
Leaf thrips	*Retithrips syriacus* (Mayet)	Thysanoptera : Thripidae	Pomegranate
Bagworm	*Clania cramerii*	Lepidoptera : Psychidae	Pomegranate
Leaf eating caterpillar	*Porasa lepida* (Cramer)	Lepidoptera : Limacodidae	Pomegranate
Castor capsule borer	*Dichocrocis punctiferalis* (Guenee)	Lepidoptera : Pyralidae	Pomegranate
Castor semilooper	*Achaea janata* (Linnaeus)	Lepidopter : Noctuidae	Pomegranate
Fruit fly	*Bactrocera zonata*	Diptera : Tephritidae	Pomegranate
Tussock caterpillar	*Euproctis scintillans* W	Lepidoptera : Lymantridae	Pomegranate
Banana scale moth	*Nacolela octasema* (Meyrick)	Lepidoptera : Pyralidae	Banana
Banana aphid	*Pentaloni nigronervosa* Coquerel	Hemiptera : Aphididae	Banana
Thrips	*Heliothrip kadaliphila* R & M, *Chaetanaphothrips signipennis*	Thysanoptera : Thripidae	Banana
Leaf eating caterpillar	*Parasa lepida* (Cramer)	Lepidoptera : Limacodidae	Banana
	Tiracola plagiata (Walker)	Lepidoptera : Noctuidae	Banana
Pseudostem borer	*Odoiporus longicollis* (Olivier)	Coleoptera : Curculionidae	Banana

References

Arora, R.K., Gupta, R.K. and Bali, K. 2009. Population dynamics of the leaf curl aphid, *Brachycaudus helichrysi* (Kalt.) and its natural enemies on subtropical peach, *Prunus persica* cv. Flordasun. *J. Entomol. and Nematol.*, **1** (3) : 36-42.

Atwal, A.S. 1964. Insect pests of citrus in Punjab. VI. Biology and control of citrus leaf miner, *Phyllocnistis citrella* Station (Lepidoptera : Phyllocnistidae). *Punjab Hort.*, **4** (2) : 100-103.

Awasthi, 2007. Pests of fruits and fruit trees. pp. : 169-190. In *Agricultural insect-pests and their control.* Scientific Publishers (India).

Batra, R.C., Sharma, D.R. and Chanana, Y.R. 1992. Screening of citrus germplasm for their resistance against citrus leaf miner, *Phyllocnistis citrella* Stainton. *J. Insect Sci.*, **5**(2) : 150-152.

Beeson, C.F.C. 1941. *The Ecology and Control of the Forest Insects of India and the Neighbouring Countries.* F.R.I., Dehra Dun.

Bhalla, O.P. and Gupta, P.R. 1993. Insect pests of temperate fruits. pp : 1557-1589. In : *Advances in Horticulture : Fruits Crops.* (eds) K.L. Chadha and O.P. Pareek. Vol. **3**, Malhotra Publishing House, New Delhi, India.

Bhalla, O.P. and Pawar, A.D. 1977. *A Survey Study of Insect and Non-insect Pests of Economic Importance in Himachal Pradesh.* Bulletin Department of Entomology and Zoology, College of Agriculture, Solan, Himachal Pradesh, India.

Bhan, R. 1994. *Studies on the bionomics and management of lemon butterfly (Papilio demoleus* Linnaeus) *on citrus.* M.Sc thesis, Sher-e-Kashmir University of Agricultural Sciences and Technology (J & K). pp : 85.

Bindra, O.S. 1957. Insect pests of citrus and their control. *Indian J. Hort.*, **14**: 89-98.

Butani, D.K. 1979. *Insect and Fruits.* Periodical Expert Book Agency, New Delhi, India.

Chadha, K.L. and Awasthi, R.P. 2005. Insect and mite pests. pp : 346-396. In : *The Apple : Improvement, Production and Post harvest Management.* (eds) K.L. Chadha and R.P. Awasthi, Malhotra Publishing House, New Delhi.

Des Raj and Nirmala Devi. 2004. Pests of citrus. pp : 325-338. In : *Pest Management in Horticulture Crops; Principles and Practices* L.R. Verma, A.K. Verma and D.C. Gautam, Asiatech Publishers Inc., New Delhi.

Dhaliwal, G.S., Singh, Ram and Chhillar, B.S. 2006. Pests of fruits. pp : 290-317. In : *Essentials of Agricultural Entomology.* (eds.) G.S. Dhaliwal, Ram Singh and B.S. Chhillar , Kalyani Publisher, New Delhi.

Faleiro, J.R. 2005. Advances in Integrated Pest Management of banana. In : *Advances in the Integrated Pest Management of Horticultural, Spices and Plantation Crops.* (eds.) B.S. Chhillar, V.K. Kalra, S.S. Sharma and Ram Singh. Centre of Advanced Studies (ICAR), Department of Entomology, CCS Haryana Agricultural University, Hisar (Haryana) India.

Fasih, M. and Srivastava, R.P. 1990. Parasites and predators of insect pests of mango. *Inter. Pest Cont.*, **32** (2) : 39-41.

Gaffar, S.A. and Bhat,A.A. 1990. Record of walnut weevil, *Alcidodes porrectirostris* (Marshall) (*Neomecyslobus porrectirostris*) : A new insect pest on walnut in Karnah, Kashmir. *Bull. Ent.*, **31** (2) : 231.

Ghosh, L.K. and Verma, K.D. 1988. A new host record of *Brachycaudus helichrysi* (Kaltenbact) in India (Homoptera : Aphididae). *J. Aphid*, **2** (1 & 2) : 66-68.

Godase, S.K., Bhole, S.R., Shivpuje, P.R. and Patil, B.P. 2004. Assessment of yield loss in mango (*Mangifera indica*) due to mango hopper (*Idioscopus niveoparvasus*) (Homoptera : Cicadellidae). *Indian J. Agric. Sci*, **74** : 370-372.

Gupta, B.P., Joshi, R. and Tripathi, G.M. 1986. Control of peach leaf curling aphid, *Brachycaudus helichrysi* (Kalt.) with granular formulations of some insecticides. *Prog. Hort.*, **18** : 129-131.

Gupta, P.R., Thakur, J.R. and Dogra, G.S. 1988. Management of peach leaf curl aphid, *Brachycaudus helichrysi* (Kalt.) infesting stone fruits in Himachal Pradesh. *J. Tree Sci.*, **7** : 26-30.

Hansen, J.D. 1993. Dynamics and control of the mango seed weevil. *Acta Horticulturae*, **341** : 415-420.

Horton, D.L. , Fuest, J. and Cravedi, P. 2008. In : *The Peach : Botany, Production and Uses* (Desmond R. Layne and Daniele Bassi. (eds) pp : 467-504.

Hussain, M.A. and Khan, M.A.W. 1949. Bionomics and control of the walnut weevil (*Alcidodes porrectirostris* Mshll.) (Alcidinae : Coleoptera). *Ind. J. Ent.*, **11** : 77-82.

Joshi, K.C. and Agarwal, S.B.1979. Tent caterpillar and its control. *Indian Horticulturist*, **24** (1) : 11-13.

Kakar, K.L., Dogra, G.S. and Nath, A. 1987. Incidence and control of pomegranate fruit borer, *Virachola Isocrates* (Fabr.) and *Deudorix epijarbas* (Moore). *Indian J. Agric. Sci.*, **57** : 749-752.

Kapoor, V.C. 1993. *Indian Fruit Flies (Insecta : Diptera: Tephritidae).* Oxford & IBH Publishing Co. Pvt. Ltd., New Delhi, pp : 228.

Khajuria, D.R. and Sharma, H.K. 1997. Use of miscible oil and insecticidal combinations in management of SanJose Scale (*Quadraspidiotus pernciosus*) on apple (*Malus domestica*). *Indian J. Agric. Sci.*, **67** : 488-489.

Khan, K.M., Radke, S.G. and Borle, M.N. 1984. Seasonal activity of citrus psylla, *Diaphorina citri* Kuwayama in the Vidarbha region of Maharashtra. *Bull. Ento.*, **25** (2) : 143-146.

Kumar, D., Roy, C.V., Yazdani, S.S., Hameed, S.F. and Khan, Z.R. 1985. Efficacy of some insecticides against hopper complex on mango (*Mangifera indica* Linn.). *Pesticides*, **19** (11) : 42-43.

Lakra, R.K. 2004. Insect pests of fruits of arid and semi-arid regions of India. pp : 347-386. In : *Pest Management in Horticulture Crops; Principles and Practices* (eds.) L.R. Verma, A.K. Verma and D.C. Gautam, Asiatech Publishers Inc., New Delhi.

Lakra, R.K., Gupta, D.S. and Bhanot, J.P. 1980. Nature and extent of damage and varietal susceptibility to peach borer, *Sphenoptera lafertei* Thompson. *Indian J. Ent.*, **42** : 233-239.

Lakra, R.K., Singh, Z. and Kharub, W.S. 1983. Population dynamics of citrus psylla, *Diaphorina citri* Kuwayama in Haryana. *Indian J. Ent.*, **45** (3) : 301-310.

Lal, K.B. and Singh, R.N. 1947. Seasonal history and field ecology of the woolly aphids in the Kumaun Hills. *Indian J. Agric. Sci.*, **17** : 211-218.

Lindegren, J. and Vail, P. 1986. Susceptibility of Mediterranena fruit fly, melon fly and oriental fruit fly (Diptera : Tephritidae) to the entomogenous nematode, *Steinernema feltiae* in laboratory tests. *Environ. Entomol.*, **4** : 465-468.

Maison, P., Massonie, G. 1982. First observation on the specificity of the resistance of peach to aphid transmission of plum pox virus. *Agronomie*, **2** (7) : 681-683.

Mani, M. and Krishnamoorthy, A. 1996. First record of *Ooencyrtis papilionis* on pomegranate butterfly, *Deudoris Isocrates* Fabr. (Lycaenidae : Lepidoptera). *Entomon*, **21** : 275-276.

Mani, M. and Krishnamoorthy, A. 1999. First record of *Trichogramma chilotreae* Nagaraja and Nagarakatti on pomegranate butterfly, *Deudorix* Isocrates (Fabr.). *Entomon*, **24** : (in press).

Masoodi, M.A. and Trali, A.R. 1987. Seasonal history and biological control of san jose scale, *Q. perniciosus* (Comstock). *J. Biol. Control*, **1** : 3-6.

Masoodi, M.A., Trali, A.R., Bhat, A.M., Tikku, R.K. and Nehru, R.K. 1989. Establishment of *Aphytis* sp. *Proclia* of san jose scale in Kashmir. *Indian. J. Pl. Protec.*, **17** : 71-73.

Mohyuddin, A.I. and Mahmood, R. 1993. Integrated control of mango pests in Pakistan. *Acta Horticulturae*, **341** : 467-483.

Nanjan, K. and Kumar, S. 1983. Notes on the control of fruit borer in pomegranate. *South Indi. Hort.*, **31** : 104-105.

Patel, J.J., Patel, J.R., Valand, V.M., Patel, B.H. and Patel, M.J. 1998. Bioefficacy of some of the new insecticides against leaf miner, *Phyllocnistis citrella* and Psylla, *Diaphorina citri* infesting citrus. *Indian J. Ent.*, **60** (1) : 50-56.

Pena, J.E., Aluja, M. and Wysoki, M. 2009. Pests. pp. : 317-366. In : *The Mango, 2nd Edition Botany, Production and Uses*. Richard E. Litz. (ed.), Tropical Research and Education Centre and Centre for Tropical Agriculture, University of Florida, USA.

Pruthi, H.S. and Batra, H.N. 1960. *Important fruit pests of North-West India*. Bulletin No. **80.** Indian Council of Agricultural Research, New Delhi.

Pruthi, H.S. and Mani, M.S. 1945. Our knowledge of insect and mite pests of citrus in India and their control. *Scientific Monograph.* Imperial Council of Agricultural Research, New Delhi. pp : 16-42.

Radke, S.G. and Kandalker, H.G. 1990. Chemical control of citrus leaf miner. *Indian J. Ent.*, **52** (3) : 397-400.

Radke, S.G., Borle, M.N. and Vawalkar, K.S. 1984. *National Seminar Pest Management Citrus, Cotton, Sugarcane and Sorghum. Progress and Problem,* PKVK (MS), Nagpur, 5-7 February, 1984. pp : 112.

Rai, K.M. and Tripathi, G.M. 1984. Resistance of various rootstocsk to the attack of apple wooly aphid, *Erisoma lanigerum* (Hausmann). *Progress Horti.*, **16** : 359-362.

Rajput, C.B.S. and Hari Babu, R. Sri. 1985. *Citriculture*. Kalyani Publishers, New Delhi-Ludhiana. pp. : 364.

Ramachandra Rao, Y. 1930. Mango hopper problem in South India. *Agricultural J. India*, **25** : 17-25.

Rawat, U.S., Pawar, A.D. and Singh, R. 1989. Establishment of *Aphelinus mali* (Hald.) in cold and dry zones of Kinnaur district in Himachal Pradesh. *J. Insect Sci.*, **2** : 109-113.

Rawat, U.S., Thakur, J.N. and Pawar, A.D. 1988. Role of natural enemies in integrated control of San Jose scale and woolly aphid in Himachal Pradesh. pp. : 182-187. In : *Proceedings of National Symposium on Integrated Pest Control; Progress and Perspectives.* Association for advancement of Entomology, Trivandrum, Kerala, India.

Reddy Parvatha, P. 2009. In : *Advances in Integrated Pest and Disease Management in Horticultural Crops,* Vol. 1, *Fruit Crops.* Studium Press (India) Pvt. Ltd., New Delhi.

Rohilla, H.R. 2005. Advances in Integrated Pest Management of guava and fig. In: *Advances in the Integrated Pest Management of Horticultural, Spices and Plantation Crops.* (eds.) B.S. Chhillar, V.K. Kalra, S.S. Sharma and Ram Singh Centre of Advanced Studies (ICAR), Department of Entomology, CCS Haryana Agricultural University, Hisar (Haryana) India.

Sahai, B. and Joshi, L.D. 1965. Bionomics and biological control of San Jose Scale (*Quadraspidiotus perniciosus* Comstock) in UP. *Punjab Hort. J.*, **5** : 37-43.

Sharma, D.R. 2005a. Advances in Integrated Pest Management of mango, sapota and grapevines. In: *Advances in the Integrated Pest Management of Horticultural, Spices and Plantation Crops.* (eds.) B.S. Chhillar, V.K. Kalra, S.S. Sharma and Ram Singh Centre of Advanced Studies (ICAR), Department of Entomology, CCS Haryana Agricultural University, Hisar (Haryana) India.

Sharma, D.R. 2005b. Advances in Integrated Pest Management of citrus. In: *Advances in the Integrated Pest Management of Horticultural, Spices and Plantation Crops.* (eds.) B.S. Chhillar, V.K. Kalra, S.S. Sharma and Ram Singh Centre of Advanced Studies (ICAR), Department of Entomology, CCS Haryana Agricultural University, Hisar (Haryana) India.

Sharma, H.K. and Chander, R. 1992. Effectiveness of different methods for the control of flat headed peach tree borer *Sphenoptera lafertei* Thomson on stone fruits. pp. : 281-283. In : *Emerging Trends in Temperate Fruit Production in India.* (eds.) K.L. Chadha, D.K. Uppal, R.N. Pal, R.P. Awasthi and S.A. Ananda. NHB Technical Communications I, Gurgaon.

Sharma, J.P. and Khajuria, D.R. 2004. Pests of almond, cherry and walnut. pp : 295-309. In : *Pest Management in Horticulture Crops; Principles and Practices* (eds.) L.R. Verma, A.K. Verma and D.C. Gautam, Asiatech Publishers Inc., New Delhi.

Sharma, J.P. and Thakur, J.R. 1996. Integrated management of insect pests of apple. pp. : 223-238. In : *Plant Protection and Environment.* D.V.R. Reddy, H.C. Sharma, T.B. Gour and B.J. Divakar (eds). Plant Protection Association of India.

Sharma, J.P. and Thakur, J.R. 2004. Insect and mite pests of apple and pear. pp : 237-258. In : *Pest Management in Horticulture Crops; Principles and Practices* Eds. L.R. Verma, A.K. Verma and D.C. Gautam, Asiatech Publishers Inc., New Delhi.

Shivankar, V.J. and Singh, S. 1998a. Citrus leaf miner – *Phyllocnistis citrella* – A pest of citrus to recon with (Abstr.). pp. 11-13. In: *Int. Conf. on Pest & Pesticide Mangmt. For Sustainable Agric.*, CS Azad Univ. Agric. & Technol., Kanpur, India,

Shivankar, V.J. and Singh, S. 1998b. *Management of Insect Pests in Citrus.* Tech. Bull. No. **3,** National Res. Centre for Citrus, Nagpur, India, P. 99.

Shivnakar, V.J., Rao, C.J. and Singh, S. 2008. *Citrus Entomology a Decade at NRCC.* Tech. Bull. No. **11**, National Res. Centre for Citrus, Nagpur, Maharashtra, India, P. 78.

Shukla, R.P., Tandon, P.L. and Singh, S.J. 1984. Baculovirus- a new pathogen of mango nut weevil, *Sternochetus mangiferae* (Fabricius) (Coleoptera : Curculionidae). *Curr. Sci.*, **53** : 593-599.

Shukla, R.P., Tandon, P.L. and Suman, C.L. 1988. Intra-tree distribution of eggs and diapausing adults of the stone weevil. *Acta Horticulturae*, **231** : 566-570.

Singh, H.P. and Palaniswami, M.S. 2010. Tropical fruits. In : *Horticulture in different agro-climatic conditions- Four decades of coordinated research (Vol. 1 : Fruits, plantation crops, Spices, Medicinal and Aromatic Plants).* Westville Publishing House, New Delhi.

Singh, K. and Sharma, V.K. 1990. Chemical control of citrus leaf miner (*Phyllocnistis citrella* Stainton). *Adv. in Horti. and Forestry.* **1** : 92-94.

Singh, S.P. 1995. *Technology for Production of Natural Enemies.* Tech. Bull. No. **4**, Project Directorate of Biological Control, Bangalore, India. P. 221.

Sohi, A.S. and Sohi, A.S. 1990. Mango leafhoppers (Homoptera : Cicadellidae) – a review. *J. Insect Sci.*, **3** : 1-12.

Srivastava, R.P. 2004. Pests of mango. pp : 310-324. In : *Pest Management in Horticulture Crops; Principles and Practices* (eds. L.R. Verma, A.K. Verma and D.C. Gautam) Asiatech Publishers Inc., New Delhi.

Srivastava, R.P. and Verghese, A. 1985. Fogging as a new method of hopper control in mango. *Indian J. Ent.*, **47** : 113-115.

Tandon, P.L. 1993. Insect and mite pests of tropical fruits. In : *Advances in Horticulture-Fruit crops.* Malhotra Publishing House, New Delhi. **3** (3) : 1530-1550.

Tandon, P.L. and Verghese, A. 1994. Present status of insect and mite pests of grapes in India. pp. : 149-157. *Drakshavrita Souvenir.* Maharashtra Growers Association, Pune, India.

Thakur, J.N. and Gupta, P.R. 2004. Insect pests of peach, plum and apricot. pp : 279-294. In : *Pest Management in Horticulture Crops; Principles and Practices* (eds. L.R. Verma, A.K. Verma and D.C. Gautam) Asiatech Publishers Inc., New Delhi.

Thakur, J.N., Pawar, A.D. and Rawat, U.S. 1988. Observations on the correlation between population density of apple woolly aphid and its natural enemies and their effectiveness in Kullu valley (HP). *Plant Protec. Bull.*, **40** (2) : 13-15.

Thakur, J.R., Chauhan, U., Dogra, G.S. 1995. Autumn population and setting behaviour of peach leaf curl aphid, *Brachycaudus helichrysi* (Kalt.)on early varieties in Himachal Pradesh. *J. Aphid*, **5 (1 & 2) :** 77-85.

Thakur, J.R. and Dogra, G.S. 1980. Woolly apple aphid, *Erisoma lanigerum*, research in India. *Trop. Pest Manag.*, **26** : 8-12.

Thomas,P., Kannan, A., Degwekar, V.H. and Ramamurthy, M.S.1995. Non-destructive detection of seed weevil-infested mango fruits by x-ray imaging. *Post Harvest Bio. and Techno.*, **5** : 161-165.

Verghese, A. 1998. Status of stone weevil and fruit fly with reference to management for export of mango (Abstr.). In: *National Symp. On Mango Prodn. And Export,* Central Inst. Of Subtropical Hort., Lucknow.

Verghese, A. 2004. Pests of grapes. pp : 339-346. In : *Pest Management in Horticulture Crops; Principles and Practices* (eds. L.R. Verma, A.K. Verma and D.C. Gautam) Asiatech Publishers Inc., New Delhi.

Verghese, A., Jayanthi, P.D.K. and Sudha Devi, K. 2000. Developing an IPM based export protocol and forewarning model for mango fruit fly, *Bactrocera dorsalis* Hendel. *Nat. Seminar Hitech Hort.*, Bangalore.

Verma, A.N. and Khurana, A.D. 1974. Further new host record of *Indarbela* sp. (Lepidoptera : Metarbelidae). *HAU J. Res.*, **4** (3) : 253-254.

Wadhi, S.R. and Batra, H.N. 1964. Pests of tropical and sub-tropical fruit trees. pp. : 251-252. In : *Entomology in India.* N.C. Pant (ed.). Entomology Society of India, New Delhi.

Waterhouse, D.F. 1993. *The Major Arthropod Pests and Weeds of Agriculture in Southeast Asia.* Australian Centre for International Agricultural Research, Canberra, pp. : 141.

Whalon, M.E. and Croft, B.A. 1984. Apple IPM implementation in North-America. *Ann. Revi. of Ent.*, **29** : 435-470.

Chapter-5

Rodent Damage and Management in Horticultural Crops

V.P. Sabhlok
Department of Zoology
CCS Haryana Agricultural Univeristy, Hisar

Rodents are well recognized pests and are considered the number one enemy of man. These animals, though considered primitive, have been so successful and abundant worldwide that they alone account for more than half of the known mammalian fauna. They can subsist on a variety of food and even without water for a long time. Due to their prowess of adaptability, they have successfully adapted to varied ecological and physiological niches, and therefore are inhabitants of practically each and every latitude and altitude. They multiply at super fast speed because they become sexually mature when they are just 6-8 months old. They then pose a challenge to man in all respects. According to Prakash (1988) in India there are 135 species and about 300 sub-species of rodents belonging to 46 genera. At least 15 of these rodent species are known to be the serious pest of public health hazard (Jain and Tripathi, 1988). Rodents not only damage the crops in the fields, where the infestation occurs at all stages of food production, processing, storage and distribution, they are also responsible for a number of diseases in the human beings.

Rodents attack the-crops shortly after sowing, during the vegetative growth and during the ripening of the seed heads. They damage the various crops of the *kharif* season *viz.*, jowar, maize, ground nut and cotton as well as the crops of the *rabi* season *viz.*, wheat and barley etc. They damage other crops also which include paddy, sugarcane, soyabean and Bengal gram etc. Hence some well planned rodent management is the need of the time and before going for the planning of any control method in a given situation it is essential to have an estimation of the population and the species involved.

Characters of Rodents

1. They are omnivorous i.e. they are able to eat anything and everything.
2. They can not vomit and it is because of this character that they always try to sample the various diets before actually consuming them.
3. They can be diurnal, noctural, arboreal or terrestrial.
4. They show a bimodal feeding activity and show one peak in the morning and the other during the evening.
5. They become sexually mature when they are 6-8 months old.
6. The gestation period is 18-21 days. Gestation period is the time interval between the conception and the delivery of young ones.
7. The litter size varies from 1 to 22. Litter size means the number of young ones at the time of delivery.
8. They have ever growing incisors. If the growth of these ever growing incisors is not checked, they may grow so much that they may pierce the forehead of the rodent and it may die. In order to avoid this, the rodents always nibble something or the other because of which they keep a check upon the growth of incisors.
9. They do not have canines. The space between the incisors and the premolars is called diastema.
10. They are fossorial i.e. they are adapted for digging and they make their burrows inside the field.
11. They have a life span of 1-2 years.

Rodent Species of Economic Importance in India

Sr.No.	Common name	Scientific Name	Habitat
1	Five striped palm squirrel	*Funambulus pennanti*	Orchards, gardens, farm houses, roadside tree plantations
2	Three striped palm squirrel	*Funambulus palmarum*	Orchards
3	House rat	*Rattus rattus*	Residential premises, stores, go downs, poultry farms and plantation crops
4	House mouse	*Mus musculus*	Residential premises, stores, go- downs, poultry farms
5	Indian mole rat	*Bandicota bengalensis*	Fields, go downs, poultry farms
6	Large bandicoot rat	*Bandicota indica*	Fields, residential premises, poultry farms
7	Soft-furred field rat	*Rattus maltada*	Crop fields, poultry farms
8	Norway rat	*Rattus norvegicus*	In drains at sea ports, waste lands, godowns, poultry farms
9	Short-tailed mole rat	*Nesokia indica*	Crop fields
10	Indian gerbil	*Tatera indica*	Crop fields, waste lands. poultry farms
11	Desert gerbil	*Meriones hurrianae*	Fields, waste lands
12	Indian bush-rat	*Golunda ellioti*	Crop fields
13	Common Indian field mouse	*Mus boduga*	Fields
14	Brown spiny mouse	*Mus platythrrix*	Fields
15	Indian crested porcupine	*Hystrix indica*	Rocky habitat

Damage

The rodents have diverse food habits and can subsist practically on anything and everything, though they are basically herbivores. They have attained the status of major vertebrate pests in all ecosystems as they cause severe damage to cereal crops, stored grains, orchards, vegetable crops, grasslands, forests as well as residential premises and poultry farms. Rodents damage plants during their fossorial feeding and gnawing activities, and the damage is spread to all parts of the plants. The rodent damage is restricted primarily to root system and to underground rhizome and stem. Above ground damage may either be restricted to some specific part or widely spread, i.e. to stem, leaves, flowers, fruits, etc. Depending upon the severity and nature of damage, the plant may be uprooted and killed (if the roots are completely destroyed), or the vitality/ vigour of the plant may be adversely affected which results in slowed growth or reduced yield of the plant. The extent of rodent damage is directly related to the density of the rodent population in particular area. The species responsible for damage also vary in diverse niches and continents.

Sheikher and Jain (1991) emphasized that the damage was more pronounced in trees which were 1.5 m or less away from the surface opening of active burrow of rodents (*B. bengalensis* and *R. meltada*) in the pecan nut orchards of Himachal Pradesh (India). The age of the plant also influences the severity of the damage as it was more in plants which were less than 10 years of age; the incidence of death being as high as 17.4%. Apparently the older trees could withstand the damage due to their more elaborately and strongly developed root system. A similar relationship was recorded by Swihart and Picone (1994) in apple orchards. The trees growing nearer to burrows recorded a reduced growth in terms of height, maximum canopy, width, stem diameter as well as apple size and weight. Rodent damage is wide spread in apple, citrus orchards, banana cultivation, pineapple, avocado, pear and mango crop in India, Latin America (Argentina, Brazil and other countries), and southern Africa (Fiedler, 1988). Rodent damage in guava orchards was also recorded in India (Malhi and Sheikher, 1989). Patel *et al.* (1993) recorded 4.6% damage in pomegranate crop in the arid region of Rajasthan (India). Plants of *Zizyphus mauritiana* (ber), a cash crop in tropics in India, face rodent damage (Jhala *et al.*, 1989). *Meriones shawi* attack lowers the fruit yield in olive, almonds and pistachio nut orchards due to debarking and consumption of the nuts (Fiedler, 1988).

Rodent damage to fruit trees is not restricted to only establish orchards but also in the nurseries. Parshad *et al.* (1984) reported that 26.8% plants in almond nursery were damaged 3 days after sprouting and an additional 14.9% suffered damage after 15 days of sprouting due to the activity of ground rodents. The nurseries located near fallow land or some *gorge* suffer relatively higher damage probably due to the fact that rodents find better refuge and chance to build up their population since human interference is almost negligible here as against the area where well-kept environment prevails.

Coconut suffers heavily 'from rodent activity; up to 77% damage in Colombia mainly by *Rattus* and *Sigmodon* species, resulting in total failure of coconut industry was reported by Elias and Fall (1988). Rodent damage in this crop starts from nursery till trees are fully grown, and from tender nut stage till storage of nuts. In India, *R. rattus* and *Vandeleuria* sp., *R. rattus wroughtoni* and *R. rattus rufescens* damage coconut on crown, while *B. bengalensis* inflicts damage in nurseries and along with *R. rattus* in storage; the damage varying from 9 to 30% to tender and 8 to 50% to mature nuts, besides 6 to 8% in nursery (Sridhara, 1992; Kapadia, 1995; Rangareddy, 1995). Oil palm plantations are severely damaged by rodents in India. Rodents kill young plants (less than 3 years old) by damaging their collar. In mature and bunch bearing palms, the rats gnaw at the female flower bearing branches and also the tender and ripe fruits (Subiah, 1983). Valencia *et al.* (1994) reported that 19% of the palms in oil palm plantations in eastern plains of Colombia were damaged annually by rodents. In Himachal Pradesh, tea plantations are also infested by rodents, especially *B. bengalensis* and *M. musculus* (Sheikher and Jain, 1990). They affect the quality and quantity of the produce.

In vegetable crops, damage is initiated at sowing of the seed itself as the rodents dig and consume the seeds thereby resulting into thinning of the crop, and spreads to early nursery/growth phase. However, the damage is not usually alarming at this stage as the crop fields are devoid of adequate cover and the rodents appear not to venture deep in the field due to danger from predators. Normally in vegetable crops the damage becomes discernible at the maturity or a little earlier at the time of flowering or fruit setting. Rodents inflict heavy damage to tomato crop everywhere, showing special preference for green, unripe and solid berries, consuming their endocarp portion. The extent of damage varies from 8.99% (Ramesh and Katiyar, 1985) to 12.56% (Singh and Saxena, 1984). *B. bengalensis* and *T. indica indica* have been shown to inflict 2.62-35.58% damage to tomato in Gujarat (Katodia *et al.*, 1993). Advani and Mathur (1982) have reported 19% damage in Rajasthan to this crop. In Meghalaya, 4.93% damage is caused due to the activity of *B. bengalensis* and *Rattus indica* (Singh, 1996).

Advani and Mathur (1982) had recorded 5.5% damage to cauliflower and 7.1 % to cabbage in Rajasthan. Saxena and Sahai (1990) observed that the cereals (*viz.*, wheat and barley) harboured more active rodent burrows as compared to those in pea during *Rabi* season. Rodent damage to cucumber (8.96%), sponge gourd (9.85%), muskmelon (11.4%) and bottle gourd (14.32%) in Haryana (Kumar *et al.*, 1995), to brinjal (4.5%), carrot (15.8%), radish (19.8%), onion (8.8%), garlic (6.6%), spinach (4.7%), bottle gourd (4.1%) and okra (4.4%) in Rajasthan (Adavani and Mathur, 1982), and watermelon (9.94-19.79%) in Punjab (Chopra and Parshad, 1986) has also been recorded.

Rodent Management

Among vertebrate pests, rodents are the most serious pest posing a serious challenge to human economy as their infestation occurs at all the stages of food production, processing, storage and distribution. Various methods of the rodent management include:

i) Mechanical methods- Environmental control

Rodent proofing is very important. It is made possible by excluding the rodent from a particular area especially in residential premises, stores, warehouses and godowns (Fitzwater and Prakash, 1973). Improving the sanitary conditions, designing new and repairing the old buildings in a rat proof manner help in reducing rodent activity effectively (Barnnett and Prakash, 1980). In fields number of rodents can be reduced by habitat alteration. Rodents prefer the fields with weeds. One could achieve up to 90 per cent reduction in population of rodents by merely removing the-weeds. Cultivation of 20-30 meter strip of an unpalatable crop like guar (*Cyamopsis tetragonaloba*) around the cash crop may act as a barrier to the migrating rodents, thus, saving the main crop. Paddy crop may be prevented from the rodents by planting the ginger crop around the paddy crop (Prakash and Mathur, 1987).

ii) Trapping

It is an effective method of the pest management. Trapping is possible by using the live traps of single catch e.g. wooden or wire traps and Sherman traps on a cumulative catch e.g. wonder traps (Jain *et al.*, 1992). Trapping is a useful method for rodent control in the residential premises, stores, godowns and poultry farms. It is also very effective in small infestations. Trapping can be enhanced by a pre-baiting of two days. Trapping can also be enhanced by changing the bait material and by adding sugar or ground nut oil (Chopra, 1985). Trapped rodents can be killed by dipping the traps in water.

iii) Chemical method

It is one of the easiest and most effective methods for the rodent control. Chemicals used for killing rodents are called rodenticides. The rodenticides are applied in solid, gas or in the liquid form. These can be of different types.

I. ACUTE TOXICANTS

These chemicals are applied in a single dose. Rodents die within few hours. Surviving rodents may develop bait shyness which may last for 3-4 months. Disadvantage of these acute toxicants is their toxicity to a wide variety of animals. There is usually neither a specific antidote nor the time to administer treatments. Commonly used acute toxicants are

i) Zinc phosphide

It is the most common rodenticide. Its toxication is because of the release of phosphine gas in the stomach and there by causing death within hours. It is usually applied at 2 per cent concentration (Prakash and Mathur, 1987; 1992).

ii) RH-787 (VACOR)

It is also the single dose rodenticide which is also very toxic. It acts very slowly thereby taking several hours to kill the rodents (Marsh and Howard, 1975). Rodents die from paralysis and pulmonary arrest. It is effective at 0.5 and 1.0 per cent concentration. It has been banned in many countries.

iii) Scilliroside (Red squill)

It is the oldest rodenticide still in use. It is supposed to be safe for non-target species. It is effective against *R. rattus, R. norvegicus, B. bengalensis. M. booduga* and *T. indica*. It is used in concentration of 0.05-0.2 per cent in the bait (Chopra *et al.,* 1984). It has widely been used in USA.

iv) Gophacide

It is an acute rodenticide effective against *R. norvegious* and *M. musculus* in India (Muktabai, 1980).

v) Barium carbonate

It was formerly widely used rodenticide. A high dose is required and is generally used at 20 per cent concentration (Chopra and Sood, 1982). Of the various acute rodenticides, zinc phosphide is the only recommended rodenticide which is available in the market in India.

II. ANTICOAGULANTS

These are the chemicals that interfere with blood clotting. Rodents after eating substantial anticoagulant bait die due to internal on the external bleeding. These are slow acting rodenticides. There is no bait shyness development because of these chemicals. These are safe for the wide use because of the availability of antidotes. These chemicals are classified as:

Hydroxycommarius e.g. Warfarin, Ferrnarin, Difenacoum, Flococuafen, Bromadiolone, Brodifacoum etc.

Indantiones; e.g. chlorofacinone, Diphacinone Anticoagulants are categorized as multi feed and single feed/dose anticoagulants.

MULTIPLE FEED ANTICOAGULANTS (1st generation anti-coagulants)

These include:

i) Warfarin

It is used for managing rodents in many countries. It is used at 0.025% concentration. The number of days required for the feeding of this anti-coagulant varies and a minimum of 6-14 days are needed to kill the rodents (Mathur *et al.*, 1992).

ii) Fumarin

It is widely used to control rodents. The number of days required for the feeding of this anticoagulant varies from 4-21 days.

iii) Coumatetralyl

It is sold under the trade name 'Racumin'. It requires a feeding of 5-10 days at 0.0375 per cent concentration.

iv) Chlorophacinone

It is effective at a concentration of 0.0075 per ce.nt. It reduces the population of *M. hurrianae* and *T. indica* by 50 per cent in the fields (Mathur and Prakash, 1984).

v) Diphacinone

It is used at 0.025 per cent concentration. It kills all the individuals of *R. rattus* up to 17th day.

vi) Calli Feral

It is a vitamin activated engosterol. It is effective against rats and mice which are resistant to warfarin. A seven days feeding at 0.1 per cent concentration kills all the house mice (Muktabai, 1979).

Saturation Baiting

It exploits the fact that there is slow appearance of the symptoms of poisoning which helps in avoidance of the development of poison bait shyness. By the time rodents stop feeding they have already ingested a lethal dose. This allows/results in 100 per cent control of rodents without pre-baiting. Bait points of 250 g each are continuously supplied so that the excess of bait is available until the rodents stop feeding. In India only Warfacin is being recommended and being extensively used specially in residential premises. There is the development of resistance to this anticoagulant in some parts of the world (Jackson *et al.*, 1975). In India this resistance has been seen in the *R. rattus* and *B. bengalensis* (Arora and Lal, 1979).

III. SINGLE DOSE OF IIND GENERATION ANTICOAGULANTS

These are the anticoagulants having the properties of acute rodenticides. A single feed is sufficient to kill the rodents. These include:

i) Difenacoum 3

It is more effective than warfarin against *B. bengalensis* and *T. indica*. A concentration of 0.005 per cent is required. The mortality is delayed by 8-20 days (Mathur, 1982).

ii) Brodifacoum

It has shown exceptional results in all the rodent species. A single day exposure of 0.005 per cent is sufficient for the complete kill (Saxena and Sharma, 1981). In coconut plantation, killing of 74.5 per cent rodents with 0.005 per cent brodifacum has been observed. In Poultry farms when offered in protected bait containers, it is highly acceptable and toxic against rodents (Prashad *et al.*, 1987).

iii) Bromadiolone 3-3

It is highly effective against rodents in India and elsewhere (Chopra, 1988). Burred baiting of rodents for 1-2 days with 0.005 per cent bromadiolone bait has been very effective and results in more than 90 per cent kill. It has been found to be effective against rodents infesting coconut, wheat, groundnut, gram and the oil palm. Pulse baiting in 8-10 bait stations/acre at 60-100 days after sowing of whet effectively kills the rodents and thereby resulting in 92.72 per cent reduction in pre-harvest killer cut damage. In poultry farms, single application of 0.005 per cent bromadiolone bait is effective in killing most of the rodents (Singh, 1986).

iv) Bromethalin

A concentration of 0.005 per cent is highly effective against residential and field rodents (Jackson, 1985).

v) WL-I08366

It is anti-coagulant 'snell' rodenticide. It is effective like brodifacaum and other single dose anti-coagulants.

Pulse Baiting

It involves the repeating of the same poison after specific periods to deal with the residual populations of the rodents. In pulse baiting small bait at sufficient places are laid and the bait is replenished only after a specific period.

Mechanism of Action of Anti-coagullants

Anticoagulants act by reducing the ability of blood to clot and the death occurs a few days after the ingestion of a required dose. Anticoagulants have an effect similar to that of vitamin K deficiency. Vitamin K deficiency causes damage to the work of the blood vessels resulting in a severe hemorrhage followed by death.

Fumigants

The fumigants are the chemicals which exist in a gaseous state having sufficient concentration and fatal to the given organisms. Fumigants used in India include carbon tetrachloride, methyl bromide and aluminium phosphide etc.

Hydrogen Cyanid (Cyanogas)

It was earlier used by pumping into the burrows of the rodents. It is highly poisonous to the domestic animals and the human beings. It has been replaced by aluminium phosphide. The pellets of aluminium phosphide release the phosphine gas in the humid air. Rodents can effectively be killed with fumigation with aluminium phosphide. There is no bait shyness. Fumigants should never be used in houses.

Repellants

Repellants are the substances which repel or deter the pests. Endrin has been used as the repellant. Malathion has also the properties of a repellant can be used to protect the food in storage and the seeds during germination. Ultrasound can also be a repellant. High intensity ultrasound contributes to the rodent control.

Chemosterilants

A chemosterilant is a chemical which can cause a permanent on a temporary sterility in either sex or both the sexes directly or through some other physiological mechanism thereby reducing the number of off springs and also altering the fecundity of the off-springs produced. Colchicine is valuable chemosterilant and its use along with nitrofuran has been advocated for the bandicoots (Srivastava, 1966b). Male anti-fertility compound, alpha-chlorohydrn is highly effective (Kassa and Jackson, 1984). An integrated approach with the use of rodenticides for the initial reduction of rodent populations followed by chemosterilant treatment may help to keep the population low.

BIOLOGICAL CONTROL

This includes the use of predators or the parasites.

1. Predators

Rodents are subjected to the predation by lizards, snakes (Minton, 1966) and birds such as owls, hawks, kites etc. (Ali, 1977; Chopra, 1980). These predators are important nature's check on rodents.

2. Parasites and Disease

It involves the manipulations of a dangerous nature while attempts are made to introduce parasites or the diseases of the vertebrate animals.

Pathogen salmonella was used to kill rodents. Rodents immediately develop a resistance to salmonella.

Methods for Population Census and Baiting Techniques

Population census method

In an area with virtually no permanent boundaries it is rarely possible to count all the animals. Direct or indirect methods of estimating the rodent population are

i) Absolute Population Estimates

Absolute population is by counting of the individuals in a given geographical area. Rodents being small, nocturnal and fussorial in habit it is rarely possible to count them. Estimation of the population is done by marking method popularly known as catch, mark and release method. The basic idea is marking the animals so that they can be released unharmed and unaffected into the area and upon recapture are recognized again. Proper sized traps are placed at an interval of 10-15 meters. Pre-baiting is done. Rodents are captured for one night. They are marked and released. Recapturing interval of few days is advocated between first and second trapping. The second trapping include both marked and unmarked individuals. Population can be calculated by Lincoler index.

$$N = \frac{A \times n}{M}$$

Where N= Population size; a = Number of animals marked; n =Total number of rodents (marked and unmarked) captured in second trapping; m =Number of marked rodents captured in second trapping.

Trap Index

This method involves trapping of rodents using a particular type of traps. Trap index is calculated as

$$Z = N\frac{N}{M \times n}$$

Where, Z=Trap index/number of trape/day; N= Total number of animals trapped; m = Number of traps used in trap time/grid; n = Number of days for which trapping is done.

Live Burrow Count

In agricultural and barren lands, live burrow count is another useful index of rodent populations. It involves the blocking of burrows in the afternoon and the counting the freshly opened burrows in next morning. This gives the rough indication of rodent's abundance. Burrows of many species have more than one opening. Some rodents are solitary while others are colonial. Some plug their burrows and are not easily recognized.

Census Method

In census method food intake is the criterion of abundance. Sufficient quantities of food are kept at many points in an area. Food consumed is recorded and replenished. It is repeated for a number of days. Mean daily intake is calculated to estimate the rodents present in the area. Daily intake of food per g body weight of the rodents must be known. Where many species co-exist, it is not possible to do the calculations. Census baiting is an index used to compare the pre-control and the post-controls of rodents in the poison baiting trials.

Activity Marker Census

It involves the placement of markers containing thin layer of the marking ink for checking the sings of rodents. Rodent activity can be used as an index to depict the relative abundance of rodents.

Baiting Techniques

Burrow baiting, poison baiting in grid at 10 meter interval and the baiting in bait stations may be adopted. Burrow baiting leads to high percent control. It is not possible to apply burrows baiting where the vegetation cover is thick e.g. in sugarcane crop. In such cases poison baits at regular and definite intervals or in bait stations are the effective ways. In the residential premises, godowns, stores. poultry farms etc. baits should be placed at places most frequented by the rodents (Prakash, 1976). Baiting must be carried-out to knock out all the species in a single operation. To avoid the accidental poisoning of non-target animals, bait containers are used. These include hollow, bamboo. tin boxes, coconut shells, broken pitchers.

References

Advani, R. and Mathur, R.P. 1982. Experimental reduction of rodent damage of vegetable crops in Indian villages. *Agro-Ecosystems*, **8:** 39-45.

Fiedler, L.A. 1988. Rodent pest problems and management in eastern Africa. *FAO Plant Protec. Bull.*, **36**(3): 125-134.

Jhala, Re., Patel, Z.P. and Patel, C.B. 1989. Rodent damage to ber and cashewnut trees. *Rodent News Letter*, **13:** 4-5.

Kapadia, M.N. 1995. Rodent damage to coconut plantations in Gujarat. *Rodent Newsletter,* **19:** 6-7.

Malhi, E.S., and Sheikher, E. 1989. Rodent damage and control in guava (*Psidium guava*) orchards at Srinagar (Garhwal Himalayas). *J. Tree Sci.,* **8:** 27-30.

Patel, N., Tripathi, R.S., Parveen, F., Jain, A.P. and Mathur, N. 1993. Vertebrate pest damage to pomegranate in arid horticulture. *Rodent Newsletter,* **17** (1-2): 6-7.

Ramesh, P. and Katiyar, RN. 1985. Short tailed bandicoot rat, *Nesokia indica* G. damage in tomato, *Rodent Newsletter,* **9**(1) : 6-9

Rangareddy, A 1995. Species composition of rodent pests and their management in coconut in Godavari delta. *Rodent Newsletter,* **19:** 10.

Sheikher, E. and Jain, S.D. 1991. Rodegt damage and control in pecan orchards. Proc. Indian *National Science Àcademy,* **B57:** 391-396.

Sridhara, S. 1992. Rodent pest management in rice, plantation crops, horticultural crops and poultry. *Rodent Newsletter,* **16**(1) : 12-14.

Subiah, S. 1992. Rodent problem in oil palm plantation in Hut Bay, Little Andaman and suggested control measures. *Indian J Farm Chemicals,* **1** : 32-40.

Swihart, RK and Picone, P.M. 1994. Damage to apple trees associated with wood chuck burrows in orchards. *J. Wild life Management,* **58:** 357-360.

Valencia, D., Elias, D.I. and Ospina, I.A. 1994. Rodent pests in Colombian agriculture. pp. 92-94. In: *Proc. 16th Vertebrate Pest Conf. (*eds. W.S. Halverson and AC Crabb*).* University of California, Davis.

❑❑❑

Part - II

Chapter-6

Effect of Pesticides on Soil Microbes and their Activities

K. Kukreja, S. Suneja, R. Gera and R.C. Anand
Department of Microbiology, CCS HAU, Hisar

The modern improved crop varieties are responsive to fertilizers and demand higher fertilizer application rates to sustain high grain yields. With higher fertilizer inputs, the pest problems have become prominent and endangered the yields necessitating the large scale use of agrochemicals. Some of these chemicals are purposefully released or they may be accidentally released into the environment as waste water or residues from industrial manufacturing and processing of fields.

Soil is one of the most dynamic sites of biological interactions in nature, serving as a growth medium for vegetation and as a habitat for microorganisms and thus has a major role in determining the overall quality of our environment. The physicochemical properties of soil strongly influence the growth, activities and population dynamics of microorganisms in soil and also modify the effects of chemicals on the microbial community. Microorganisms are scavengers in soil. Due to physiological variability, they degrade a variety of chemical substances including insecticides, fungicides and herbicides in soil to derive energy and other nutrients for their growth and metabolism (El- Shahaat *et al.*, 1987; Bhuyan *et al.*, 1993). As a result, the population density of the active microorganisms increases which favorably influences the biological transformation of nutrient elements in the soil (Rangaswamy and Venkateswarlu, 1993; Jana *et al.*, 1998). On the other hand, there are many pesticides which exert adverse effect on growth of soil microorganisms (Martinez-Toledo *et al.*, 1992). However, no definite conclusions can be made on the effects of different pesticides on the growth and activities of microorganisms in soil since different groups of insecticides exhibit manifold variations in toxicity. The degree to which a pesticide affects microbial population and their activities is largely dependent upon the chemical, its dosage and the particular physicochemical parameters of the environment such as soil type, temperature, water content, pH, method of application and other factors. Soil physicochemical factors are particularly important and probably account for the variations in toxic effects often seen with the same compound. Some of the soil microorganisms and soil microbial processes appear to be more sensitive to pesticides than the others. For example nitrification appears to be highly sensitive whereas N- mineralization is relatively resistant. These results reflect the diversity of microorganisms mediating these processes. The present effort has been made to review briefly the effects of pesticides on microorganisms and their activities.

1.0 MICROORGANISMS AND THEIR ACTIVITIES

A diverse range of microorganisms exist in soil. These include bacteria, fungi, actinomycetes, algae and protozoa. Number of microorganisms are normally higher in soil than other habitats. All these organisms have an important role in the recycling of nutrients. Soil microorganisms have a primary catabolic role in the environment through degradation of plant and animal residues. The activities of microorganisms are essential to the global cycling of carbon, nitrogen, sulfur, phosphorus and other elements because many substances can not be degraded by organisms other than microbes.

The application of pesticides may affect the soil microbial populations like bacteria, fungi, actinomycetes and algae etc. leading to elimination, modification or decrease in soil biological processes which are essential for maintaining the fertility of the soil. Therefore, monitoring the changes in number and metabolic activities of these microorganisms enables the assessment of the effects on specific species by pesticides.

2.0 EFFECT OF PESTICIDES ON SOIL MICROBIAL POPULATION

The microbial balance of soil is disturbed by the addition of toxic agrochemicals like insecticides, fungicides, nematicides and herbicides etc. to the soil. The addition of these chemicals exerts inhibitory or stimulatory effects on the growth and metabolic activities of microorganisms.

Misra and Gaur (1977) studied the influence of gamma-BHC (lindane) on soil microbial population and found that at low concentrations (1 and 10 ppm) the compound was toxic to bacteria and to a lesser extent to actinomycetes but stimulated the growth of *Azotobacter.* At 100 ppm, lindane was toxic to all the organisms tested. Herbicides like propanil, nitrofen, 2, 4-D, butachlor and bentazon alone or in combination with insecticides like thimet and furadon in rice crop at their recommended doses did not have any adverse effect on soil microflora (Mukhopadhyay, 1980). Helmeezi *et al.* (1984) reported the temporary inhibition of soil bacteria and fungi by the application of herbicides ethalfluralin + chlorbromuron and ethalfluralin + atrazine which ceased at the end of growing season. Mandal *et al.* (1987) observed initial depressive effect on total microbial population particularly bacterial flora of soil with pre-emergence application of herbicides like goal, tok E-25, saturn and machete which however, recovered within few days. Singh (1990) reported initial decrease in population of bacteria, fungi and actinomycetes with butachlor application in rice crop and these adverse effects gradually reduced with passage of time.

Rajendran *et al.* (1990) studied the effect of organochlorine pesticides on bacterial population of a tropical estuary and found that DDT was found to be highly toxic to bacterial population followed by lindane and endosulfan. Edwards *et al.* (1992) studied the short term changes in soil bacterial population by carbofuran insecticide in field and laboratory. Largest increase in bacterial population was observed at 7 and 15 days after treatment. Significant increase in total bacterial populations and presumed carbofuran-degraders due to carbofuran treatment were associated with increased populations of *Pseudomonas sp.* and *Flavobacterium sp.* Application of carbofuran appeared to provide a competitive advantage to these species over actinomycetes persisting beyond 20 days after treatment.

Continuous use of pre-emergence herbicides pre-tilachlor and butachlor in rice-rice pulse system showed a little suppression of soil bacteria (Ali and Asokraja, 1994). Kumar and Kandaswamy (1994) revealed that initial suppression of soil microflora up to 30 days after the application of herbicide recovered later on. Gopalswamy *et al.* (1994) reported a decrease in bacterial, fungal and actinomycetes population up to 20 days after application of pre-emergence herbicides in rice which recovered after 30 days. Ghinea *et al.* (1998) found that among soil microorganisms, bacteria were most sensitive to herbicides.

Das and Mukherjee (2000) observed that application of all the insecticides in general and HCH and phorate in particular significantly increased the population of microorganisms in soil. The most predominant genera of microorganisms such as *Bacillus, Micrococcus* and *Aspergillus* were not affected by most of the insecticides. However, some of the insecticides stimulated the growth of *Bacillus, Proteus, Trichoderma* and *Fusarium* etc. Among the tested insecticides, stimulations were more pronounced with HCH followed by phorate and fenvalerate. Stimulation of bacterial population with recommended doses of fluchloralin and reduction with higher doses was observed under pot culture conditions (Shukla *et al.* 2001). Omar and Abdel- Sater (2001) studied the effect of soil treatment with brassinal (herbicide) and selecron (insecticide) on counts of bacteria and observed stimulation of bacterial population at field application rate and reduction in number at higher levels.

Ekundayo (2003) studied the effect of eleven pesticides at their recommended rates on the population of bacteria, actinomycetes, fungi and protozoa. In general, pentachloronitrobenzene, hexachlorocyclohexane and phenylmercuric acetate were toxic particularly to soil microorganisms, whereas the herbicides dacthal, preforan and dual were quite harmless in soil. Protozoa and fungi were more susceptible to fungicides than bacteria and actinomycetes. Fliessbach and Mader (2004) evaluated short and long-term effects of two potato pesticide spraying sequences on soil microorganisms. In short term, cumulative pesticide side effects on microbial biomass and microbial activities averaged 19% with glufosinate and 45% with dinoseb. After 135 days, (long term) values returned to normal. Jha and Mishra (2005) investigated the side effects of HCH, DDT and endosulfan pesticides on viable counts of bacteria, fungi, actinomycetes, *Cyanobacteria* and most probable number of *Rhizobium, Azotobacter Azospirillum, Nitrosomonas* and *Nitrobacter.* Highest microbial population was recorded in pulse growing soils which contained the least pesticide residues. An additional experiment was conducted to evaluate the effect of endosulfan, cypermethrin and chlorpyriphos on soil microorganisms. Pesticide application drastically reduced the microbial population. Cypermethrin had least effect on microbial population, while chlorpyriphos had the major effect.

Jiang *et al.* (2005) showed that application of atrazine stimulated the growth of soil microorganisms. Also the respiration intensity of soil increased greatly and specially bacteria and fungi also increased greatly. In our laboratory also, Suneja *et al.* (2008) studied the effect of carbofuran and carbosulfan alone and in combination on soil microflora. The total bacterial population and free living diazotrophs population reduced initially by nematicide application (Table 1). With progress of time, population increased in nematicide treatments till the end of the experiment. Population of fungi and actinomycetes remained higher in nematicides treated soils till the end of the experiment. Suneja *et al.* (2010) also observed that pesticides used for control of termites in cotton crop had some effect but of low importance with regard to microbial population.

Table1 : Effect of nematicides on soil microflora in cotton crop under pot culture conditions

Treatments	Bacterial population ($x10^5$)g^{-1} dry soil			Fungal population ($x10^3$)g^{-1} dry soil			Actinomycetes ($x10^4$)g^{-1} dry soil			Free living diazotrophs ($x10^4$)g^{-1} dry soil		
	Sampling time (Days after nematicides application)											
	10	30	60	10	30	60	10	30	60	10	30	60
T1 control (untreated)	112.0	45.3	35.0	4.0	8.3	17.3	13.6	4.3	20.6	25.3	12.3	6.3
T2 carbofuran 3% 1Kg ha^{-1}	24.3	60.3	32.3	9.6	17.0	19.6	47.6	27.6	35.3	9.6	17.0	10.0
T3 carbosulfan 3% seed treatment	42.3	226.0	63.6	7.0	10.6	27.0	22.3	22.0	27.0	16.0	27.3	7.6
T4(T2+T3)	24.3	118.0	57.0	9.9	11.4	20.0	23.0	27.6	32.3	8.3	15.0	8.0
CD 5%	18.831	29.332	14.522	2.768	4.585	5.710	7.090	4.167	NS	9.843	7.806	NS

2.1 Effect of Pesticides on Diazotrophs

Diazotrophs are abundantly present in cultivated fertile soils and play important role in nitrogen economy of the soil. Nitrogen fixed by various legumes in association with rhizobia and free living diazotrophs results in increased soil fertility (Singh, 2005). Scanty information is available on the effects of pesticides on the diazotrophs.

Bollen (1960) found that 2, 4-D at rates above 500 ppm decreased nitrogen fixation by diazotrophs. Mendoza (1973) studied the effect of five pesticides on *Azotobacter chrococcum* and *Rhizobium trifolii*. DDT and menazon added to the culture medium at more than 500 ppm favored the increase of *A. chrococcum* but did not affect *R. trifolii* even up to 1000 ppm. Although carbaryl, dimethoate and melathion reduced the growth of both species, at high concentrations, there was inverse relationship between the rate of increase of the bacteria and the concentration of insecticide. Gomah *et al.* (1974) found that mephosfolan and phosfolan (insecticides) significantly increased the number of nitrogen fixing bacteria (*Azotobacter* and *Clostridium*). Misra and Gaur (1977) also observed that lindane application stimulated the growth of *Azotobacter*. De Felipe *et al.* (1987) observed decrease in nitrogen fixation rate by the addition of herbicides in soil. El-Sayed *et al.* (1993) found that carbamate insecticides sevin (carbaryl), larvin (Thiodicarb) and lannate (methomyl) significantly increased the population of *Azotobacter* at recommended or 10 fold the recommended dose.

Mallikarjunaiah (1995) studied the effect of fungicides and insecticides on growth and nitrogen fixation in *Azotobacter* and observed that brassicol, benlate, thiophamine and vitavax stimulated the growth and nitrogen fixing ability of *Azotobacter*. The rest of the fungicides except sulfur were toxic to all the *Azotobacter* cultures tested. Phorate also stimulated the growth of few cultures. Nitrogen fixing capacity also increased in few cultures. Allievi *et al.* (1996) reported that herbicide bentazone applied at 10 and 100 ppm, significantly reduced the number of anaerobic nitrogen fixing bacteria after 30 weeks. Allievi and Gigliotti (2001) found negligible effects of cinosulfuron (at field rates) on *Azotobacter* population. RuiFu *et al.* (2004) studied the long term effect of methylparathion on soil microflora and found decrease in total count of *Azotobacter* in the contaminated soil compared with control. Adeleye *et al.* (2004) investigated the effect of atrazine, paraquat dichloride and 2,4-D on *Azotobacter vinelandii*, *Rhizobium phaseoli* and *Bacillus subtilis* under laboratory conditions and found that 2, 4-D was most toxic of three herbicides tested and *Azotobacter* was found to be most sensitive.

2.2 Effect of Pesticides on Nitrifying Bacteria

Nitrifying bacteria play a major role in N-cycle and also in maintaining the soil fertility. Nitrification has been used as a index of soil fertility (Paul and Clark, 1989) and probably this is the most sensitive process to pesticides (Lata *et al.*, 2004). The continuous use of pesticides may affect the population of *Nitrosomonas* and *Nitrobacter* resulting in decreased crop yield (Pandey and Rai, 1993). Reduction in their population may result in inhibition of plant growth due to non availability of nitrogen.

Kukreja and Mishra (1987) studied the effect of three herbicides namely emisan, furadon and tribunil on the activity of *Nitrosomonas* and *Nitrobacter* in liquid culture. Emisan inhibited *Nitrosomonas* and *Nitrobacter* even at a concentration of 1 $\mu g\ ml^{-1}$ while MPN count of *Nitrosomonas* and *Nitrobacter* taken after 12 days showed complete absence of these bacteria in the presence of emisan which showed that emisan is bactericidal. Pandey and Rai (1993) studied the effect of different doses of aldrin, 2, 4-D, BHC, carbofuran, basalin and machete on nitrifying bacteria and found reduction in population with higher doses which was temporary. Gonzalez-Lopez *et al.* (1993) studied the effect of various concentrations of insecticides phorate and melathion on soil microorganisms and found that nitrifying bacteria were not affected by the insecticides.

Shukla and Mishra (1997) found that nitrification process was inhibited more in acidic soil than neutral pH soil by butachlor application in rice and inferred that bacteria like *Nitrosomonas* and *Nitrobacter* may be affected more in acidic conditions. Cernakova *et al.* (1992) observed that insecticide astellic containing 50% pirimiphos had a negative effect on nitrification. Nagaraja *et al.* (1998) observed inhibition of nitrification process by atrazine application and inhibition increased with increase in concentration of atrazine. Alleivi and Gigliotti (2001) observed the inhibition of nitrification by cinosulfuron after one week of incubation even at the field rate (42 μg kg^{-1} soil). Gigliotti *et al.* (1998) studied the effect of bensulfuron methyl at 16 and 160 μg kg^{-1} soil in two non-flooded soils and found that higher doses substantially inhibited nitrification. The nitrifying bacterial population significantly increased in the soil contaminated with methylparathion compared with control as a result of long term application of the herbicide. In our laboratory, Saini *et al.* (2009) studied the effect of long term use of clodinafop herbicide in wheat under rice-wheat system on different microbial populations and found inhibition of nitrifying bacteria and free living diazotrophs at early stage after the application. However, bacterial population significantly reduced only in rotational herbicide plot as compared to weedy control at early stages after application of herbicide.

2.3 Effect of Pesticides on Microbial Activities

Available data indicates that pesticides not only influence the population of various groups of soil microbes, but also affect their physiological activities. The activities of microorganisms in soil are essential for transformation of various elements. The delicate balance between microbes and their enzyme activities in soil is very important for nutrient release. Activities of soil enzymes may predict the potential capacity of the soil to perform certain biological transformations of importance to soil fertility. Reports on the impact of pesticides on soil enzymes show considerable diversity depending upon type of chemical used, rate and mode of application, climatic factors, composition of soil and organic matter content. The variations in the effect are due to the fact that in soil, numerous microorganisms exist and each microorganism may respond differently to the pesticide treatment. These chemicals may either enhance or inhibit soil enzymatic activities.

Krezel and Musial (1969) found that dehydrogenase and urease respond to herbicides but concentration required for inhibition tends to be high especially in the field where effects were often negligible. Tate (1974) studied the effect of four pesticides viz., carbofuran, DDT, fenithrothion and fensulfothion in commercial formulations at the recommended rate of 2.24 kg ha^{-1} and found no significant effect on soil respiratory activity. Fenitrothion at 100 ppm and all insecticides at 500 ppm except carbofuran significantly inhibited soil respiratory activity. Malkomes (1979) evaluated the effect of dichlorprop and methabenzthiazuron on soil dehydrogenase activity in wheat and found that dehydrogenase activity was stimulated by dichlorprop but remained unaffected by methabenzthiazuron. Herbicide isoproturon when applied alone or in combination with insecticides like dinoseb acetate and bifenox in wheat caused limited transient stimulation and inhibition of dehydrogenase activity. Palaniappan and Balasubramanian (1985) studied the influence of two pestcides namely carbofuran and fluchloralin on soil enzymes invertase, dehydrogenase, amylase and urease under laboratory conditions. Carbofuran had a definite influence on soil microbial activities especially at the higher doses, stimulating amylase activity and inhibiting invertase and dehydrogenase. Fluchloralin inhibited the activity of all four enzymes at 5 and 10 ppm but at recommended rate, herbicide had no significant effect.

Pozo *et al.* (1995) found that ie. chlorpyriphos treatment at concentration of 2.0 -10.0 kg ha^{-1} decreased activities of acid and alkaline phosphatases and dehydrogenase significantly, but recovered after 14 days to levels similar to those in control soil without chlorpyriphos. Ismail *et al.* (1996) reported reduction in urease activity for the entire four week study in the presence of metachlor at 10 µg g^{-1} in Malaysian soil. Later on Ismail *et al.* (1998) evaluated the effect of metasulfuron methyl on urease activity in loamy sand and clay loam soils up to 28 days and found reduction in urease activity for the entire period of study in both the soils. Nagaraja *et al.* (1998) found the inhibition of phosphatase and dehydrogenase activity by the application of atrazine in acidic, saline-alkaline and neutral soils. Shukla *et al.* (2001) studied the effect of fluchloralin on dehydrogenase activity in sandy loam soil and found gradual decline in dehydrogenase activity up to 15 days on exposure to herbicide at all concentrations from 1 to 10 ppm.

Omar and Abdel-Sater (2001) studied the effect of soil treatment with herbicide (bromonol) and the insecticide (profenofos) on soil alkaline phosphatase activity. Alkaline phosphatase activity in treated soil was accelerated with both pesticides even at higher application rates suggesting a direct role of alkaline soil pH in increasing resistance of alkaline phosphatase to pesticides. Das *et al.* (2002) studied the effect of two herbicides alachlor and atrazine on dehydrogenase activity in three different soils and found initial enhancement in dehydrogenase activity over control by alachlor application. The application of mefenacet at concentration of 0.067, 0.100, 0.133, 0.200 and 0.267 $\mu g\ g^{-1}$ soil stimulated the dehydrogenase activity in rice (Yang Fang *et al.*, 2004). LiJun *et al.* (2005) evaluated the effects of four pesticides (thiophanate-methyl, mancozeb, fenvalerate and abamectin) on soil catalase activity and soil respiration. Catalase activity was inhibited at first, then stimulated and finally recovered. The stimulation and inhibition was stronger with higher concentration. Soil respiration was slightly stimulated during initial period, inhibited after 5 days and then recovered to control level after 12 days. In our laboratory also, we studied the effect of long term use of clodinafop herbicide in wheat under rice-wheat cropping system on soil dehydrogense activity (Saini, 2006). It was slightly higher in herbicide treated plots than weedy control under green manuring conditions at 30 and 60 days after treatment (DAT) while in without green manuring treatments, activity was significantly less in herbicide treated plots than weedy control (Table 2).

Table 2 : Effect of long term use of herbicide applied in wheat under rice-wheat rotation system on soil dehydrogenase activity.

	mg TPF kg^{-1} soil $24h^{-1}$				
Treatments	**BT***	**15 DAT**	**30 DAT**	**60 DAT**	**90 DAT**
Green manuring					
Permanent herbicide**	38.44	20.92	27.56	44.63	29.64
Rotational herbicide***	43.76	27.25	15.58	40.27	7.49
Weed free	32.22	26.04	15.59	38.42	13.13
Weedy	32.96	24.56	12.03	34.75	13.85
Without green manuring					
Permanent herbicide**	23.83	18.85	22.11	33.03	13.25
Rotational herbicide***	14.76	13.27	10.58	22.35	6.17
Weed free	24.45	19.96	18.43	42.41	18.93
Weedy	22.69	28.70	32.77	63.50	7.85
CD(P=0.05)	4.32	3.87	3.09	9.50	1.76

*Before the application of herbicide

** Clodinafop 60 g ha^{-1}

***Clodinafop 60 g ha^{-1}

3.0 CONCLUSIONS

The intensive use of chemicals in agriculture and generation of chemical wastes from industry has aroused public concern as to how these chemicals might affect humans and the environment. Soil is the main sink for most of these chemicals and has a major role in determining their fate. The biological activity of soil is the result of the activities of resident microorganisms, invertebrates and plants. The microbial component mainly bacteria, fungi, algae and protozoa are important to the fertility of soils through their role in the degradation of plant and animal remains and recycling of nutrients in the soil. Among the most important processes mediated by microorganisms are the transformations of carbon, nitrogen, sulfur and phosphorus. Anything that disrupts the activity of the microorganisms could be expected to affect the nutrient status of soil and would therefore have serious ecological consequences. Because of this, it is important to understand the effects of various pesticides on soil microorganisms and their activities. The degree to which a chemical affects the microbial activity depends upon the chemical, its dosage and method of application and also the physicochemical properties of the soil, such as pH, moisture content, soil type and temperature. Some microbial processes are more sensitive than others, for example, nitrification process appears to be more sensitive than nitrogen mineralization.

Adverse effect of pesticidal chemicals on soil microorganisms becomes a foreign chemicals major issue. Soil microorganisms show an early warning about soil disturbances by foreign chemicals than any other parameters. Different studies have shown that agricultural chemicals do not appear to have long-term harmful effects on soil microorganisms and their activities when applied at recommended field levels. However when applied at higher levels, these chemicals do appear to have an adverse short-term effects on microorganisms and their activities. Contradictory results are often observed for the same chemical applied at same level and these contradictory results are explained on the basis of differences in soil types, method of application and laboratory procedures. So the basic problems have to be solved before adequate predictive environmental risk assessment can be developed for a particular chemical on microorganisms in soil.

References

Adeleye I . A, Okorodudy, E. and Lawal, O. 2004. Effect of some herbicides used in Nigeria on *R. phaseoli, A. vinelandii and B. subtilis. J. of Environ. Biol.,* **25** (2): 151-156.

Ali, A and Asokraja, N. 1994. VI *Biennial Conference*. ISWS. pp142. (c.f. *Agricultural Reviews*, **20(1):** 48-52).

Allievi, L and Gigliotti, C. 2001. Response of bacteria and fungi of two soils to sulfonylurea herbicide cinosulfuron. *J. Environ. Sci. Health*, **36** (2): 161-175.

Allievi, L. Gigliotti, C. Salardi, C. Vaisecchi, G. Brusa, T. and Ferrari, A. 1996. Influence of herbicide bentazon on soil microbial community. *Microbio. Res.*, **151** (1): 105-111.

Bhuyan, S. Sreedharan, B. Adhya T.K. and Sethunathan, N. 1993. Enhanced biodegradation of r-hexachlorocyclohexane (r-HCH) in HCH (commercial) acclimatized flooded soil: factors affecting its development and persistence. *Pesticides Sci.*, **38**: 49-55.

Bollen, W.B. 1960. Iinfluence of insecticides on soil microorganisms and their biochemical activities. *Pesticides*, 16-23.

Cernakova, M. Kurucova, N. and Fuchsova, D. 1992. Effect of the insecticide Actellic on soil microorganisms and their activity. *Folia Biologica*, **37:** 219-222.

Das, A.C. and Mukherjee, D. 2000. Soil application of insecticides influences microorganisms and plant nutrients. *Appl. Soil Eco.*, **14**: 55-62.

Das, S.K. Pahwa, M.R. and Yadava, R.B. 2002. Soil dehydrogenase activity in relation to atrazine and alachlor application. *Range Manage. Agroforestry*, **23(2):** 126-129.

De Felipe, M.K. Ferenadez-Pascual M. and Pozeuls, M. 1987. Effects of the herbicide lindex and simazine on chloroplast and nodule development, nodule activity and grain yield in *Lupinus albus. Plant and Soil*, **101**: 99-105.

Edwards, D.E. Kremer, R.J. and Keaster A.J. 1992. Characterization and growth response of bacteria in soil following application of carbofuran. *J. of Environ. Sci. and Health*, **27**: 139-154.

Ekundayo, E.O. 2003. Effect of common pesticides used in the Niger Delta basin of Southern Nigeria on soil microbial populations. *Environ. Monitor. Assessment*, **89** (1): 35-41.

El-Sayed, M.A. El-Shanshoury, A.R., Swelim M.A. and Abd-El-Salam, S.S. 1993. Microbiological and physicochemical properties of carbamate insecticides treated soil. *J. of Agro. Crop Sci.*, **170**: 217-223.

El-Shahaat, M.S., Othman, M.A.S. Halfawym, E. and Marei A.S. 1987. Effect of carbamate and synthetic pyrethroid pesticides on some soil microbial activities. *Alexandria J. Agri. Res.*, **32**: 427-438.

Fliessbach, A and Mader, P. 2004. Short and long term effects on soil microorganisms of two potato pesticide spraying sequences with either glufosinate of dinoseb as defoliants. *Bio. and Fertility Soils*, **40(4):** 268-276.

Ghinea, L. Iancu, M. Turcu, M. and Stefanic, C. 1998. The impact of sulfonyl urea and non selective herbicide on biological activity of sandy soil. *Weed Technology*, **12**: 55-57.

Gigliotti, C. Allievi, L. Salardi, C. Ferrari, F. and Ferrari, A. 1998. Microbial ecotoxicity and persistence in soil of the herbicide benslfuron methyl. *J. Environ. Sci. Health*, **33(4):** 381-398.

Gomah, A.N. Abdellatif, M.A. and Elbasuony E.N. 1974. Effect of systemic insecticides on growth of different plants and soil microorganisms. *Alexandria J. Agri. Res.*, **22:** 443-452.

Gonzalez-Lopez, J. Martinez-Toledo, M.V. Rodelas, B. and Salmeron, V. 1993. Studies on the effects of the insecticides phorate and melathion on soil microorganisms. *Environ. Toxico. Chem.*, **12:** 1209-1214.

Gopalswamy, G. Raj S.A. and Kareem, A.A. 1994. Interaction of herbicides with *Azolla* and soil microbes. *Indian J. Weed Science*, **26**: 28-34.

Helmeezi, B. Nagy, M. Katai, J. Bessenyei, M. and Szegi, J. 1984. Effect of herbicide combinations containing ethafluralin on the soil microflora and the nitrogen fixing capacity of *Azotobacter. Soil Bio. and Conser. of the Biosphere*, **1**: 239-250.

Ismail, B.S. David I and Omar, O. 1996. Effects of metachlor on activities of enzymes in a Malaysian soil. *Environ. Sci. Health B*, **30**: 485-497.

Ismail, B.S. Yapp, K and Omar, O. 1998. Effects of metsulfuron -methyl on amylase, urease and protease activities in two soils. *Australian J. of Soil Res.*, **36:** 449-456.

Jana, T.K. Debnath, N.C. and Bask, R.K. 1998. Effect of insecticides on decomposition of organic matter, ammonification and nitrification in a Fluventic Ustochrept. *J. Indian Soc. Soil Sci.*, **46**: 133-134.

Jha, M.N. and Mishra, S.K. 2005. Decrease in microbial biomass due to pesticide application/residues in soils under different cropping systems. *Bull. Environ. Contam. Toxi.*, **75(2):** 316-323.

Jiang, H. Xianzhu, D. and Shunpeng, L. 2005. Effects of atrazine and its degrader *Exiguobacterium sp* BTAHI on soil microbial community. *Yingyong-Shengtai-Xuebao*, **16(8):** 1518-1522.

Krezel, Z and Musial, M. 1969. The effect of herbicides on soil microflora. II. The effect of herbicides on enzymatic activities of the soil. *Acta Microbiogica. Poland.(Ser.B)*, **1:** 93-97.

Kukreja, K. and Mishra, M.M. 1987. Evaluation of some pesticides as nitrification inhibitors. *Ann. of Bio.*, **3(1):** 71-76.

Kumar, K and Kandaswamy, O.S. 1994. VI *Biennial Conference* ISWS. pp. 144(c.f. *Agricultural Reviews*, **20(1):** 48-52).

Lata, Pandey, A.K. and Narayan, K.P. 2004. Effect of insecticides on nitrifying bacteria in soil. *Ann. of Agri. Res. New Series*, **25(1):** 92-94.

LiJun, F. ShiXi, Z. and Hai, W. and WenJin, Y. 2005. Effects of four pesticides on catalase activity in soil and soil respiration. *J. Fujian Agri. Forestry Uni. Nat. Sci.*, **34(4)**: 441-445.

Mallikarjunaiah, R.R. 1995. Effect of fungicides and an insecticide on growth and nitrogen fixation in *Azotobacter. Mysore J. Agri. Sci.*, **29:** 36-42.

Malkomes, H.P.1979. The effect of post-emergence application of two herbicides in winter wheat on soil microflora. *Zentralblatt-furr-Bakteriologi-parasitenkunde-Infektionskrankheiten-und-Hygiene-II*, , **134(7)**: 573-586.

Mandal, B.B. Bandyopadhyay, P. Bandyopadhyay, S. and Maity, S.K. 1987. Effect of some pre-emrgence herbicides on soil microflora in direct seeded rice. *Indian Agri.*, **31(1):** 19-23.

Martinez-Toledo, M.V. Salmiron, V. and Gonzalez-Lopez J. 1992. Effect of an organophosphorus insecticide, phenophos on agricultural soil microflora. *Chemosphere*, **24**: 71-80.

Mendoza, N.C. 1973. The effects of five insecticides on the growth of *Azotobacter chroococcum* and *Rhizobium trifolii. Anales-del-Instituto-Nacional-de-Investigaciones-Agrarias-General*, **2**: 21-35.

Misra, K.C. and Gaur, AC. 1977. Influence of the interactions of lindane application and ecological factors on soil microbial population. *J. of Entomological Res*, **1:** 132-135.

Mukhopadhyay, S.K. 1980. Effect of herbicides and insecticides alone and their combinations on soil microflora. *Indian J. Weed Sci.*, **12:** 53-60.

Nagaraja, M.S. Ramakrisana, P. and Siddaramappa, R. 1998. Effect of atrazine on urea-N mineralization and activity of some soil enzymes. *J. Indian Soc. of Soil Sci.*, **46(2):** 189-192.

Omar, S.A. and Abdel-Sater, M.A. 2001. Microbial populations and enzyme activities in soil treated with pesticides. *Water, Air Soil Pollu.*, **127**: 49-63.

Palaniappan, S.P. and Balasubramanian A.1985. Influence of two pesticides on certain soil enzymes. *Agri. Res. J. of Kerala*, **23:** 189-192.

Pandey, R.K. and Rai, S.N. 1993. Effect of varying doses of pesticides on nirtrifying bacteria and nitrogen transformation. *Indian J. Agri. Chem.*, **26:** 123-132.

Paul, E.A. and Clark, F.E. 1989. *Soil Microbiology and Biochemistry*, Academic Press, NewYork, pp. 133-148.

Pozo, C. Martinez-Toledo, M.V. Salmiron, V. Rodelas B. and Gonzalez-Lopez, J.1995. Effect of chlorpyrifos on soil microbial activity. *Environ. Tox. Chem.*, **14**: 187-192.

Rajendran, R. Rajendran, N. and Venugopalan, V.K. 1990. Effect of organochlorine pesticides on the bacterial populations of a tropical estuary. *Microbiological Letters*, **44:** 57-63.

Rangaswamy, V. and Venkateswarlu, K. 1993. Ammonification and nitrification in soils, and nitrogen fixation by *Azospirillum sp.* as influenced by cypermithrin and fenvalerate. *Agri. Ecosy. Environ.*, **45**: 311-317.

RuiFu, Z. Zhong, LI. C, Jian, H. Ting-Ting, H. and ShunPeng, L. 2004. Ecological effects of long term methyl parathion contamination on soil microflora. *Rural Eco. and Environ.*, **20(4)**: 48-50.

Saini, B. 2006. *Impact of clodinafop herbicide on soil microflora and their biochemical activities in wheat field soil.* M.Sc. Thesis, CCS HAU, Hisar, Haryana, India.

Saini, B. Suneja, S. and Kukreja, K. 2009. Impact of Long Term Use of Clodinafop in wheat on soil microbes. *Indian J. of Weed Sci.*, **41 (1& 2):** 50-53.

Shukla, A.K. Magu S.A. and Das, T.K. 2001. Effect of fluchloralin on soil microorganisms and nitrifiers. *Ann. of Plant Sci.*, **9:** 109-112.

Shukla, A.K. and Mishra, R.R. 1997. Effect of butachlor on nitrogen transformation and soil microbes. *J. of Indian Soci. of Soil Sci.*, **45:** 571-574.

Singh, G. 2005. Effects of herbicides on biological nitrogen fixation in grain and forage legumes- A review. *Agri. Rev.*, **26(2):** 133-140.

Singh, S.P. 1990. *Biennial Conference of Indian Society of Weed Science*, pp. 72-74. (c.f. *Agricultural Review*, **20(1):** 48-52.

Suneja, S. Kukreja, K. and Anand, R.C. 2010. Effect of pesticides on soil microbes in cotton rhizosphere. *Environ. Eco.* **28(2B):** 1235-1237.

Suneja, S. Kukreja, K. Goyal, S. and Bansal R.K. 2008. Effect of nematicides on soil microflora in cotton. *Indian J. Eco.* **35(1):** 25-27.

Tate, K.R. 1974. Influence of four pesticides formulations on microbial processes in a New Zealand pasture soil. I. Respiratory activity. *New Zealand J. Agri. Res.*, **17**: 1-7.

Yang Fang, Y. Hang, M. and XiangChi, Z. 2004. Effects of mefenacet on microbial respiration and enzyme activities in paddy soil. *Acta Pedologica Sinica*, **41(1):** 93.

Chapter-7

Compatibility between Microbial Pesticides with Synthetic Pesticides

Monika Geroh, H.D. Kaushik*, P. Bhatnagar* and Asha
Department of Zoology, *Entomology
CCS Haryana Agricultural University, Hisar

The food plants of world are damaged by more than 10,000 species of insects, 30,000 species of weeds, 1,00,000 diseases (caused by fungi, bacteria, viruses and other microorganisms and 1000 species of nematodes (Hall, 1995). The global losses due to various categories of pests vary with the crop, geographic location and weather. Total yield losses from different pests to all the crops have been estimated to be US$ 500 billion worldwide (Oerke *et al.*, 1994). In India, the current losses due to insect pests in principal field crops have been estimated to be Rs. 6, 89,400 million (Dhaliwal *et al.*, 2004). In addition, there is loss of Rs. 60,000 million due to insect pests during storage. Thus, total losses to major crops and food grains in storage caused by insect pests are estimated to be Rs. 750 billion annually (Dhaliwal and Arora, 2006). Among regions, losses in Asia and Africa almost reaches 50 percent whereas in Europe (28.2%), North America (31.2%) and Oceania (36.2%) are below average. Losses due to animal pests in Asia (18.7%) are almost double those in developed countries and losses from weed competition in Africa and Asia are approximately double than those in Europe (Oerke *et al.*, 1994). To combat such heavy losses mostly chemical pesticides are used by the farmers but indiscriminate use of pesticides has led to severe ecological consequences like destruction of natural enemy fauna, effect on non target organisms, residues in consumable products including packed pure and mineral water and ultimately resistance to the pesticides, to which we solely rely. Insecticides are always required to suppress rapidly expanding insect pest populations. Strategies should be employed to increase efficiency and accelerate insect mortality by combining microbials with sub lethal doses of chemical insecticides and botanicals. Biointensive pest management (BIPM) is the recent trend in Indian farming and attracting the farmers for higher income to their produce. This has resulted due to increased awareness among the end users and concerns about the deteriorating ecological situations among the eco-campaigners.

Microbial pesticides consist of disease causing microorganisms, which are disseminated in the pest populations in large quantities in a manner similar to application of chemical pesticides. It includes bacteria, viruses, fungi, nematodes and some protozoans. Pathogens may exert controlling effect by means of their invasive properties, toxins, enzymes and other substances (Lewis, 2006). The use of microbial insecticide has been proved to be a viable alternative to chemical insecticides, especially in management of forest insect pests. The combination of particular active ingredients and their ability of replication in host populations in the fields define the major differences between microbial insecticides and chemical insecticides. These differences have practical implications for their use in insect pest management programmes.

Microbial insecticides have some striking advantages in contrast to chemical insecticides. They tend to be host specific, safe, and have no toxic residues. They also ensure the survival of natural enemies and unlikely stimulate resistance in target pests. Some microbial insecticides are compatible with chemical insecticides and can often be used in combination with them. These are compelling reasons for increasing the efforts to improve these agents and increase their utility for IPM programmes (Cuperus *et al.*, 2004). However, microbial insecticides have some disadvantages too. Their high specificity restricts their production and marketing. Certain upper or lower limits of pH, temperature and light intensity might lead to their failure.

Fungi include some microbials which invade insects by spores, landing on the cuticle the germ tube penetrates the cuticle. *Beauveria bassiana* and *Metarhizium anisopliae* used against cabbage caterpillar, and soil pests, respectively, are most important. Several viruses, bacteria, protozoa and nematodes (ingested microbials) and rely on ingestion by the host to initiate infection. Nuclear polyhedrosis and granulosis (used against lepidopteran and hymenopteran pests) and cytoplasmic viruses (against lepidopteran and dipteran pests) are very effective. The most important spore forming bacterium of insects is *Bacillus popilliae* (infects scarab dung beetles) and *B. thuringiensis* (attack lepidoptera, mosquitos and blackfly larvae). Protozoa, especially *Nosema,* have been used against moths.

Most microbial agents are suitable for the use as microbial insecticides. The industry also has maximum interest in this approach, because the multiple applications create the best opportunity for product sales. The estimated market for bioinsecticides in 1995 was US$ 380 million including $ 3-4 million for viruses. By 2000, the market was predicted to have increased by $116-141 million, with viruses making up $5-6 million of the total (Hunter- Fujita *et al.,* 1998). Presently the microbial insecticides account for 2-5 per cent of the world pesticide market. The market is dominated by products based on *B. thuringenesis* and nematodes which occupy about 90 per cent of the total share of microbial pesticides (Hagler, 2000).

Fungi and Compatibility with Insecticides

Concerns of entomopathogenic fungi as alternative pest control agents are increasing even though chemical pesticides have been used as the main control agents for pests and diseases in crop production. Fungal biological control agents and selective insecticide may act synergistically increasing the efficiency of the control, allowing the lower doses of insecticides, preservation of natural enemies, minimizing environmental pollution and decreasing the likelihood of development of resistance to either agent (Boman, 1980; Moino and Alves, 1998; Ambethgar, 2009). However, use of incompatible insecticides may inhibit growth and reproduction of the pathogens and adversely affect integrated pest management (Duarte *et al.,* 1992; Malo, 1993).

Beauveria bassiana (Bb) has been found to be compatible with lufenuron 5EC (0.4 litres/ha), thiamethoxan 25 WG (100g/ ha), methomyl 40SP (1.0 kg/ ha) and Ethion 50EC (1.5 litres/ ha) for insect pest management in soybean. *Bacillus thuringiensis* was also found to be compatible with all the above insecticides except ethion, which inhibited *Bt* colonies completely. Hence, *Bt* and Bb formulations can be used in mixture along with these compatible insecticides and may be sprayed at flowering to control major pests. Mixture of lufenuron and *Bt* or Bb would be an ecofriendly combination for the management of major defoliatiors on soybean (Ansari and Sharma, 2005).

A two year study was conducted to determine the efficacy of *Beauveria bassiana* in combination with *Bacillus thuringiensis* and insecticides against *H. armigera* on cotton (Baraiya and Kapadia, 2005) (Table 1). The results revealed that half the dose of *B. bassiana* (1.25 kg/ ha) combined with half the dose of cypermethrin (0.0045%) or chlorpyriphos (0.025%) were found as effective and economical as the recommended synthetic insecticides (cypermethrin 0.009%, endosulphan 0.07% and chlorpyriphos 0.05%).

Table 1 : Efficacy of *Beauveria bassiana* in combination with synthetic insecticides for the management of insect pests

Treatment	Pest	Crop	Pest reduction (%)	Yield (kg/ ha)	Cost: Benefit ratio	Reference
Beauveria bassiana (2.5 kg/ ha)	*Helicoverpa armigera*	Cotton	22.34	1881	1:3.74	Barariya and Kapadia, 2005
Bb (1.25 kg/ ha) + Endosulfan (0.035%)	*H. armigera*	Cotton	38.35	1806	1:0.81	Barariya and Kapadia, 2005
Bb (1.25 kg/ ha) + Carbaryl (0.1%)	*H. armigera*	Cotton	78.87	2426	1:8.59	Barariya and Kapadia, 2005
Bb (1.25 kg/ ha) + Chlorpyriphos (0.025%)	*H. armigera*	Cotton	83.09	2328	1:3.01	Barariya and Kapadia, 2005
Bb (1.25 kg/ ha) + Cypermethrin (0.0045%)	*H. armigera*	Cotton	78.92	2317	1:8.02	Barariya and Kapadia, 2005
Bb (1,000 conidia/ mm^2) + Pyridaben (0.05 a. i. µg/ml)	*Tetranychus cinna barinus*	-	65.40	-	-	Shi *et al.*, 2005
Bb (1,000 conidia/ mm^2) + Pyridaben (1.0 a. i. µg/ml)	*T. cinnabarinus*	-	77.70	-	-	Shi *et al.*, 2005
Bb (1,000 conidia/ mm^2) + Pyridaben (2.5 a. i. µg/ml)	*T. cinnabarinus*	-	85.30	-	-	Shi *et al.*, 2005

Contd.

Bb (10^8 conidia ml^{-1})	*T. urticae*	-	91.00	-	-	Gatarayiha, 2009
Bb (10^8 conidia ml^{-1}) + azoxystrobin (1×X)	*T. urticae*	-	82.10	-	-	Gatarayiha, 2009
Bb (10^8 conidia ml^{-1}) + azoxystrobin (10^{-1}×X)	*T. urticae*	-	89.70	-	-	Gatarayiha, 2009
Bb (10^8 conidia ml^{-1}) + azoxystrobin (10^{-2}×X)	*T. urticae*	-	85.80	-	-	Gatarayiha, 2009
Bb (10^8 conidia ml^{-1}) + flutriafol (1×X)	*T. urticae*	-	4.50	-	-	Gatarayiha, 2009
Bb (10^8 conidia ml^{-1}) flutriafol (10^{-1}×X)	*T. urticae*	-	18.30	-	-	Gatarayiha, 2009
Bb (10^8 conidia ml^{-1}) + flutriafol (10^{-2}×X)	*T. urticae*	-	25.50		-	Gatarayiha, 2009

Maximum percent reduction in pest population (83.09%) was observed with *Bb* (1.25 kg/ ha) + chlorpyriphos (0.025%) treatment followed by *Bb* (1.25 kg/ ha) + cypermethrin (0.0045%) (78.92%), *Bb* (1.25 kg/ ha) + carbaryl (0.1%) (78.87%), *Bb* (1.25 kg/ ha)+ Endosulfan (0.035%) (22.34%). All treatments were better than the *Beauveria bassiana* (2.5 kg/ ha) alone treatment. However, the yield of cotton obtained and cost: benefit ratio was highest with *Bb* (1.25 kg/ ha)+ carbaryl (0.1%) treatment. Thus, *B. bassiana* can be integrated with synthetic insecticides for ecofriendly management of *H. armigera* on cotton (Baraiya and Kapadia, 2005).

Likewise Shi *et al.* (2005) recorded combination of *B. bassiana* (1,000 conidia/ mm^2) + pyridaben (1.0 a. i. µg/ml) to be the most effective treatment as compared to other against plant mite, *T. cinnabarinus* (Table 1).

When the fungicide, azoxystrobin and flutriafol were sprayed along with *Bb*, effective control of *T. urticae* was obtained (Gatarayiha, 2009). With azoxystrobin, the corrected mortality ranged between 82.1% and 89.7 %; these mortalities demonstrated that none of the levels of azoxystrobin impaired the activity of *Bb* as no significant difference was established from that caused by *Bb* sprayed alone. Where *Bb* was sprayed together with flutriafol at the three levels of concentration, the corrected mortality of mites assessed 10 days after treatment ranged between 4.5% and 25.5% (Table 1). The percentage Abbott's corrected mortalities due the fungal treatments were not far from the percentage overt mycosis. These mortality parameters proved that the fungicide significantly reduced the control efficacy of *Bb* against *T. urticae* at all the levels tested.

Asi *et al.* (2010) evaluated the influence of some selective insecticides on mycelia growth and conidial germination of *Metarhizium anisopliae* (Metsch.) Sorokin and *Paecilomyces fumosoroseus* (Wize) Brown and Smith. Chlorpyriphos was found to be the most toxic insecticide to mycelia growth and conidial germination followed by methomyl, thiodicarb and chlorfenapyr. Flufenoxuron, Lufenuron, Indexacarb and Emamectin Benzoate were comparatively less toxic to mycelial growth (36.78- 48.67% inhibition) and conidial germination (40.32-49.97% inhibition) of the fungal pathogens. Conversely, methoxyfenozide, triflumuron, abamectin and prophenophos were compatible with significantly lesser inhibition in growth (25.19-36.47%) and conidial germination (27.78- 43.66%) of the fungi. Spinosad was found safe to conidial germination and growth of the fungi. The observed variations in the inhibitory potential could be due to inherent variability of chemical insecticides to entomopathogenic fungi.

The beneficial effects of insecticides may also potentially expand the pest host range of fungal agents eg. *Paicelomycesus fumosoroseus* was not effective against greenhouse infestations of aphids, *Aphis gossypii* and *Macrociphoniella sanborni*, but when combined with azadirachtin (Margosan- O), efficacy was enhanced and good control of these aphids was attained thereby increasing the cost- effectiveness of the chemical control stratergy (Lindquist, 1993).

Entomopathogenic fungi are slow acting as compared with chemical insecticides is a particular challenge. The common route of infection is by the conidial germination followed by penetration of the insect cuticle rather than ingestion. Because these organisms do not show spectacular short term effects they get little chance to compete with and serve as alternative chemical pesticides.

Bacteria and Compatibility with Insecticides

The compatibility of bacteria with other control tactics make them ideal candidates for use in IPM programmes. *Bacillus thuringiensis* subsp. *Kurstaki* (Dipel 8L) in alteration with endosulphan resulted in a significant reduction in percent bollworm damage on cotton from 31.9

(boll basis) and 16.7 (loculi basis) in untreated check to 19.6 (boll basis) and 8.2 (loculi basis) in the alternated application (Butter *et al.*, 1995).

The efficacy of *Bacillus thuringiensis* alone and in combination with insecticides was evaluated for the management of *Spodoptera litura* (Fabricius) on castor (Jethwa and Kapadia, 2001). It was found that application of *Bt* (1.0 kg/ha) combined with chlorpyriphos (0.025%) was the most effective and was at par with *Bt* (1.0 kg/ha) combined with endosulfan (0.035%) (Table 2). However, *Bt* @ 2.0 kg/ha alone caused 38.35 percent reduction in *H. armigera* population on cotton (Barariya and Kapadia, 2005).

Table 2 : Efficacy of *Bacillus thuringiensis* in combination with synthetic insecticides for the management of insect pests

Treatment	Pest	Crop	Pest reduction (%)	Yield (kg/ ha)	Cost: Benefit ratio	Reference
Bacillus thuringiensis (1.0 kg/ ha)	*Spodoptera litura*	Castor	2.60	817	1:1.03	Jethwa and Kapadia, 2001
Bacillus thuringiensis (1.5 kg/ ha)	*S. litura*	Castor	2.44	990	1:0.77	Jethwa and Kapadia, 2001
Bacillus thuringiensis (2.0 kg/ ha)	*S. litura*	Castor	1.95	1280	1:1.34	Jethwa and Kapadia, 2001
Bt (1.0kg/ ha) + Chlorpyriphos (0.025%)	*S. litura*	Castor	1.80	1585	1:3.34	Jethwa and Kapadia, 2001
Bt (1.0 kg/ ha)+ Endosulfan (0.035%)	*S. litura*	Castor	1.88	1549	1:3.49	Jethwa and Kapadia, 2001
Bacillus thuringiensis (2.0 kg/ ha)	*H. armigera*	Cotton	38.35	1806	1:0.81	Barariya and Kapadia, 2005
Btk (500g/ ha) + Deltamethrin (0.0014%)	*H. armigera*	Tomato	63.42	-	-	Dass, 2006
Btk (500g/ ha)+ Endosulfan (0.025%)	*H. armigera*	Tomato	77.35	-	-	Dass, 2006

Dass (2006) found deltamethrin (0.0014%) and endosulfan (0.025%) to be high compatible with *Btk* as 63.42 and 77.35 percent reduction in *H. armigera* population was recorded on tomato crop (Table 2).

Bacillus cereus in combination with reduced doses of endosulphan and fenvalerate has been found to be superior as compared to alone treatment of *B. cereus* against the larvae of *Achaea janata*. The application of reduced doses of insecticides first followed by application of *B. cereus* was more effective than reversing the sequence and application of insecticide and pathogen alone (Basappa and Lingappa, 2003).

The efficacy of *Bacillus thuringiensis* alone and in combination with different insecticides for the management of *H. armigera* on chickpea was studied at Junagarh (Gujarat) (Patel and Kapadia, 2001). The treatments were applied twice, one at flowering stage and second at pod formation stage. Considering the field efficacy and economics of all the treatments, application of *Bt* @1.0 kg/ ha combined with endosulfan (0.035%) was found to be the most effective treatment followed by *Bt* @1.0 kg per ha combined with polytrin-C (0.022%).

Field studies were conducted in 1992 and 1993 in Hermiston, Oregon, to evaluate the efficacy of transgenic *Bt* potato (Newleaf®, which expresses the insecticidal protein Cry3Aa) and conventional insecticide spray programs against the important potato pest, *Leptinotarsa decemlineata* (Say), Colorado potato beetle (CPB), and their relative impact on non-target arthropods in potato ecosystems. Results from the two years of field trials demonstrated that Newleaf potato plants were highly effective in suppressing populations of CPB, and provided better CPB control than weekly sprays of a microbial *Bt*-based formulation containing Cry3Aa, bi-weekly applications of permethrin, or early- and mid-season applications of systemic insecticides (phorate and disulfoton). When compared with conventional potato plants not treated with any insecticides, the effective control of CPB by Newleaf potato plants or weekly sprays of a *Bt*-based formulation did not significantly impact the abundance of beneficial predators or secondary potato pests. In contrast to Newleaf potato plants or microbial *Bt* formulations, however, bi-weekly applications of permethrin significantly reduced the abundance of several major generalist predators such as spiders (Araneae), big-eyed bugs (*Geocorus* sp.), damsel bugs (*Nabid* sp.), and minute pirate bugs (*Orius* sp.), and resulted in significant increases in the abundance of green peach aphid (GPA), *Myzus persicae* (Sulzer) - vector of viral diseases, on the treated potato plots. While systemic insecticides appeared to have reduced the abundance of some plant sap-feeding insects such as GPA, lygus bugs, and leafhoppers, early and mid-season applications of these insecticides had no significant impact on populations of the major beneficial predators. Thus, transgenic *Bt* potato, *Bt*-based microbial formulations and systemic insecticides appeared to be compatible with the development of integrated pest management (IPM) against other potato pests such as GPA because these CPB control measures have little impact on major natural enemies. In contrast, the broad-spectrum pyrethroid insecticide (permethrin) is less compatible with IPM programs against GPA and the potato leafroll viral disease (REED Gary).

Three different microbials *Bt* var. kurstaki (*Btk*, Dipel 8L), *Saccharopolyspora spinosa* (Spinosad 45SC, Tracer) and *B. bassiana* (pure culture) were tested alone and in combination at their pre-determined $LC_{50}S$ and $LC_{25}S$ against 3rd instar larvae of *H. armigera*. Individual treatments of (0.0018 or 0.003 %) and its combinations (0.006 or 0.0013%) with *Btk* (0.02 to 0.08%) resulted in significantly higher larval mortality of 100% and 64.8% respectively. The individual treatment of *Btk* (0.05 to 0.19%) and *B. bassiana* ($1.6x10^{5 \text{ to}}$ $2.5x10^5$ spores/ ml) were at par resulting in 53.0 to 75.6 and 60.4 to 75.3% larval morality respectively. Combined treatments of microbial agents at theirs LC_{25} resulted in additive lethal effect in most of the combinations (Sridevi *et al.*, 2004).

Bajwa and Aliniazee (2001) evaluated the suceptibiltty of *Bt* based insecticides Dipel (100 milllion Intrenational units/ 100 litres) and MVP (250 ml/ 100l) in combination with reduced rates of azinphosmethyl 2.5-5.0 g and carbaryl 12 g/100 litres) were found to be quite selective to spider or they show negligible effect. However full field rates of orgnaphosphate azinphosmethyl (25 g/ 100 litres) and carbamate carbaryl (60 g/ 100 litre) were slightly to moderately toxic causing 25- 75 percent mortality.

Baculoviruses and Compatibility with Insecticides

Baculoviruses may also be combined with conventional pesticides. The NPVs of a number of important pests like *H. armigera*, *A. gemmatalis*, *H. zea*, *S. litura* and *Amastaca albistriga* (Walker) have been reported to be compatible with chemical insecticides. Application of HaNPV@ 300 LE/ ha plus endosulphan @500 ml/ha resulted in minimum number of *H. armigera* larvae (2.4) per meter row length as compared to 6.13 larvae in the standard (endosulphan 1 litre/ha) and 10 larvae in untreated check on chickpea (Kumar and Malik, 1998).

NPV of *S. litura* @ 46 x 10^6 OBs/ml combined with carbaryl (0.02%) significantly reduced the leaf damage in tobacco. A combination of SlNPV @125 LE/ha with chlorpyriphos (0.04%) resulted in cent percent reduction in larval population 7 days after spray. However, there are some reports of antagonistic response between NPVs of *S. litura* and

Spilarctia oblique Walker with selected doses of some insecticides (endosulphan, fenitrothion) (Battu *et al.*, 2002).

Two sprays of HaNPV (250 LE/ ha) at initiation of flowering and at 7 days after first day followed by one spray of endosulphan @1.25 litres per ha after 10 days of second spray of NPV gave the maximum yield (17.86 q/ ha) with minimum pod damage (8.90%) by on chickpea. Endosulfan @ 0.625 litres per ha in combination with NPV was also equally effective, resulting in 9.00 per cent damage and 17.73 q/ ha yield (Chaudhary *et al.*, 2001).

Two studies conducted consecutively for three years have evaluated the potential of sequential application of biopesticides for the management of pest complex on pigeonpea. A study on the control of pod borer complex on pigeonpea at Anand (Gujarat) has revealed that four alternate sprays of *Bacillus thuringiensis* (*Bt*) and HaNPV showed significant control of borer complex (Jani *et al.*, 2001). There was minimum seed damage (20.75%) and maximum yield (1040 kg/ ha) recorded in this treatment as compared to all other combinations (Table 3). All treatments were better than control. The number of larvae was 3.10, 3.27, 3.31 and 3.33 with endosulfan, endosulfan-*Bt*-HaNPV-*Bt,* NSKE-*Bt*-HaNPV- *Bt and Bt*-HaNPV-endosulfan-*Bt,* respectively. Similarly the yield varied from 98 to 983 kg/ ha with various treatments.

A similar study was conducted at Rajendranagar (Andhra Pradesh) (Rao *et al.*, 2001) which revealed that four sprays of *Bt*-HaNPV-endosulfan-*Bt* at 10 day interval, starting at flower initiation stage were highly effective against *H. armigera* larval population by reducing the pod damage (23.44%) and increasing the yield (559 kg/ ha) (Table 3). The application of botanical, neem seed kernel extract followed by *Bt*-HaNPV- *Bt* reduced the number of larvae to 21.88 as compared to 37.21 larvae in control. The yield of crop was also higher (550 kg/ ha) than control (293 kg/ ha).

The bioefficacy of combined effect of NPV (2.3x 10^6 PIB's per ml), *B. thuringiensis* and two insecticides, viz. endosulphan and carbaryl (0.05% each), against third instar larvae of *Spilarcita oblique* Walker on soybean was studied by Agnihotri and Khan (2006). It was found that *Bt* var. kurstaki (*Btk*) + endosulphan (0.05%) +NPV and *Bt* var. aizawani (*Bta*) + carbaryl (0.05%) + NPV provided 100 percent mortality of larvae. It was significantly higher than the mortalities observed from *Btk* + carbaryl (0.05%) + NPV and *Btk* + endosulfan+ NPV, both recording 86.6 percent mortality. *Btk* and *Bta* varieties were more effective in combination with NPV and insecticides as compared to *Btt*.

Table 3 : Effect of sequential application of biopesticides on insect pests infesting pigeonpea

Treatment	Pest	Larval population (No./10 plants)	Seed damage (%)	Yield (kg/ ha)	Reference
Bt-HaNPV-*Bt*- HaNPV	Pod borer complex	3.10	20.75	1040	Jani *et al.*, 2001
Bt-HaNPV-Endosulfan -*Bt*	Pod borer complex	3.33	23.19	983	Jani *et al.*, 2001
Endosulfan-*Bt*-HaNPV- *Bt*	Pod borer complex	3.27	21.9	930	Jani *et al.*, 2001
NSKE-*Bt*-HaNPV- *Bt*	Pod borer complex	3.31	22.31	901	Jani *et al.*, 2001
Bt-HaNPV-*Bt*- HaNPV	*Helicoverpa armigera*	12.92	23.9	558	Rao *et al.*, 2001
Bt-HaNPV-Endosulfan -*Bt*	*H. armigera*	9.08	21.20	34	Rao *et al.*, 2001
Endosulfan-*Bt*-HaNPV- *Bt*	*H. armigera*	10.17	23.44	559	Rao *et al.*, 2001
NSKE-*Bt*-HaNPV- *Bt*	*H. armigera*	29.08	21.88	550	Rao *et al.*, 2001
Endosulfan	*H. armigera*	10.58	15.01	51	Rao *et al.*, 2001

At six days following the first application of Acal- AaHIT WP (I) at 2x10^{12} PIB's per hectare averaged significantly less *H. virescence* damaged squares, blooms and bolls than untreated cotton. Baculoviruses and *Bt* treatments reduced crop damage vs untreated throughout the study. Weekly application at high rates of had no adverse effect on densities of non-target arthropods.

Combination of Acal- AaHIT plus chlorfenapyr produced control equivalent to full rate of chlorfenapyr. Another study showed that Acal-AaHIT alone and in combination with cypermethrin produced results comparable to cypermethrin standard alone. All treatments were significantly better than cypermthrin standard alone. Similar results were concluded when combined with acephate. All studies indicated that baculoviruses alone and in combination with has the potential to reduce crop damage and control permissive insect species (Gard, 1997).

Boomathi *et al.* (2006) evaluated the toxic effect of spinosad and its combination with other microbial insecticides against 2nd and 5th instar larvae of *H. armigera* on pigeon pea flowers and pods. The results were highly encouraging as spinosad and endosulfan alone 100 percent mortality of 2nd instar after 24 hours of treatment as well as of 5th instar after 72 hours of treatment. HaNPV alone was not highly effective as only 23.3 percent mortality was observed after 72 hours of treatment (5th instar) but in combination with Spinosad 100 (2nd instar) and 83.3(5th instar) percent mortality was observed after 24 and 72 hours of treatment, respectively.

Fadare and Amusa (2003) compared the efficacy of three microbial (biotrol, dipel and thuricide) and three chemical insecticides (monocrotophos, endosulfan and carbaryl) on four major lepidopterans and their natural enemies in replicated field trials at Moor Plantation, Ibadan. Thuricide was evaluated at different combinations with monocrotophos in a second trial. The results showed that the microbials caused the mortalities of destructive bollworms and leafroller but allowed the survival of their natural enemies. The chemicals on the other hand caused mortalities of both destructive and useful species. Both groups of insecticides enhanced seed cotton yields. Application of thuricide followed by monocrotophos was better than other combinations evaluated.

Similarly, the spray formulations of biotrol, dipel and thuricide gave good control of the grape leaf-folder *Desmia funeralis* (Hubn) (Pyralidae) as the chemical insecticide, carbaryl (Ali Niezee and Jensen, 1973).

The effect of microbial and chemical insecticides on cotton insects was compared by McGarr *et al.* (1970), and reported percent bollworm damages of 23.8, 14.2, 18.7, 20.2 and 39.7 for methyl parathion, *B. thuringiensis* (HD-I), Toxaphene + methyl parathion, carbaryl + methyl parathion and control, respectively. They also reported significant yield increases of seed cotton over the unsprayed control treatments. They concluded that *B. thuringiensis* was more effective than the chemicals in controlling the bollworms.

Bull *et al.* (1979) had reported a deliberate suppression of beneficial species populations with methyl parathion and subsequent rapid increases in outbreak of *Heliothis* sp. on cotton fields. Simultaneous application of both microbial and chemical insecticides though not significantly different from 'Thuricide' or monocrotophos alone, may not be advisable because of adverse reactions of emulsifiable concentrate insecticides (Morris and Armstrong, 1975; Morris, 1975). But with monocrotophos and *Bt.* there is no such risks as Fadare and Amusa (2003) has confirmed that both are compatible. However the chemical may have adverse effects on parasites and predators in the agroecosystem. Sequential application of microbial insecticides and low doses of chemical insecticides has been a major input in the implementation of integrated control of insect pests. The trials have confirmed that microbial formulations can be as effective as the commonly used chemical insecticides on lepidopterous pests. There was also superior performance of sequential application of thuricide followed by monocrotophos over all other combinations.

Conclusion

Microbial insecticides, being highly selective conserve the populations of parasites and predators as well as other beneficial species while they suppress lepidopterous populations for which they are specific. There are numerous examples where application of chemical pesticides has enhanced the efficacy of entomopathogen against insect pests. Sublethal doses of chemical pesticides can act as physiological stressor/ behavioral modifier and thereby predispose insects to disease and also, the application of sublethal doses of insecticides has substantially enhanced the efficacy of pathogen. Enhanced efficacy of entomopathogenic hyphomycetes applied in combination with imidacloprid against sugarcane rootstock borer weevil was attributed to reduced mechanical removal of conidia from surfaces of larval cuticles, imidacloprid severely impaired larval movements in soil (Quintela and McCoy, 1998).

Applications of chemical insecticides for the control of cotton pests is not advisable early in the season as they may reduce yields due to an apparent adverse reactions by the plants, and such applications may result in increased bollworm, beet armyworm or cabbage looper populations. The potential application of a microbial insecticide followed by chemical insecticide should be adopted in pest control programmes (Fadare and Amusa, 2003).

To increase the utility of microbial pathogens in IPM programmes systematic surveys should be done in different agrochemical zones to identify naturally occurring pathogens. Detailed studies are necessary on properties, mode of action and pathogenecity of insect pathogens. Ecological studies on dynamics of diseases in insect populations are necessary because environmental factors play an important role in development of diseases and ultimate control of insect pests. Efforts should be made to minimize the loss of infectivity of certain pathogens due to photo inactivation. Studies on safety of insect pathogens to higher animals, beneficial insects and plants should be undertaken. The self perpuating nature of most of pathogens in both space and time could certainly prove to be an asset in sustainable agriculture.

References

Agnihotri, M. and Khan, M. A. 2006. Bioefficacy of combined effect of biopesticides and insecticides against *Spilarctia oblique* on soyabean. *Ann. Pl. Protec. Sci.* **14(1):** 45-58.

Ali Niazee, N. J. and Jensen, F. L 1973. Microbial control of the grape leaf folder with different formulations of *Bacillus thuringiensis*. *J. Econ. Entomol.*, **66**: 151 – 158.

Ambethar, V. 2009. Potential of entomopathogenic fungi in insecticide resistance management (IRM): A review. *J. Biopestic.*, **2(2)**: 177-193.

Ansari, M. M. and Sharma, A. N. 2005. Compatibility of *Bacillus thuringenesis* and *Beauveria bassiana* with new insecticides recommended for insect pest control in soyabean. *Pestology*, **29 (9):** 18-20.

Asi, M. R., Bashir, M.H., Afzal, M., Ashfaq M. and Sahi, S.T. 2010. Compatibility of entomopathogenic fungi, *Metarhizium anisopliae* and *Paecilomyces fumosoroseus* with selective insecticides. *Pak. J. Bot.*, **42(6)**: 4207-4214.

Bajwa, W. I. and Aliniazee, M. T. 2001. Spider fauna in apple ecosystemof Western Oregon and its field susceptibility to chemical and microbial insecticides. *J. Econ. Entomol.*, **94(1):** 68-78.

Baraiya, K. P. and Kapadia, M. N. 2005. Field efficacy of *Beauveria bassiana* alone and in combination with insecticides recommended against *Helicoverpa armigera* (Hubner) in cotton. pp.82-83. In: Abstr. *Conference on biopesticides: Emerging trends* (eds. Koul, O., Dhaliwal, G. S., Shanker, A., Raj, D. and Kaul, V. K.). Nov. 11-13, Palampur (India), Society of Biopesticide Science, India, Jalandhar.

Basappa, H. and Lingappa, S. 2003. Evaluation of *Bacillus cereus* alone and in combination with certain insecticides against castor semilooper, *Acahaea janata* L. and its parasitoids. pp.147-155. In: *Biopesticides and pest management* (eds. Koul, O., Dhaliwal, G. S., Marwaha, S. S. and Arora, J. K.). Vol. 2. Campus Books International, New Delhi.

Battu, G. S., Arora, R. and Dhaliwal, G.S. 2002. Prospects of Baculoviruses in integrated pest management. pp.215-238. In: *Microbial biopesticides* (eds. Koul, O. and Dhaliwal, G. S.). Taylor and Francis, London.

Boman, H.G. 1980. Insect responses for microbial infections. pp.769-744 In: *Microbial control of pests and plant diseases.* (ed. Burges, H.D.). Academic Press, New York.

Boomathi, N., Sivasubramanian, P. and Raguraman, S. 2006. Toxic effect of spinosad and its combination with other insecticides against *Helicoverpa armigera. Ann. Pl. Protec. Sci.* **14**(1): 228-229.

Bull, D. L., House, V. S., Ables, J. R. and Morrison, R. K. 1979. Selective methods for managing insect pests of cotton. *J. Econ. Entomol.*, **72**: 841–846.

Butter, N. S., Battu, G. S., Kular, J. S., Singh, T. H. and Brar, J. S. 1995. Integrated use of *Bacillus thuringenesis* Berliner with some insecticides for the management of bollworms on cotton. *J. Entomol. Res.*, **19**(3): 255-263.

Chaudhary, H. R., Sharma, K. P., Mathur, A. K., Gopal, G. and Trivedi, S. K. 2001. Efficacy of NPV and its combination with endosulphan against gram pod borer, *Helicoverpa armigera* (Hubner) in chickpea under humid south eastern plain zone of Rajasthan. pp.207-208. In: *Biological control: Contributed papers* (eds. Singh, D., Dhaliwal, V. K., Mahal, M. S., Brar, K. S. and Singh, S. P.). Punjab Agricultural University, Ludhiana.

Cuperus, G. W., Berberet, R. C. and Noyes, R. T. 2004. The essential role of IPM in promoting sustainability of agricultural production systems for future generations. *In: Integrated Pest Management: Potential, Constraints and Challenges.* CAB International, Wallingford. 265-280 pp.

Dhaliwal, G. S. and Arora, R. 2006. *Integrated pest management: Concepts and approaches.* Kalyani Publishers, New Delhi.

Dhaliwal, G. S., Arora, R. and Koul, O. 2004. Neem research in asian continent: Present status and future outlook. pp. 65-96. In: *Neem: Today and in the new mellenium* (eds. Koul, O. and Wahab, S.). Kluwer Academic Publishers, The Netherland.

Duarte, A., J.M. Menendez and N. Trigueiro. 1992. Estudio preliminar sobre la compatibilidad de *Metarhizium anisopliae* com algunos plaguicidas quimicos. *Revista Baracoa,* **22**: 31-39.

Fadare, T. A. and Amusa, N. A. 2003. Comparative efficacy of microbial and chemical insecticides on four major lepidopterous pests of cotton and their (insect) natural enemies. *African J. Biotech.,* **2**(11): 425-428.

Gard, I.E. 1997. Field testing a genetically modified baculovirus. In: *Proceedings of BCPC Symposium: Microbial Insecticides: Novelity or Necessity?* University of Warwick, Coventry, UK.

Gatarayiha, M.C. 2009. *Biological Control of the two-spotted spider mite, Tetranychus urticae Koch (Acari: Tetranychidae).* Ph. D. Thesis. Discipline of Plant Pathology, School of Agricultural Sciences and Agribusiness, Faculty of Science and Agriculture, University of KwaZulu-Natal, Pietermaritzburg. 218 pp.

Hagler, J. R. 2000. Biological control of insects. pp. 207-241. In: *Insect pest management: Techniques for environmental protection* (eds. Recheigl, J. E. and Recheigl, N. A.). Lewis publishers, Boca Raton, Florida. 207- 241.

Hall, R. 1995. Challenges and prospects of integrated pest management. pp. 1-19. In: *Novel approaches to integrated pest management* (ed. Reuveni, R.). Lewis publishers, Florida, USA. 1-19 pp.

Hunter- Fujita, F. R., Entwisstle, P. H., Evans, H. F. and Crook, N. E. 1998. *Insect viruses and pest management.* John Wiley and Sons, Chichester, England.

Jani, J. J., Godhani, P. H., Mehta, D. M. and Yadav, D. N. 2001. Developing microbial pesticides based module for the control of pod borer complex in pigeonpea. In: *Biological control: Contributed papers* (eds. Singh, D., Dhaliwal, V. K., Mahal, M. S., Brar, K. S. and Singh, S. P.). Punjab Agricultural University, Ludhiana. 151 pp.

Jethwa, D. M. and Kapadia, M. N. 2001. Field efficacy and economics of *Bacillus thuringenesis* var. kurstaki alone and in combination with insecticides against *Spodoptera litura* on castor. pp. 174-175. In: *Biological control: Contributed papers* (eds. Singh, D., Dhaliwal, V. K., Mahal, M. S., Brar, K. S. and Singh, S. P.). Punjab Agricultural University, Ludhiana.

Kumar, S. and Malik, V. S. 1998. Management of gram pod borer, *Helicovera armigera* (Hubner) by nuclear polyhedrosis virus. pp. 329-333. In: *Ecological agriculture and sustainable development* (eds. Dhaliwal, G. S., Randhawa, N. S., Arora, R. and Dhawan, A. K.). Indian Ecological Society for Research in Rural and Industrial Development, Chandigarh.

Lewis, L. C. 2006 Ecological considerations for the use of entomopathogens in integrated pest management. pp. 249-268. In : *Ecologically based integrated pest management* (eds. Koul, O. and Cuperus, G. W.). CAB International, Wallingford, UK.

Lindquistt, R.K. 1993. Integrated insect, mite and disease management programs on greenhouse crops: Pesticides and application methods. pp. 38-44. In: *Proceedings of Ninth Conference on Insect and disease management in ornamentals* (eds. Robb, K. and Hall, J.), Del mar, California. Society of American Florists, Alexandria, Virginia.

Malo, A.R. 1993. Estudio sobre la compatibilidad del hongo *Beauveria bassaina* (Bals.) Vuill. conformulaciones comerciales de fungicidas e insecticidas. *Revista Colombiana de Entomologia*, **19**: 151-158.

Mcgarr, R. L., Dulmage, H. T. and Wolfenbarger, D. A. 1970. The delta endotoxin of *Bacillus thuringiensis* H.D. - 1 and chemical insecticides for the control of tobacco budworm and the bollworm. *J. Econ. Entomol.*, **63**: 1357 - 1358.

Moino Jr., A.R. and Alves, S.B. 1998. Efeito de Imidacloprid e Fipronil sobre *Beauveria bassiana* (Bals.) Vuill. E *Metharhizium anisopliae* (Metsch.) Sorok. e no comportamento de limpeza de *Heterotermes tenuis* (Hagem). *Anais da Sociedade Entomológica do Brasil*, **27**: 611-619.

Morris, O. N. 1975. Effect of some chemical insecticides on the germination and replication of commercial *Bacillus thuringiensis*. *J. Invertebr. Pathol.*, **26**: 198 - 204.

Morris, O. N. and Armstrong, J. A. 1975. Preliminary field trials with *Bacillus thuringiensis*, chemical insecticide combinations in the integrated control of the spruce/ budworm *Choristomeura fumiferana*. *Can. Ent.*, **107**: 1281 - 1288.

Oerke, E. C., Dehne, H. W., Schonbeck, F. and Weber, A. 1994. *Crop production and crop protection*. Elsevier Science B. V., Amesterdum.

Patel, N. M. and Kapadia, M. N. 2001. Field efficacy of *Bacillus thuringenesis* var. kurstaki alone and with insecticides against *Helicoverpa armigera* (Hubner). pp. 189-190. In: *Biological control: Contributed papers* (eds. Singh, D., Dhaliwal, V. K., Mahal, M. S., Brar, K. S. and Singh, S. P.). Punjab Agricultural University, Ludhiana.

Quintela, E.D. and McCoy, C.W. 1998. Synergistic effect of imidacloprid and two entomogenous fungi on behaviour and survival of *Diaprepes abbreviatus* (Coleoptera: Curculionidae) in soil. *J. Econ. Entomol.*, **91**(1): 110-122.

Rao, A. G., Rahman, S. J., Swarnarsree, P. and Saxena, R. 2001. Evaluation of sequential application of biopesticides for the management of *Helicoverpa armigera* (Hubner) in pigeonpea. In: *Biological control: Contributed papers* (eds. Singh, D., Dhaliwal, V. K., Mahal, M. S., Brar, K. S. and Singh, S. P.). Punjab Agricultural University, Ludhiana. 169 p.

Shi, W.B., Jiang, Y. and Feng, M.G., 2005. Compatibility of ten acaricides with *Beauveria bassiana* and enhancement of fungal infection to *Tetranychus cinnabarinus* (Acari: Tetranychidae) eggs by sub-lethal application rates of pyridaben. *Appl. Entomol. Zool.*, **40**: 659-666.

Sridevi, T., Krishnayya. P. and Arjuna Rao, P. 2004. Efficacy of mocrobials alone and in combination on larval mortality of *Helicoverpa armigera* (Hubner). *Ann. Pl. Protec. Sci.*, **12**(2): 243-247.

Chapter-8

Effect of Pesticides on Parasites and Predators in Horticultural Crops

Sunita Yadav*, P. Bhatnagar and Manmeet Brar Bhullar***
*Department of Entomology, Punjab Agricultural University, Ludhiana
**KVK, Kaithal, CCS Haryana Agricultural University, Hisar

The availability of wide range of pesticides though has been responsible for increasing crop production through regulating arthropod (insect and mite) pest populations in agricultural and horticultural crop production systems but unfortunately, improper and excessive use of pesticides has resulted into a number of potential ecological problems including pest resistance, resurgence, residues in food, feed and fodder, environmental contamination, direct hazards to the users and mainly destruction of beneficial organisms including natural enemies. The most important group of natural enemies is entomophagous insects i.e. insect predators and parasites. Predatory insects are usually much larger than their prey and are generally voracious feeders that kill and eat a wide variety of insects as they grow and reproduce. Many predators are active in both their immature as well as adult stage. Insect parasites/ parasitoids are often tiny, non-stinging wasps. Parasitic wasps are free-living in the adult stage, but in the larval stage are parasitic on specific insects. The parasitic larvae eat their hosts from within, ultimately resulting in the death of the host insect. The compatibility of natural enemies with pesticides is essential as both these management strategies are important part of Integrated Pest Management programs designed to regulate arthropod pest populations and minimize plant damage.

Effect of Pesticides on Natural Enemies

Natural enemies come in contact with pesticides through a variety of means, including direct exposure to the chemical, through contact with the pesticide residue, through predation and during host feeding by adult or immature parasitoids. They may also ingest the toxic material while feeding on plant material to obtain nutrients or water. Natural enemies are usually more susceptible to the effects of pesticides than their plant-feeding hosts or prey owing to their generally smaller size, searching habits and usually less-developed enzyme based detoxification systems. The pesticides may affect the natural enemies directly or indirectly by affecting physiological and/or behavioral parameter of natural enemies. Direct effects include i) short (immediate mortality) and ii) long term (sub lethal) impacts on natural enemies due to direct exposure to spray or due to coming in contact with residues while walking over the treated surface. The immediate mortality caused by pesticides in natural enemies is influenced by many biological factors such as their weight, size and sex, developmental stage, starvation and nutritional effects, diapause state, circadian rhythm, searching behavior. Predators are usually more tolerant of pesticides than parasitoids. Relatively little attention has been directed towards long-term, sub lethal effects of pesticides on natural enemies. Sub lethal effects of pesticides include reduced daily fecundity, reduced total progeny production, decreased viability, altered predation or parasitism behavior; loss of the ability to recognize hosts; loss of coordination, reduction of predation efficiency, temporary paralysis or knockdown, termination of feeding, and repellency from treated hosts/prey, increased developmental times, production of deformed F1 progeny, decreased production of F1 female progeny, reduced survivial of F1 progeny (Johnston and Tabashink, 1999)

Indirect effects are those in which the impact of pesticide is mediated through natural enemies host or prey (Waage, 1989). Indirect effects may be caused by reduction of host or prey populations that serve as food sources for natural enemies (Powell *et al.*, 1985), a change in the host or prey distribution (Waage, 1989), and ingestion of pesticide-contaminated prey or hosts (Goos, 1973). The indirect effects are sometimes more subtle or chronic compared to direct effects as they inhibit the ability of natural enemies to establish populations, suppress the capacity of natural enemies to utilize prey, impact parasitism (for parasitoids) or consumption (for predators) rates, decrease female reproduction, reduce prey availability, inhibit ability of natural enemies to recognize prey, influence the sex ratio (females: males), and reduce mobility, which could impact prey-finding. The indirect effects of pesticides on natural enemies may vary depending upon the type of natural enemy (parasitoid or predator), their life stages exposed to pesticides (immature or adult), age and sex (male or female). Furthermore, natural enemies, particularly parasitoids, may be indirectly affected by feeding on contaminated honeydew excreted by phloem-feeding insect prey, which could significantly affect their performance.

Natural Enemies of importance in Horticultural Crops and their Susceptibility to Pesticides

Pesticides vary in their activity, which not only impacts how they kill arthropod pests but also how they indirectly influence natural enemy populations. The type of pesticides (contact, stomach poison and systemic) and the application method (foliar, drench and granular) may determine the pesticide mode of action and the extent of any indirect effects on natural enemies. For example, broad-spectrum, nerve toxin pesticides (organophosphate, carbamate, pyrethroids etc.) may be both directly and indirectly more harmful to natural enemies than non-nerve or selective pesticides (Bt, NPV, insect growth regulators, insecticidal soaps etc.). Numerous reviews and articles have been published on the impact of pesticides on biological control agents (Metcalf, 1986; Croft, 1990; Messing and Croft, 1990; Hardin *et al.*, 1995; van Emden and Peakall, 1996), assay techniques to evaluate pesticide effects on natural enemies (Hassan, 1985; 1989; Croft, 1990; Wright and Verkerk, 1995), and ecological and physiological pesticide selectivity (Mullin and Croft, 1985; Poehling, 1989; Croft, 1990).

Coccinellids, popularly known as ladybird beetles (Coleoptera: Coccinellidae) are of great economic importance because of its potential to control many soft-bodied insect pests particularly the aphid on which it feeds voraciously in the immature as wells as mature stages (Samal and Misra, 1982). Due to the use of synthetic organic insecticides, this predatory fauna is largely being eliminated. Vostrel (1991) stated that most of times tested insecticides (carbamates and synthetic pyrethroids) exert higher mortality of all stages of *Coccinella septempunctata* followed by fungicide, acaricides. Dimethoate, fenthion and parathion (Stacherska, 1963), Phosphamidon (Sarup *et al.*, 1965); fenitrothion and trichlorfon (Singh and Malhotra, 1975) and monocrotophos, ethion, fenthion, methyl demeton, malathion and fenitrothion (Sharma and

Adalakha, 1981) were reported to be toxic to *C. septumpunctata*. Swaran, 1999 determined the safety margin of twenty-four insecticides for the adults of *C. septempunctata* by bioassay technique. The order of toxicity on the basis of LC_{50} in decreasing order was lambdacyhalothrin, decamethrin, alphamethrin, dimethoate, cypermethrin, monoctotophos, phosphamidon, fenitrothion, fenpropatbrin, methyl parathion, dichlorovos, malathion, carbaryl, polytrin, fenvalerate, quinalphos, profenofos, pyrethrin, trebon, methyl demeton, lindane, endosulfan, aphidan and menazon. Shukla *et al.*, 1990 found oxydemeton-methyl 25 EC (0.040%) as most toxic and endosulfan 35 EC (0.070%) least toxic to both larvae and adults of the coccinellid. Gour and Pareek, 2005 evaluated contact toxicity of 9 insecticides against grubs and adults of *C. septempunctata* cypermethrin and dimethoate were rated as highly toxic; Ethofenprox, malathion and imidacloprid were rated as moderate toxic and acephate, cartap hydrochloride, endosulfan and neem extract were rated as less toxic. Thus, diazinon, trichlorfon, dimethoate, phosphamidon, formothion and carbaryl should not be applied in the field when the coccinellids are active.

Lady bird beetles, *Menochilus sexmaculata* Fab and *Cheilomenes sexmaculata* (Fab) are also effective aphid predators. Endosulfan was reported to be safer to the larvae of predatory coccinellid by many workers (Makar and Jadhav, 1981; Choudhary and Ghosh, 1982; Babu, 1988; Sonkar and Desai, 1998). Makar and Jadhav (1981) found monocrotophos as less toxic, while quinalphos as highly toxic to the larvae of *M. sexmaculatas*. Tewari and Krishnamoorthy (1985) showed toxic nature of synthetic pyrethroids against the larvae of *M. Sexmaculatus*. Tank *et al.*, (2007) reported that dichlorvos caused maximum (97.16%) mortality of eggs of *C. sexmaculata* followed by cypermethrin, fenvalerate and phosphamidon which were at par causing 55.26 to 63.52 % egg mortality and significantly least (10.55%) mortality was caused by acetamiprid and endosulfan. Similarly dichlorvos ranked first in its toxic action against grubs and adults of *C. sexmaculata* whereas, acetamiprid and endosulfan proved least toxic. Rests of the insecticides were found to be moderate to highly toxic.

The green lacewing, *Chrysoperla carnea* is one of the most common predators in agroecosystem with a wide range of prey including aphids, eggs and young lepidopterous larvae, whiteflies, mites and other soft bodied insects. The susceptibility of *C. carnea* to pesticides vary, for example, synthetic pyrethroids were reported to have, comparatively low toxicity against *C. carnea* (Rajakulendran and Plapp 1982; Sterk *et al.*, 1999), while organophosphate insecticides were usually more toxic to *C. carnea* (Bigler and Waldburger, 1994). Hassan *et al.* (1985) reported that the insecticides namely, pirimicarb, bromophos, heptenophos and trichloroform were harmless to slightly harmful, while diflubenzuron, fenvalerate, dimethoate, methamidophos, chlorpyriphos and methidathion were harmful to the larvae of *C. carnea.* Krishnamoorthy (1985) tested the level of toxicity of several pesticides to eggs, larvae and adults of the green lacewing under laboratory conditions. The egg stage was little affected by insecticidal sprays. The endosulfan, dicofol, monocrotophos, phosalone, methyl demeton, phosphomidan, dimethoate, sulphur and dithane were found to be totally harmless to both laervae and adults while quinalphos, chlorpyriphos, malathion and dichlorvos were highly toxic. Singh and Verma (1986) tested relative toxicity of insecticides against neonate larvae of *C. carnra* for 24 hrs at recommended dose level. Endosalfan, quinalphos, monocrotophos, phenthoate and fenitrothion caused 74-89% larval mortality over a 72-h period under laboratory. Phosalone, carbaryl and cypermethrin were moderately toxic (34.1-38.1% mortality), while fenvalerate was the least toxic (19.2% mortality).The effect of different botanical insecticides viz. neemark, repelin, wellgro, seed kernel extract, nicotine sulphate and neemrich on oviposition of *C. carnea* as well as their ovicidal action was studied under laboratory condition. These botanicals except nicotine did not affect the hatching of the *C. carnea* eggs (Yadav and Patel, 1990). Kapadia and Puri (1991) found that in laboratory, triazophos, monocrotophos and fenpropathrin had greater residual toxicity to larvae of *C. carnea* than endosulfan, cypermethrin and fenvalerate. Vogt (1994) found insect growth regulators moderately harmful to *C. carnea* larvae. Jakob (1996) investigated the effect of neem extracts on apple pests and their natural enemies in the laboratory and field. He found that neem extracts had no adverse side effect on the predator *C. carnea.* Sarode and Sonalkar (1999) studied the biosafety of nine insecticides to *C. carnea* in the laboratory. None of the insecticides were found safer to the predator. However, neem seed extract proved comparatively safe which resulted in maximum hatching of eggs, followed by monocrotophos and phosalone. Chlorpyriphos, deltamethrin and cypermethrin were found highly toxic to the predator eggs.

Meyerdirk, *et al*, (1982) tested toxic residual activities of four pesticides, phosmet, carbaryl, diazinon and dimethoate against natural enemies of the citrus mealybug, *Planococcus citri* (Risso). Species tested included four parasites, *Pauridia peregrina* Timberlake, *Leptomastidea abnormis*, *Leptomastix dactylopii*, and *Anagyrus pseudococci*, and two predators, *Cryptolaemus montrouzieri* and *Sympherobius barberi*. Phosmet and carbaryl had significantly high toxic residual activity up to 30 days post treatment against the majority of natural enemies tested whereas diazinon and dimethoate toxic residue activity decreased significantly in 9 days. Therefore, in the insect-pest management programme, the use of phosmet and carbaryl should be avoided. Mani (1993) found dichlorvos, dicofol and few botanicals to be safe to exotic parasitoid, *Leptomastix dactylopii* in citrus orchards. Mani *et al.*, 1995 reported that mean longevity and progeny production of both male and female parasitoids (*L. dactylopii*) of *P. citri* was adversely affected with all the pesticidal treatments (dichlorvos 0.01%, neem seed kernel extract 5%, copper oxychloride 0.20%, dicofol 0.05% and diflubenzuron 200 p.p.m.). It is concluded that these selective chemicals should be avoided when *L. dactylopii* is released in the field.

Tewari and Krishnamoorthy (1983) found endosulfan to be safer to *Podiobuis foveolatus* as compared to synthetic pyrethroids in brinjal. The application of NSKE for control of *Spodoptera litura* did not affect the emergence of egg parasitoid, *Telenomus remus* Nixon. The longevity of the parasitoid was reduced when oviposition was made on pretreated egg masses. Topical application of seed of neem oil and custard apple showed slightly toxic effect to the predators of mirid bug, *Cyrtorhinus lividipennis* Reuter and showed no adverse effect on predatory spider, *Lycosa pseudoannulata* (Rosenberg and Strand). Neem formulations were safe to the predatory coccinelid, *Menochilus sexmaculatus* (Fabricious) but highly toxic to its hyperparasite, *Terrastichus coccinella*. These formulations were also found to be safe to the predatory mites and did not show ovicidal activity against the eggs of lacewing.

Walton and Pringle (1999) studied the effects of regularly used table grape insecticides and fungicides on 1-day-old adults of the parasitoid *Coccidoxenoides peregrinus* (Timberlake) of vine mealybug, *Planococcus ficus* (Signoret) in the laboratory. The insecticides chlorpyrifos, endosulfan and cypermethrin were highly toxic to the parasitoid, while the fungicides penconazole and mancozeb were not toxic. These results suggest that the insecticides may be detrimental to a biological control system using *C. peregrinus* while the two fungicides tested can be used in augmentative releases of *C. peregrinus*.

Wakgari and Giliomee (2003) tested through Laboratory bioassay the effect of contact insecticides (methidathion, methomyl, methyl-parathion, parathion, profenofos and prothiofos) and IGRs against C. *peregrines* in citrus orchards. All insecticides were proved highly toxic, causing 98-100% mortality in less than 6 h of treatment. The IGR's fenoxycarb and triflumuron did not cause significant parasitoid mortality. However, a mixture of pyriproxyfen and mineral oil caused a marginally significant mortality.

Moura *et al.* (2006) studied the effects of the insecticides abamectin, acetamiprid, cartap and chlorpyrifos on larvae, pupae (within the host egg) and adults of the egg parasitoid *T. pretiosum* Riley under laboratory conditions. Cartap and chlorpyriphos proved to be the most harmful insecticides, affecting both the emergence success and parasitism capacity of this parasitoid, whereas abamectin and acetamiprid were selective.

Prasad and Nath (2011) reported that neem based and biological origin pesticides are less harmful to the natural enemies than newer and conventional synthetic organic insecticide like imidacloprid and cypermethrin on the population of natural enemies i.e., braconid wasp, coccinellid beetle and predatory spider in brinjal ecosystem.

Nemade *et al.* (2009) found that imidacloprid seed treatments were safer for natural enemies of *Earias vittella* in okra field. Among the foliar spray endosulfan spray was safe to natural enemies whereas fenvalerate was found most toxic. Foliar spray of imidacloprid was found safer to spiders but proved least safer to other natural enemies.

Suma *et al.* (2009) tested the side-effects of the insecticides chlorpyrifos-methyl, buprofezin, pyriproxifen, spinosad and a narrow range mineral oil on *Aphytis melinus* DeBach, *Coccophagus lycimnia* Walker and *L. dactylopii*. Chlorpyrifos-methyl and spinosad caused 100% mortality on all tested parasitoids just 24 h after the treatment. According to the IOBC classification of toxicity, the mineral oil was slightly harmful on *L. dactylopii*; buprofezin was harmful on *C. lycimnia*, and slightly harmful on *L. dactylopii* and *A. melinus*; pyriproxifen resulted moderately harmful on *C. lycimnia* and slightly harmful on *L. dactylopii* and *A. melinus*. The progeny production of *A. melinus* surviving female treated with buprofezin was significantly reduced *C. lycimnia* females treated with pyriproxifen did not produce any progeny.

Bayram *et al.* (2010) investigated the inhibitory effects of a pyrethroid insecticide (permethrin), Adion 20% EC on the flight responses, host-searching behaviour and foraging behaviour of *Cotesia vestalis* (Hymenoptera: Braconidae), a larval parasitoid of the diamondback moth, *Plutella xylostella* (Lepidoptera: Plutellidae), under laboratory conditions and reported that the application of the insecticide had an inhibitory effect on the wasps' searching behaviour and the mortality of *C. vestalis* adults on the insecticide-treated plants significantly higher than in the control plants treated with distilled water, consequently reduced the effectiveness of *C. vestalis* as a biological control agent against *P. xylostella*.

Preetha *et al.* (2010) compared the toxicity of imidacloprid, thiamethoxam and methyl demeton on *Trichogramma chilonis* Ishii, *Chelonus blackburni* Cameron and *Bracon hebetor* under laboratory conditions. Imidacloprid 17.8 SL did not cause any adverse effects on the adult emergence and parasitization of *T. chilonis*. The recom-mended dose of imidacloprid (25 g a.i./ha) caused 56 per cent mortality and was found to have moderate impact on the adults of *C. blackburni*. On the other hand, it was found to be toxic to the parasitoid *B. hebetor*, causing 70 per cent mortality at 48 hours after treatment.

Sharma and Kaushik (2010) evaluated effect of spinosad 45 SC alongwith six chemical insecticides *viz.*, emamectin benzoate 5 WSG, cypermethrin 10 EC quinalphos 25 EC, endosulfan 35 EC, lambda cyhalothrin 5 EC, chlorpyrifos 20 EC was against pest complex of eggplant and their natural enemies (*Encarsia lutea, Chrysoperla carnea* and ladybird beetles). Spinosad @ 162.5 ml/h was found safe to natural enemies whereas the chemical insecticides proved toxic to them.

Regupathy and Ayyasamy (2011) observed that avoidance of insecticide spraying resulted in the appearance of notable number of biocontrol organisms in papaya fields in Tamilnadu. Six species of spiders *viz. Clubiona*spp. crab spider, *Thomisus* spp., Jumping spider *Phidippus* sp., *Plexippus* sp., *Araneus* sp. wolf spider, *Lycosa pseudoannulata* were found and the most predominant one was *Araneus* sp. Nine species of coccinellids, *viz. Brumoides suturalis* (Fab.), *Cheilomenes sexmaculata* (Fab.), *Coccinella septumpunctata* Linnaeus, *Coccinella nigrita* Fab, *Cryptolaemus montrouzieri* Mulsant, *Hippodamia variegata* (Goeze), *Hyperaspis maindroni* Sicard, *Nephus regularis* Sicard and *Scymnus coccivora* Ayyar recorded from the fields. Two species of chrysopids, *Chrysoperla zastrowi silemi (carnea)* and *Mallada* sp. were observed. Parasitoids *A. papayae* and *Torymus* sp. (Torymidae) only were found. Four species of ants *viz., Camponotus compressus* (Fabricius), *Camponotus sericeus* (Fabricius), *Camponotusparius* Emery and *Tapinoma melanocephalum* (Fabricius) were found to be associated with mealybug.

Stara *et al.* (2011) tested the side effects of methoxyfenozide, indoxacarb, pyridaben, acetamiprid, azadirachtin A, spinosad, and propargite on *Aphidius colemani, Aphidoletes aphidimyza*, and *Neoseiulus cucumeris* under laboratory conditions. Methoxyfenozide had low toxic effect on all three species, causing mortality after 24 h in 4.4, 11.4, and 29.3% of *N. cucumeris, A. colemani, and A. aphidimyza*, respectively. Similarly, indoxacarb caused mortality after 24 h in 11.9, 20.0, and 24.9% of *A. aphidimyza, N. cucumeris, and A. colemani*, respectively.

Methoxyfenozide was shown to significantly reduce fecundity of *A. aphidimyza*. In contrast, there was no effect of pure azadirachtin A on *A. colemani* fecundity. Thus both methoxyfenozide and indoxacarb would be suitable for use in the integrated pest management (IPM).

Simmons and Shaaban (2011) evaluated eight biorational insecticides (based on oil, plant derivatives, insect growth regulator and fungus) in the field for their influence on populations of six natural enemies of *B. tabaci*. Natural populations of two predators (*Chrysoperla carnea* Stephen and *Orius* spp). and two genera of parasitoids (*Encarsia* spp. and *Eretmocerus* spp.) were evaluated in eggplant, *Solanum melongena* L. Also, augmented field populations of three predators [*C. carnea, Coccinella undecimpunctata* L. and *Macrolophus caliginosus* (Wagner)] were evaluated in cabbage, cucumber and squash. Regardless of natural enemy or crop, jojoba oil, Biovar and Neemix had the least effect on abundance of the natural enemies in comparison with the otherinsecticides during a 14 day evaluation period. Conversely, Admiral, KZ oil, Mesrona oil, Mesrona oil + sulfur and natural oil had a high detrimental effect on abundance of the natural enemies.

Mohan and Chauhan (2012) evaluated toxicity of Fenazaquin, HMO and Neem Baan on predatory mite, *Neoseiulus longispinosus* (Evans) which is an important predatory mite of *Tetranychus urticae* on apple. Fenazaquin was found very toxic and showed corrected mortality of 84.4 and 97.8 per cent under field and laboratory conditions respectively after 72h and 96h of treatment. HMO was categorized non-toxic to slightly toxic showed corrected mortality of 20.5 and 26.0 per cent under field and laboratory conditions respectively. Neem Baan was found non-toxic as corrected mortality varied from 4,4 arid 13.4 per cent in 72h and 96h of treatment.

Moura *et al.* (2010) evaluated the effects of abamectin 18 CE (0.02 g a.i. L^{-1}), carbaryl 480 SC (1.73 g a.i. L^{-1}), sulfur 800 GrDA (4.8 g a.i. L^{-1}), fenitrothion 500 CE (0.75 g a.i. L^{-1}), methidathion 400 CE (0.4 g a.i. L^{-1}), and trichlorfon 500 SC (1.5 g a.i. L^{-1}) as applied in integrated apple production in Brazil on the survival, oviposition capacity, and egg viability of the lacewing, *Chrysoperla externa* (Hagen) (Neuroptera: Chrysopidae). First- and second-instar larvae were exposed to pesticide residues sprayed on glass plates. Carbaryl, fenitrothion and methidathion were found harmful to *C. externa*. Trichlorfon was harmful to first-instar larvae and slightly harmful to second-instar larvae. Abamectin and sulfur were slightly harmful to first-instar larvae and harmless to second-instar larvae.

Procedure for testing the Efficacy of Pesticides on Natural Enemies

The International Organization of Biological Control (IOBC) Working Group on Pesticides and Beneficial Organisms, together with other workers, has made significant progress in identifying available compounds that can be used in integrated programs. Registration of pesticides should include tests on their impact on natural enemies commonly found in the crop systems where materials will be applied. Resulting information should be provided on pesticide labels. Without this information, one cannot make intelligent pesticide selections. In addition to the type of information that the IOBC/WPRS Working Group has provided, data are needed on the impact of pesticides on natural enemy populations from different areas. Studies have shown that natural enemy responses to pesticides vary with locality. This information would allow a grower to select the least disruptive pesticide for use in integrated programs and would also provide information on the ability of natural enemies to develop pesticide resistance. Pattern techniques to test the physiological selectivity of insecticides to natural enemies were developed by the IOBC. In IPM programmes, incorporation of natural enemies is possible only when pesticides of low toxicity to natural enemies as compared to pests are used. Therefore, evaluation of the effect of pesticides on beneficial organisms has become a necessity. Realizing the importance of standard test methods, the IOBC, West Palaearctic Regional Section (WPRS) formed a working group "Pesticides and Beneficial Organisms" in 1974 to develop standard procedures for testing the side effects of pesticides on natural enemies using laboratory, semi field and field test methods(Hassan,1994). No single test method could provide sufficient information on the side effects of pesticides on any natural enemy, IOBCIWPRS working group had suggested a combination of tests to be carried out in a particular sequence (Hassan *et al.*, 1985), Normally or testing the effects of pesticides on natural enemies, commercial preparations are preferred in view of the harmful effects, the adjuvants may have on the natural enemies in addition to the actual toxicants.

Choice of Natural Enemies for the Test

The natural enemies chosen for the tests should be relevant to the crop on which the particular pesticides are to be used. Also the pests that attack the crop at different stages need to be considered for selecting representative natural enemies attacking different pests and their stages. The quality of natural enemies used in such tests is of at most importance in order to obtain reliable information on the susceptibility or otherwise. Weak parasitoids which may be the result of super parasitism in the previous generation or very old individuals should not be used. Uniform age/stage individuals are always preferable for such toxicity tests.

Protocol for Assessment of Toxicity to Parasites and Predators in India (David, 1985)

1.	Bio-assay method:	Dry film / topical application method
2.	Species:	Any two important insect parasites of major crop pests and anyone insect predator species of crop pests
3.	Treatment dosages:	A minimum of 5 treatment dosages giving mortality 20 to 80 per cent
4.	Number of test insects:	10 to 20, laboratory reared, one day old adult females per replicate in case of parasites and 10 to 12 field collected predators per replicate
5.	Number of replications:	3 or more 1 untreated control
6.	Exposure period:	6 to 8 hours in case of dry film method
7.	Post-treatment period :	1 day in case of parasites and 8 to 12 hours in the case of predators
8.	Test conditions:	Test insects should be conditioned for 6 to 12 hours at a temp. of 28±2° C. Same temp. should be maintained for exposure and post treatment period
9.	Mortality counts:	Moribund insects should be treated as dead. If mortality is observed in control population Abbott's (1925) formula should be used to correct mortalities in treatments Corrected per cent mortality = $\frac{T-C}{100-C}$ Where, T = Per cent mortality in treatment C = Per cent mortality in control

10. Calculation: LC_{50} should be calculated statistically & equation regression should be provided (Finney, 1947)

Standard Characteristics Test Methods

A. Laboratory, initial toxicity tests

Parasitoids and predators of uniform age may be exposed to a fresh dry pesticide film applied on glass plates, plant leaves or soil depending upon the behavior of the natural enemy. Pesticides may be tested at recommended concentrations or at different concentrations to arrive at the LC50 values. Tests are carried out under conditions of controlled temperature and humidity favourable to the parasitoid. Normally 27±2°C and 65±5% R.H. are maintained. Mortality in water treated controls should not exceed 10 per cent. Forced ventilation to avoid accumulation of pesticide fumes is to be provided when closed cells or cages are used. Calculated quantity of the pesticide solution is sprayed with a Potter's Tower or a sprayer to provide an even and reproducible film on the experimental surface. Adult natural enemies of uniform age (0-24 h old) are to be exposed to the residues for a defined period (6, 12 or 24 hrs.) and the mortality recorded at the end of the exposure period. The natural enemies are to be provided with adequate food and hosts during the exposure period to avoid undue mortality due to absence of hosts and adult food. Simultaneously, the effect of insecticides on the beneficial capacity of the natural enemy can be evaluated by recording the parasitism/predatism observed in the hosts that were offered.

a) exposure to fresh dry pesticide film; b) recommended concentration of pesticide; c) application on glass plate, leaf or sand (soil); d) even film of pesticide, standard amount of 1-2 mg fluid cm^2 on glass or leaf and 6 mg fluid cm^2 on sand (soil); e) laboratory-reared organisms uniform in age; f) adequate exposure period before evaluation; g) adequate ventilation; h) water-treated controls; j) reduction in beneficial capacity/mortality; k) four evaluation categories:

1= harmless (50%), 2= slightly harmful (50-79%), 3= moderately harmful (80-99%), 4=harmful (>99%).

Test of Immature Stages

Hosts containing the immature stages of the natural enemy could be sprayed with the insecticides in a Potter's Tower. Percentage of individuals completed normal development and emergence and subsequent parasitism/predatism could be observed for determining the harmful effect of the pesticides.

Semi-Field, Initial Toxicity Tests

In this type of test, the laboratory-reared beneficials are exposed to a fresh dry pesticide film applied on natural material (i.e. plant or soil) and the test units are placed in the field under rain cover and partial shade or under field-stimulated environment conditions. Plants should have dense foliage to provide sufficient leaf surface for the pesticide treatment. Cages should be designed to suit the size and shape of the test plants. The plants, but not the experimental cages, are sprayed with the pesticide at recommended concentration to the point of run-off. After he pesticide has dried, the plants are caged, the beneficial are released and food is provided. The test units are placed in the field under transparent rain cover. Excessive heat and sunlight are excluded by providing partial shading. After an adequate exposure period, host or prey animals are placed among the treated plants and the performance of the beneficials is compared with control units treated with water.

a) exposure to fresh dry pesticide film; b) recommended concentration of pesticide; c) wet spraying of plants (point of run-off); d) field cages under field or field-simulated conditions; e) water treated controls; f) laboratory-reared organisms, uniform in age; g) adequate contact through dense foliage; h) food+ host/prey near the centre of treated foliage; j) adequate exposure period before evaluation; k) four evaluation categories; 1= harmless (25%), 2= slightly harmful 25-50%), 3= moderately harmful (51-75%), 4= harmful (>75%).

Semi-field, Persistence Test

Pesticides which are found to be harmless to the natural enemies in the laboratories are likely to be harmless to the same organisms in the field. In such cases further testing in semi-field and field experiments is not advocated. The technique used to test the persistence (duration of harmful activity) of pesticide residues involves the treatment of plants or soil, maintaining them under field 'or field-simulated environment conditions, and exposing the beneficials to the treated substratum, taken at different time intervals after application. The duration of harmful activity is the time required for the pesticide residue to lose effectiveness so that a reduction in parasitism of less than 30%. a) exposure to pesticide residues; b) recommended concentration of pesticide; c) wet spraying of plants (point of run-oft); d) weathering under field or field-simulated environment; c) experiments up to one month after treatment; t) laboratory-reared arthropods, uniform in age; g) water-treated controls; h) four evaluation categories: 1= short-lived «5 days), 2= slightly persistent (5-15 days), 3= moderately persistent (16- 30 days), 4= persistent (>30 days).

Field Test

In this type of test, crops inhabited by naturally occurring or laboratory-reared beneficials are directly treated with pesticide to be tested. The time and number of pesticide treatments, as well as the dose used, should be taken according to common and good agricultural practice. Water-treated controls should be set up. The design of the trials, the number of plots, the number and size of samples should be made according to recognized statistical methods. Dead and/or living individuals are collected and the number of individuals has to exceed a certain limit to allow statistical analysis.

a) crops inhabited by beneficials are directly treated: b) laboratory-reared or naturally occurring arthropods; c) sampling at intervals before and after treatment(s); d) recommended dose and number of treatments (good agricultural practice); e) water- treated control; t) dead and/or living individuals collected; g) number of individuals to exceed a certain limit to allow statistical analysis; h) four evaluation categories; 1= harmless (25%), 2= slightly harmful (25-50%), 3= moderately harmful (51-75%), 4= harmful (>75%).

Techniques/methods to reduce the negative Impact of Chemicals on Natural Enemies

- Apply pesticides only when pest population reaches above Economic threshold Levels (ETL). This will give time to the natural enemies to work and maintain pests at low levels. This requires development of ETL and sampling programs along with evaluation techniques to judge biological control effectiveness. Economic thresholds should be a cornerstone of every IPM program (Higley and Pedigo, 1996a). However, relatively few thresholds exist compared with the number of crop pests for which chemicals are applied (Peterson, 1996).
- Implementation of economic thresholds also requires monitoring programs that estimate population densities of both the pest and its natural enemies. The impact of pesticide

treatments on natural enemies of secondary pests should also be considered when decisions are made to treat for primary pests.

- To use pesticides in most efficient and least disruptive manner first define the *biological target* by i) Identification of all pests and associated natural enemies to be potentially affected by pesticide applications ii) Knowledge of the behavior and micro-habitats of targeted pests and associated natural enemies iii) Knowledge of the dosage responses of pests and associated natural enemies to available pesticides iv) Knowledge of all feasible application methods for each compound. This information will provide the basic foundation for maximizing benefits derived from pesticide selectivity.
- After the *biological target* is defined, use *Pesticide selectivity* for the most efficient use of a pesticide. *Pesticide selectivity* is the capacity of a pesticide treatment to spare natural enemies while destroying the target pest. Selectivity is a relative measure expressed as the natural enemy/pest ratio produced by one pesticide treatment compared with the ratio produced by a reference pesticide treatment. Pesticide selectivity may be divided into *physiological selectivity* and *ecological selectivity* based on the mechanism by which a pesticide treatment is preferentially toxic to pests versus their natural enemies.

The ecological selectivity is the use of nonselective insecticides in a selective way as a means to minimize exposure of natural enemies to the insecticide. This selectivity is usually accomplished through

i. Temporal discrimination: application of pesticides at hours of the day when temperatures are mild, because that is when there is less movement of natural enemies and other organisms.

ii. Modification of application methods: Use smaller nozzles for applying lower volumes and finer spray. The smaller droplets will penetrate deep in the foliage and their better retention will reduce the level of run off to the soil where the pesticide can affect soil-living natural enemies. Target the pest by using V lance to improve under leaf spray cover with reduced doses.

iii. Granular applications are generally safer than sprays. Sprays are safer than dusts. Water soluble concentrates and emulsifiable concentrates are generally safer. Aerial applications should be totally discouraged due to large scale mortality of natural enemies. Ground application is comparatively safer. Fine spray particles cause less mortality than coarse particles. Use only those insecticides which have been recommended as safe to the natural enemies.

iv. Reduce the application rates: Follow the basic principle of IPM i.e. needs based applications of insecticides, in right quantities with recommended dilutions as higher doses affect the natural enemies more than target pest. Doses can even be reduced below recommended levels, provided the application quality is good. Test reduced doses first on small scale to be sure that they are still effective.

v. Use non-persistent pesticides

vi. Habitat discrimination - application of pesticides to the parts of the habitat where the pest is more frequently found than the natural enemy.

vii. Habitat partitionment- localizes the application by spot or stratified spraying to spray only part of the plant or field. This allows natural enemies to survive in the unsprayed areas and to re-enter the sprayed area when pesticide residues have diminished.

On the other hand, the *physiological selectivity* employs insecticides with low toxicity to the natural enemies or those which are more toxic to pests than to natural enemies (Bacci *et al.*, 2006). When pest resurgence is the primary consideration, the major objective of physiological selectivity is that a significant portion of the pest population survives the treatment. When secondary pest outbreaks are the foremost concern it is preferable to use compounds with high *specificity* to control the primary pest. This tactic avoids destruction of the secondary pest's natural enemies and reduces further development of resistance in the secondary pest. Insecticides such as cyromazine (Weintraub and Horowitz 1996), abamectin, cartap and phenthoate were safer, in other words, besides presenting high efficiency in pest control, a small increase in the concentration of the insecticide does not produce a substantial increase in the mortality of the natural enemy, even when mixed with mineral oil (Leite *et al.*, 1998). Such effect occurs because these products are physiological. The cyromazine inhibits the larval development and does not inhibit the formation of chitin nor acts directly on adults. In addition, the abamectin, of the avermectin group, besides killing moth caterpillars and adults by the action of contact, may interfere in the female reproductive organs, leading to the laying of infertile eggs (Nauen and Bretschneider, 2002). Selective toxicity of some of the important insecticides was evaluated against different natural enemies which is summarized and given in Table 1.

Table 1: Comparatively safer insecticides to different natural enemies used against various pests of fruit crops

S No.	Pest	Natural Enemy	Safer Insecticide	Reference
1	Mealybugs, *Plannococcus citri*	*Coccidoxenoides peregrines*	Deltamethrin and Botanicals (Safe), Endosulfan Fenvalerate, Methyl demeton and Dichlorvos (Less toxic)	Mani *et. al.*, 2002
		Leptomastix dactylopii	Dichlorvos, Dicofol and Plant products (Safe), Fluvalinate, Endosulfan Quinalphos Methyl parathion, Chlorpyriphos and Carbaryl (Less toxic)	Mani *et. al.*, 1993
2	Oriental mealybug, *Planococcus lilacinus*	*Tetracnemoidea indica*	Fenvalerate and several botanicals (Very safe), Dichlorvos, Deltamethrin, Methyl parathion, Quinalphos and Chlorpyriphos (Less toxic)	Mani and Krishnamoorthy, *1996*
3	Psylla, *Diaphorina citri*	*Tamarixia (=Tetrastichus) radiata*	aluminum tris, copper hydroxide, diflubenzuron, and kaolin clay (Surround WP).	Hall and Nguyen, 2010
4	Citrus butterflies, *Paplio demoleus* and *P. polites*	*Trichogramma chilonis, Telenomus incommodus* and *Distratrix papilionis*	Dicofol, Copper oxychloride Neem seed extract and Neemark (Less harmful)	Mani and Krishnamoorthy, 1997
5	Pink mealy bug, *Maconellicoccus hirsutus*	*Cryptolaemus montrouzieri*	Dichlorvos, Chlorpyriphos, Fish oil resin soap	Babu, 1986; Mani, 1988

Contd.

6	Mealybug, *Ferrisia virgata*	*Aenasius advena*	Diazinon, all fungicides and Phosalone, Dichlorvos and Endosulfan (Less toxic)	Mani, 1992
7	Ash whitefly, *Siphonius phillyrea*	*Encarsia azimi*	All acaricides, fungicides, botanicals Dichlorvos and Endosulfan	Mani and Krishnamoorthy, 1996
8	Fruit borer, *Helicoverpa armigera*	*Campoletis chloridae*	Acephate and Neem (Totally safe)	Mani, 1994
		pupal stage of *Trichogramma pretiosum*	Cyazypyr @ 90 and 60 g a.i./ha	Mandal, 1980
10	White fly, *Bemisia tabaci*	*Discodon sp.*	Abamectin 18 CE	Bacci *et al.*, 2007
11	DBM, Plutella xylostella	*Cotesia plutellae*	NSKE, Fluvalinate, Carbarlyl, Acephate, Methyl demeton and Dichlovos (Safe)	Mani and Krishnamoorthy, 1984; Mani, 1995
		Diadegma semiclausum	chlorfluazuron, flufenoxuron and teflubenzuron, coppernonylphenol sulfonate, kasugamycin and oxolinic acid)	Haseeb *et al. 1995*
12	*Spodoptera litura*	*Telenomus remus*	All fungicides and Acaricides (Safe)	Mani and Krishnamoorthy, 1986
13	*Henosepilachna vigintioctopunctata*	*Pediobius foveolatus*	Endosulfan and NSKE	Terwari and Krishnamoorthy, 1983
14	Aphids, *Myzus persicae*	*Aphelinus sp.*	Methyl demeton, Deltamethrin and Fenvalerate (Non toxic)	Mani and Krishnamoorthy, 1994
		Coccinella septempunctata.	Endosulfan and menazon	Sarup *et al.*, 1965; Singh and Malhotra, 1975

Conclusion

One major purpose of Integrated Pest Management strategies is to unify the safe and sustainable use of chemical and biological control methods. The successful combination of pesticide use and biological control is dependent more on knowledge of the system, the ecology and the behavior of pests and natural enemies than on the availablility of tools and techniques. The best approach for preserving effective biological control of natural enemies is the combination of tactics including an understanding of the biology and behavior of arthropods, detailed monitoring of life history and population dynamics of pests and natural enemies, employment of selective pesticides, use of the least disruptive formulation of the pesticide, application only when absolutely necessary, basing chemical control on established economic injury levels, and application at the least injurious time. IPM in horticultural crops is facing a series of problems to maintain its effectiveness due to the side-effects of pesticides on biocontrol agents. Therefore, there is need to generate more information by screening of all the commonly used pesticides for their safety to the key parasitoids and predators.

Suggested Readings

Abbott, W. S. 1925. A method of computing the effectiveness of an insecticide. *J. Eco. Entomol.*, **18**: 265-267

Bacci, L., Crespo, A. L. B., Galvan, T. L., Pereira, E. J. G., Picanço, M. C., Silva, G. A. and Chediak M. 2007. Toxicity of insecticides tothe sweetpotato whitefly (Homoptera: Aleyrodidae) and its natural enemies. *Pest Manage. Sci.*, **63**: 699-706.

Croft, B. A. 1990. *Arthropod biological control agents and pesticides*. New York: John Wiley & Sons. 723 pp.

Fernandes, F. L., Bacci, L. and Fernandes, M. S. 2010. Impact and Selectivity of Insecticides to Predators and Parasitoids. *Entomo Brasilis*, **3**(1): 01-10.

Finney, D. J. 1947. *Probit Analysis*. Cambridge University Press, London, 333 pp.

Goos, M. 1973. Influence of aphicides used in sugar-beet plantations on arthropods. II. Studies on arachnids-Arachnoidea. *Polski Pismo Entomologiczne*, **43**:851-859.

Hardin, M. R., Benrey, B., Coil, M., Lamp, W.O., Roderick, G. K., and Barbosa, P. 1995. Arthropod pest resurgences: An overview of potential mechanisms. *Crop Protec.*, **14**:3-18

Hassan, A. S. 1994. Comparison of three different laboratory methods and one semi-field test method to assess the side effects of pesticides on *Trichogramma cacoeciae*. *Bulletin-OILB-SROP*. **17**: 133-141.

Hassan, S. A. (1989). Testing methodology and the concept of the *IOBC*/WPRS working group. In *Pesticides and non-target invertebrates* (Ed. P. C. Jepson.), (pp. 1-18). Wimborne, Dorset, United Kingdom: Intercept.

Hassan, S. A. 1985. Standard methods to test the side-effects of pesticides on natural enemies of insects and mites developed by the *IOBC*/ WPRS Working Group Pesticides and Beneficial Organisms. *EPPO Bulletin*, **15**:214-255.

Hassan, S.A., Bigler, F., Blaisinger, P. *et al.*, 1985.Standard methods to test the side-effects of pesticides on natural enemies of insects and mites developed by the IOBC/WPRS-Working Group (Pesticides and Beneficial Organisms). *Bulletin OEPPIEPPO*, **15**: 214-255.

Higley, L. G. and Pedigo, L. P. 1996. The EIL concept. In: *Economic thresholds for integrated pest management* (Eds. L. G. Higley & L. P. Pedigo), (pp. 9-21). Lincoln, NE: University of Nebraska Press.

Johnston and Tabashink, 1999. Enhanced biological control through pesticide selectivity. In: *Handbook of Biological Control: Principles and Applications* (Eds T.S. Bellows, Jr. and T. W. Fisher). Academic Press, San Diego, California, U.S.A. 1046 pp.

Leite, G. L. D., Picanço M.C., Guedes R.N.C. and Gusmão M.R. 1998. Selectivity of insecticides with and without mineral oil to *Brachygastra lecheguana* (Hymenoptera: Vespidae), a predator of *Tuta absoluta* (Lepidoptera: Gelechiidae). *Ceiba*, **39**: 191-194.

Mani, M., Sushil, S. N. and Krishnamoorthy, A. 1995. Influence of some selective pesticides on the longevity and progeny production of *Leptomastix dactylopii* How., a parasitoid of citrus mealybug, *Planococcus citri* (Risso). *Pest Manage. Horti. Ecosys.*, **1**(2): 81-86

Mani, M., Krishnamoorthy, A. and Rao, M.S. 1993, Toxicity of different pesticides to the exotic parasitoid *Leptomastix dactylopii* How. *Indian J. Plant Protec.*, **21**(1):98-99

Mani, M. and Krishnamoorthy, A. 1997. Safety of plant products and conventional pesticides to *Distratrix papilionis* (Vireck)(Hymenoptera:Braconidae), a parasitoid of citrus butterfly. Paper presented in National Symposium on Citriculture. November 17-19, 1999, Nagpur.

Mani, M. and Krishnamoorthy, A.1996. Response of the Encyrtid Parasitoid, *Tetracnemoidea indica* of the Oriental Mealybug *Planococcus lilacinus* to Different Pesticides, *Indian J. of Plant Protec.* **24**(1): 80-85.

Mani, M., Krishnamoorthy, A. and Sreenivasa R. M. 2002.Selective toxicity of different pesticides to citrus mealybug parasitoid, *Coccidoxenoides peregrima* (Timberlake), *J. Insect Sci.*, **15**: 49-52.

Messing, R., and Croft, B.A. 1990. Sublethal influences.pp. 157-183. In: *Arthropod biological control agents and pesticides* (Ed. B. A. Croft.). New York: John Wiley & Sons.

Meyerdirk, D.E., French, J.V. and Hart, W.G. 1982. Effect of pesticide residues on the natural enemies of citrus mealybug. *Environ. Entomo.*, **11**: 134–136.

Mohan, P. and Chauhan, U. 2012. Direct Effects of Some Acaricides on the Performance of *Neoseiulus longispinosus* (Evans) against *Tetranychus urticae* Koch on Apple. In: *Abstracts of International conference on Entomology*, February 17-19, 2012 organized by Department of Zoology and Enviornmental sciences, Punjabi University, Patiala. p: 262

Moura, A. P., Carvalho, G. A., Moscardini, V. F., Lasmar, O., Rezende, D. T. and Marques, M. C. 2010. Selectivity of pesticides used in integrated apple production to the lacewing, *Chrysoperla externa*. *J. Insect Sci.*, **10**(121):1-20

Moura, A.P., Carvalho G.A., Pereira A.E. and Rocha L.C.D. 2006. Selectivity evaluation of insecticides used to control tomato pests to *Trichogramma pretiosum*. *BioControl*, **51**: 769-778.

Mullin, C. A., and Croft, B. A. 1985. An update on development of selective pesticides favoring arthropod natural enemies. pp. 123-150. In: *Biological control in agricultural IPM systems* (Eds. M. A. Hoy & D. C. Herzog), Orlando, FL: Academic Press.

Nauen, R. and Bretschneider, T. 2002. New modes of action of insecticides. *Pesticide Outlook*, **13**: 241-245.

Nemade, P. W., Wadnerkar, D. W., Bansod, R. S., Kulkarni C. G. and Mali, A. K. 2009. Effect of newer insecticides on natural enemies of *Earias viitella* in okra field. *Indian J. Agri. Res.*, **43** (2): 124-128.

Peterson, R K. D. 1996. The status of economic-decision-Ievel development. pp. 151-178. In: *Economic thresholds for integrated pest management* (Eds. L.G. Higley and L.P. Pedigo). Lincoln, NE: University of Nebraska Press.

Poehling, H. M. 1989. Selective application strategies for insecticides in agricultural crops. pp. 151-175. In: *Pesticides and non-target invertebrates* (Ed P. C. Jepson). Wimborne, Dorset, United Kingdom.

Powell, W., Dean, G. J., and Bardner, R. 1985. Effects of pirimicarb, dimethoate and benomyl on natural enemies of cereal aphids in winter wheat. *Ann. App. Bio.*, **106**:235-242.

Prasad, T. G. and Nath Lok, C. S. 2011. Effect of insecticides, bio-pesticides and botanicals on the population of natural enemies in brinjal ecosystem. *Vegetos*, **24**(2):40:44.

Preetha, G., Manoharan, T., Stanley, J. and Kuttalam, S. 2010. Impact of chlornicotinyl insecticide, imidacloprid on egg, egg-larval and larval parasitoids under laboratory conditions. *J. Plant Protec. Res.* **50**: 535-540.

Sharma, S. S. and Kaushik, H. D. 2010. Effect of Spinosad (a bioinsecticide) and other insecticides against pest complex and natural enemies on eggplant (*Solanum melongena* L.). *J. Entomo. Res.*, 34 (1): 39-44

Tank, B.D., Korat, D.M. and Board, P.K. 2007. Relative toxicity of some insecticides against *Cheilomenes sexmaculata* (Fab.) in laboratory. *Karnataka J. Agri. Sci.*, **20**(3): 639-641.

Waage, J. K. 1989. The population ecology of pest-pesticide-natural enemy interactions. pp. 81-93. In : *Pesticides and non-target invertebrates* (Ed. P. C. Jepson), Wimborne, Dorset, United Kingdom: Intercept.

Walton V.M. and Pringle K.L. 1999. Effects of pesticides used on table grapes on the mealybug parasitoid, *Coccidoxenoides peregrinus* (Timberlake) (Hymenoptera: Encyrtidae). *South African J. of Eco. Viticulture*, **20**(1), 31-34.

Weintraub, P.G. and R. Horowitz. 1996. Spatial and diel activity of the pea leafminer (Diptera: Agromyzidae) in potatoes,*Solanum tuberosum*. *Environ. Ento*, **25**: 722-726.

Wright, D.J., and Verkerk, R. H. J. 1995. Integration of chemical and biological control systems for arthropods: Evaluation in a multitrophic context. *Pesticide Sci.*, **44**:207-218.

Chapter-9

Bioremediation of Chlorpyrifos Contaminated Soil-Potential and Prospects

Neeru Kadian, Santosh Satya, Anushree Malik and Prem Dureja*
Center for Rural Development and Technology, Indian Institute of Technology, New Delhi
*Division of Agricultural Chemicals, Indian Agricultural Research Institute, New Delhi

1. INTRODUCTION

Chemical pesticides are designed to control or eliminate pests such as insects, rodents, weeds, bacteria, and fungus. Although pesticides have played a significant role in increasing food production and controlling various diseases in the agriculture system but indiscriminate use of pesticides has resulted in contamination of soil, water bodies' etc. raising deep concern about food safety and human health. Excessive use of these harmful chemicals affects kidneys, developing fetus, and liver immuno-suppression and increased incidence of breast cancer. Some of these pesticides are also reported as mutagenic. Other health related problems such as impaired memory and concentration, severe depression, headache, etc have also been reported.

Through field survey (Kadian *et al.*, 2010) it has been found that chlorpyrifos is a widely used pesticide in household application as well as in horticulture crop in National capital region Delhi. This tempted us to understand the nature of chlorpyrifos (CPF) and its metabolites and assess the suitability of available techniques for detoxification of contaminated soil.

2. CHLORPYRIFOS AND ITS METABOLITES : SOME FACTS

Chlorpyrifos is typical among the organophosphate pesticide class, because it possesses a high degree of chlorination. This chlorination imparts certain properties not typically noted for this general class of compounds; making chlorpyrifos somewhat more difficult to degrade in the environment, and more lipids soluble. Reports from the Environmental Protection Agency (EPA) (1997) indicate that a wide range of water sources and terrestrial ecosystems may be contaminated with chlorpyrifos. In fact chlorpyrifos being poorly water soluble, readily gets adsorbed on to the soil and sediment, and thus it is more persistent than most other pesticides of this general group.

Due to its high toxicity, chlorpyrifos may affect the central nerves system, the cardiovascular system, and the respiratory system as well as cause skin and eye irritation (Oliver *et al.*, 2000, Serrano *et al.*, 1997). More recently, a study in Australia had shown that chlorpyrifos was present in the meconium (first bowel discharge) of new-born babies. Nearly 60% of babies in a study had chlorpyrifos in their bodies at the time of birth (Deuble *et al.*, 1999). The US EPA review of chlorpyrifos acknowledged that the insecticide and its breakdown products had also been found in the urine of 89% of children tested in one US study (Spitzer, 2000). In fact, the chemical manufacturer's (Dow Agro Sciences) own data showed that the breakdown products of chlorpyrifos i.e., TCP (3, 5, 6-trichloro-2-pyridinol) had been detected in all 416 children (aged from 0-6 years) tested in USA in 1998 (FQPA, 2000). Not only chlorpyrifos, but its breakdown product (metabolites) are also reported harmful and have great threat to living being as discussed below:

Oxon (0, 0-diethyl-0-(3, 5, 6-tri-chloro-2-pyridinyl) phosphate): It is oxygen analogue of a phosphorothioate parent chemical. These compounds belong to a special sub-category of parent chemicals and their transformation products. Phosphorothioates are organo-phosphorus compounds containing a P=S bond. This bond hydrolyzes in aqueous solution (or is metabolized in the liver) to form oxon, an analogous compound where the P=S bond has been replaced by a P=O bond. Metabolism of the phosphorothioates results in the formation of the oxygen analogs, which are more toxic than the parent phosphorothioates themselves. The cholinesterase-inhibiting effects of the parent compound are largely caused by the oxygen analog (the P=O bonded compound) because it is more physiologically active than the parent.

TCP (3, 5, 6-trichloro-2-pyridinol): The metabolite TCP, the major degradation product has been reported frequently in water (Liu *et al.*, 2001), vegetables (Zayed *et al.*, 2003), fruits (Velasco-Arjona *et al.*, 1997) and soil (Robertson *et al.*, 1998; Singh *et al.*, 2003). It has also been reported in bioremediation experiments using pure microbial cultures (Mallick *et al.*, 1999).

From the above discussion it is very clear that chlorpyrifos and its metabolites enter into different parts of ecosystem and can target each one of us easily. This type of pesticide contamination can enter into soil as a result from bulk handling of pesticide in the agronomic practices, and rinsing of containers and accidental release may occasionally lead to the contamination of surface and groundwater. Conventional methods of pesticide disposal such as incineration, burial, encapsulation target the bulk products or pesticide contaminated containers etc. Out of all the methods for toxic waste reduction, bioremediation focuses on detoxification rather than waste translocation, and is thus recognized as a potential technique (Dzantor, 1999). In this type of in-situ remediation method, most appealing observations are that they generally do less surface damage, require a minimal amount of facilities and are less expensive. It involves the use of indigenous micro flora or addition of specific microorganisms to enhance the biodegradation of pollutants.

3. BIOREMEDIATION-IN SEARCH OF A SUSTAINABLE SOLUTION

Bioremediation involves establishing suitable condition in contaminated environment so that appropriate microorganisms flourish and carry out the metabolic activities to detoxify the contaminants (Latha, 2001). Bioremediation techniques are typically more economical than conventional methods such as incineration, and some pollutants can be treated on the site itself, thus reducing exposure risks to clean-up personnel, or potentially wider exposure as a result of transportation accidents. There are basically three types of bioremediation technique; bioaugmentation, biostimulation and phytoremediation.

Researchers (Singh *et al.*, 1984; Brigante and Barbieri, 1996) reported the importance of microbial flora (biological degradation) in soil, as unsterilized soil degrades contaminants faster as compared to sterilized soil. Because many of these pollutants serve as food to some specific microbes; hence these microbes can eliminate or neutralize many toxic compounds in the environment (Grigg *et al.*, 1997). In a study, degradation of chlorpyrifos was studied in sterilized and non-sterilized soil. Half-life of chlorpyrifos in sterilized soil was 3 or 4 times longer than that in non-sterilized soil (Wu and Zhu, 2003). It showed that presence of microorganism was a critical factor in degradation. Other than the micro flora group, the initial concentration of chlorpyrifos is also a deciding factor for the rate of degradation. The half-lives were reported with a huge difference i.e., 79.2, 91.8 and 278 days, respectively, when the soil was treated with three different initial concentrations of chlorpyrifos i.e. 10, 100 and 1000 ppm chlorpyrifos.

Two of the most common enzymes involved in chlorpyrifos breakdown, organophosphate hydrolase and organophosphate acid anhydrolase, have been reported to be substrate and stereo-selective. Even by organophosphorus hydrolase (OPH), an enzyme that can degrade a broad range of OP pesticides, chlorpyrifos hydrolyzed almost 1,000-fold slower than the preferred substrate, paraoxon (Cho *et al.*, 2004) and reflect the fact that chlorpyrifos is a very inefficient substrate for OPH.

3.1 Bioaugmentation of Chlorpyrifos

Bioaugmentation is the addition of nutrients and microorganisms to a contaminated environment to increase the rate of biodegradation (Dzantor, 1999). Normally the microorganisms applied are pure culture or a consortia of specially selected microorganisms that have been chosen for their ability to ingest the contaminate. Several attempts to isolate a chlorpyrifos-degrading microbial system by repeated treatments or enrichment of soils and other media with chlorpyrifos have not been successful till 1999 (Mallick *et al.*, 1999 and Racke *et al.*, 1990). This indicates that chlorpyrifos is not a preferred or utilizable substrate for microorganisms. But chlorpyrifos has been reported to be degraded co-metabolically by bacteria, which needs extra carbon sources (Richins *et al.*, 1997; Mallick *et al.*, 1999). A large number of microbial species were reported to degrade chlorpyrifos co-metabolically in liquid medium. However, these microorganisms do not utilize chlorpyrifos as a source of carbon. Chlorpyrifos has been reported to be degraded co-metabolically in liquid media by *Flavobacterium* sp. (Sethunathan and Yoshida, 1973), *Pseudomonas diminuta* (Serdar *et al.*, 1982), *Arthrobacter* sp. (Mallick *et al.*, 1999), and also by an *Escherichia coli* clone with an opd gene which were initially isolated to degrade other organophosphate compounds. Xu *et al.* (2007) also reported a co-culture of *Serratia* and *Trichosporon* spp. could mineralize chlorpyrifos.

Singh *et al.* (2003) and Yang *et al.* (2005) isolated six chlorpyrifos-degrading bacteria and *Alcaligenes faecalis* DSP3, respectively, which were capable of degrading chlorpyrifos in liquid media as well as in soil. Singh *et al.*,(2004) isolated the first CP-degrading bacterium, *Enterobacter* B-14, which hydrolyzed CPF to diethylthiophosphate (DETP) and TCP, and utilized DETP for growth and energy. Enterobacter Strain B-14 utilized chlorpyrifos as the sole source of carbon and phosphorus. However, this strain can degrade only chlorpyrifos and not TCP. After this two other CP degrading bacteria, *Stenotrophomonas sp.*, and *Sphingomonas sp.*, were also isolated which could utilize chlorpyrifos as the sole source of carbon and phosphorus, but they did not degrade TCP (Yang *et al.*, 2006; Li *et al.*, 2007). Ghanem *et al.*, 2007 also isolated a chlorpyrifos - degrading bacterial strain from an activated sludge sample collected from the Damascus Wastewater Treatment Plant, Syria. Within 4 days the isolated *Klebsiella sp.* was found to break down 92% of chlorpyrifos when co-incubated in a poor mineral medium in which chlorpyrifos was the sole carbon source.

Already reported degraders, barely showed any diversity among the isolated microorganisms in those studies. Then recently Lakshmi *et al.*, 2008 reported a diverse enriched class of microorganisms, with capability of chlorpyrifos degradation in soil such as *Pseudomonas* fluorescence, *Brucella melitensis, Bacillus subtilis, Bacillus cereus, Klebsiella species, Serratia marcescens* and *Pseudomonas aeroginosa*, showed 75-87% degradation of chlorpyrifos as compared to 18% in control after 20 days of incubation. These strains utilize chlorpyrifos as a source of energy. Out of all these, P. aeroginosa was the most efficient, showing degradation of 62% chlorpyrifos after 10 days and 40% uptake of TCP after 6 h. It indicates TCP utilization as energy source by P. aeroginosa. Previously Feng *et al.*, (1997) showed utilization of TCP as energy source by Pseudomonas sp. Such a diverse group of degraders has not been reported previously. Other than these only a few introduced organisms like strain YC-1 (Yang *et al.*, 2006) are known to survive well in soil.

A bacterium, isolated from activated sludge and named strain TRP from genus Paracoccus, could biodegrade chlorpyrifos and 3, 5, 6-trichloro-2-pyridinol (Xu *et al.*, 2008). Strain TRP could also degrade pyridine, methyl parathion and carbofuran when provided as sole carbon and energy sources. This is the first reported bacterium that could completely mineralize chlorpyrifos and TCP.

Mineralization of TCP by a Pseudomonas sp. has been reported previously (Feng *et al.*, 1997), and also a strain of *Alcaligenes* species was found capable of biodegrading both chlorpyrifos and TCP (Yang *et al.*, 2005).

As already discussed above, *Enterobacter* sp. B-14 (which uses chlorpyriphos as the source of carbon and phosphorus) stopped degrading chlorpyriphos in the presence of other carbon sources (Singh *et al.*, 2004). However, strain TRP shows a more rapid degradation in the presence of an additional carbon source. This indicates that chlorpyriphos could also be degraded co-metabolically by strain TRP, which might signify the environmental adaptation of this bacterium.

From the above discussion it is clear that, out of all the introduced isolates some were stable and active in liquid medium and others were in soil. But out of all, very few isolated species were capable of degrading chlorpyriphos as well as its main metabolites. Also, the performance in soil is slow and even doubtful in the presence of other carbon source. Hence there is rare chance of their optimal performance in harsh open environment of contaminated sites.

3.2 Biostimulation of Chlorpyrifos

Biostimulation involves the addition of nutrients that are deficient but required for biodegradation of contaminate and is comparable to fertilizing a field (Vidali, 2001). The addition of nutrient source causes an increase in microbial population, thereby, increasing the number of indigenous microorganisms capable of degrading contaminate. In this regard, bioprocessed materials (BPMs) such as biogas slurry, mushroom spent compost, vermin compost and heap manure etc. are cheap and easily available nutrient sources, which can enhance the microbial population. In addition to this, bioprocessed material amendments improve the physical and chemical properties of soil. It is also effective in controlling disease caused by soil borne pathogens, such as *Pythium, Phytophthora, Fusarium spp.* or *Rhizoctonia solani* both in fields and in pot mixture in greenhouse. Also it helps in plant growth.

Soil has the natural capacity to clear the contaminants, but at a slow pace. With the biostimulating agents amendment we simply want to accelerate this specific process of clean up. In a specific Australian sugarcane field, microbial degradation is considered to have contributed to rapid loss of chlorpyrifos from a controlled-release formulation (where spray could not control of grey back cane grub). Addition of 10% non-fumigated soil to fumigated soil resulted in low concentrations of chlorpyrifos and TCP, with levels similar to those in non-fumigated soil (Robertson *et al.*, 1998).

In another study, reactors filled with differ mixtures of biomass (vine-branch, citrus peel, urban waste and public green compost) enhanced the degradation of chlorpyriphos in the reactors. The half-life was reported less than 14 days in reactors, compared to literature values of 60-70 days in soil (Vischetti *et al.*, 2004). Their study showed the rapid participation of active microbiological component in the presence of different biomass mixture in the degradation of chlorpyriphos. Recently, Romyen *et al.* (2007) studied the potential of different agricultural by-products for the chlorpyriphos degradation. The results indicate that the chlorpyriphos was degraded more rapidly in coconut husk than peanut shell, rice husk and peat moss. Many reports had shown that the range of half-life of chlorpyrifos in soil is wide than that in the biomass. Generally, microorganisms use carbon as a source of energy and nitrogen for building cell structure. Biomass is normally reported to have higher amount of nutrients as compared to soil. It is another important factor that promotes microbial growth in the contaminated environment in the presence of BPMs. As a result, chlorpyrifos degraded more rapidly as compared to the soil.

Sewage sludge addition resulted in an increase in faunal activity as indicated by the abundance of excrements and total porosity of the soil, but a reduction in the size of the soil microbial biomass. In the sludge amendment, there was no significant effect of biocide (chlorpyrifos) on soil porosity, microbial biomass, or microbial respiration (Adesodun *et al.*, 2005). The effect of swine, cow, and poultry derived lagoon effluents on the transformation of chlorpyrifos to TCP and subsequent transport was studied using soil/effluent incubation and leaching studies. Chlorpyrifos degradation was significantly enhanced during incubation with lagoon effluents; however, in the presence of soil, lagoon effluents don't exert any effect on CPF degradation (Huang and Linda, 1998).

Above discussion clearly reveal that these organic materials seems to be more cost effective and easy to handle as compared to bioaugmentation method, where a lots of sophisticated techniques are needed for the isolation and maintenance of pure microbial cultures.

3.3 Phytoremediation

Phytoremediation is based on the use of plants for detoxifying environmental sites contaminated with organic and inorganic pollutants. This technology exploits the ability of plants to extract and/or mineralize xenobiotics in the surrounding environment, as well as the tolerance of these plants to the contaminants.

It is generally recognized that plants can remediate organic pollutants by

i) Direct root uptake of contaminants and subsequent accumulation of nonphytotoxic metabolites in plant tissue

ii) Direct - foliar uptake of volatile contaminants from the surrounding air by foliage

iii) Release of exudates and enzymes that enhance biochemical transformations and/or mineralization due to mycorrhizal fungi and microbial activity in the rhizosphere (Anderson *et al.*, 1993; Anderson and Coats, 1994; Schnoor *et al.*, 1995).

As a well known fact, plants are great enhancer of biological activity in the surrounding soil. Previous studies have shown that biological activity leads to a greater rate of the decomposition of some contaminants. This increase in biological activity is caused by root exudates, such as sugars and organic acids upon which bacteria and fungi can feed. Plant root also penetrate further than grass roots (up to 1 meter) and improve soil aeration because of their high water use.

The uptake and metabolism of organophosphorous (OP) pesticides was investigated using aquatic plants such as parrot feather (*Myriophyllum aquaticum*), duckweed (*Spirodela oligorrhiza*), and elodea (*Elodea canadensis*) (Gao *et al.*, 2000). The decay kinetics of these pesticides from the aqueous medium followed first-order kinetics. However, rate and extent of decay depended on both the physiochemical properties of the pesticide compounds and the nature of the plant species (Gao *et al.*, 2000). This study indicated that OP pesticides could be taken up by aquatic plants and subsequently transformed via the plant-mediated transformation attributed to enzymatic reactions. Rapid uptake of organochlorine and organophosphate pesticides into plants containing high fatty acid contents is a very common phenomenon. Residues of aldrin and heptachlor in seeds were reported to be directly related to the oil content of the seeds (Nash *et al.*, 1970).

A number of studies also elucidated the importance of interacting microbial communities in the degradation of compounds. Synergism probably can be found not only among members of microbial communities, but also between higher plants and microorganisms. Data from studies of plant-microbe interactions in the rhizosphere implicate its microbial community as an important exogenous line of defense for plants against potentially harmful organic compounds in soil (Walton *et al.*, 1994). This is the defense mechanism that can be used for phytoremediation of harmful pesticides. In this context, Sardar and Kole (2005) conducted a lab scale study on the availability of plant nutrients (N, P and K) at pesticide application rate of 1, 10 and 100 ppm. They reported that the presence of chlorpyriphos -metabolites (TCP and TMP) rather than chlorpyriphos is responsible for inhibitory affect on nutrient availability.

The use of plants to remove toxic compounds from soil (phytoremediation) is emerging as a potential strategy for cost effective and environmentally sound remediation of contaminated soils especially those with low level defused persistent pesticide. Like other organic pollutants, chlorpyrifos may be absorbed by plants, and thus enter the food chain. On the other hand, plants growing in soils that have been contaminated by pesticides may be able to biodegrade the pesticide residues by means of their rhizosphere influence. Recently Wang et al., 2007 studied the effect of chlorpyrifos residues in the soil and its uptake by crops. Addition of chlorpyrifos (1-10 μg /g) in a single irrigation with distilled water resulted in absorption of chlorpyrifos by wheat (0.257-4.50 μg/g) and also oilseed rape seedlings (0.249-2.02 μg/g) during 20 d of plant growth. However, little information is available to confirm these assumptions. Other plants also transfer chlorpyriphos from soil to the plant parts as studied by Putnam *et al.* (2003) in cranberry bog. Chlorpyrifos was detected in fruit at the time of harvest (62 days, post-chlorpyrifos application), but no metabolites were found. Chlorpyrifos-oxon and 3, 5, 6-trichloro-2-pyridinol, however, were detected in earlier fruit samples and in foliage and soil samples.

Very few studies have been reported on the uptake of chlorpyrifos by plants, and no study was observed on the phytoremediation aspect of chlorpyrifos in soil system. But this technique look very promising as beside decontamination of soil, this can also enhance the aesthetic value of the contaminated site near pesticide production unit.

4. VISION AND FUTURE PROSPECTS

A close look at these techniques (i.e., phytoremediation, biostimulation and bioaugmentation) indicates that biostimulation technique is a low cost, requiring less effort and time and can be accomplished with the help of easily available materials as shown earlier by Kadian *et al.* (2008). It is clear that biostimulation of native micro flora through addition of organic materials holds great promise for degradation of chlorpyrifos avoiding the huge expenditure on extra set ups as in case of bioaugmentation. Moreover, being compatible with normal agricultural practices it is feasible at the field level and takes care of long term ecological implications of using pure microbial culture.

Thus, keeping in view that agriculture is a complex, non linear, open system and a way of life not just agribusiness, strong need is felt to design and develop holistic sustainable bioremediation practices based on judicious integration of phytoremediation, bioaugmentation and biostimulation. The abundance of microorganisms in the rhizosphere vicinity has been reported widely. In fact it is a normal synergistic relationship between plants and microorganisms which can be used in the phytoremediation of these harmful pesticides from the contaminated soil. This strategy would prove very useful in case of low level defused persistent pesticide in agricultural soil. Biostimulating agents (such as bioprocessed materials) can be integrated with the above discussed phytoremediation process, bringing several benefits, as biostimulating agents provide a extra support for the survival of phytoremediators, and in terms of nutrient support to microbial life in the contaminated environment. Further research and development (R & D) work on integration of cropping pattern, phytoremediation and bioprocessed material amendments under field conditions is warranted to recommend a technology package for detoxification of the contaminated sites.

Acknowledgement

One of the authors, Neeru Kadian, gratefully acknowledges the financial support from DST as Women Scientist Project.

References

Adesodun, J.K., Davidson, D.A. and Hopkins, D.W. 2005. Micromorphological evidence for changes in soil faunal activity following application of sewage sludge and biocide. *App. Soil Ecol.*, **29**:39-45.

Anderson, T., and Coats, J. (ed.) 1994. Bioremediation through rhizo- sphere technology. *Am. Chem. Soc. Symp. Ser.* 563. ACS, Washington, DC.

Anderson, T., Guthrie, E., and Walton, B. 1993. Bioremediation in the rhizosphere. *Environ. Sci. Technol.*, **27**:2630-2636.

Briagante, J., and Barbieri, S. M. 1996. Biodegradation and biodeterioration in latin america-biodegradation studies of atrazine in a tropical soil. *Source: www.bioline.org.br/abstract Lb96004.*

Cho, C. M., Mulchandani, A., and Chen, W. 2004. Altering the substrate specificity of organophosphorus hydrolase for enhanced hydrolysis of chlorpyrifos. *App. Environ.Microbio.*, 4681-4685.

Deuble, L., Whitehall, J. F., Bolisetty, S., Patole, S. K., Ostrea, E. M., and Whitehall, J. S. 1999. Environmental pollutants in meconium in Townsville, Australia. Department of Neonatology, Kirwan Hospital for Women, Townsville. Deparment of Pediatrics, Wayne State University, Michigan. (Unpublished).

Dzantor, E.K. and Allan, S.F. 1999. Combination of landfarming and biostimulation as a waste remediation practice. Source: *www.heidlberg.edu/depts/chm/atrazine.html.*

Environmental Protection Agency. 1997. Review of chlorpyrifos poisoning data. Environmental Protection Agency (EPA), Washington, D.C.

Feng, Y., Racke, K. D. and Bollag, J.M. 1997. Use of immobilized bacteria to treat industrial wastewater containing a chlorinated pyridinol. *App. Microbiol Biotechnol.,* **47:** 73-77.

FQPA report. 2000, MEMORANDUM, SUBJECT: CHLORPYRIFOS - Re-evaluation Report of the FQPA Safety Factor, Brenda Tarplee, Executive Secretary, FQPA Safety Factor Committee Health Effects Division (7509C). HED DOC. NO. 014077.

Gao, J., Garrison, A. W., Hoehamer, C., Mazur, C. S., and Wolfe. N. L., 2000. Uptake and phytotransformation of o, p'-DDT and p, p'-DDT by axenically cultivated aquatic plants. *J.Agric. Fd Chem.,* **48** (12): 6121-6127.

Ghanem I, Orfi M, Shamma M. 2007. Biodegradation of chlorpyrifos by Klebsiella sp. isolated from an activated sludge sample of waste water treatment plant in *Damascus. Folia Microbiol (Praha).,* **52** (4): 423-427.

Grigg, B. C., Assaf, N. A., and Turco, R. F. 1997. Removal of atrazine contamination in soil and liquid systems using bioaugmentation. *Pestic. Sci.,* **50:** 211-220.

Huang, X. L., and Linda, S. 1998. Impact of animal derived lagoon effluents on the fate of chlorpyrifos and its metabolite in soils. *Book of Abstracts, 215th ACS National Meeting, Dallas,* Publisher: American Chemical Society, Washington, D. C.

Kadian, N., Malik, A. and Satya, S. 2010. *Bioremediation of chlorpyrifos contaminated soil using biostimulation and phytoremediation techniques.* PhD Thesis, IIT Delhi.

Kadian, N., Malik, A., Satya, S. Dureja, P. 2008. Potential of organic bioprocessed materials for bioremediation (biostimulation) of chlorpyrifos contaminated soil for food safety. pp-82. Abstract In: Proc. Third Int. Meeting Environ. Biotech. and Engi. (3IMEBE), Eds., Conde *et al.,* Grafiques Terrasa, Palma de Mallorca, Spain.

Lakshmi, C. V., Kumar, M., and Khanna. S. 2008. Biotransformation of chlorpyrifos and bioremediation of contaminated soil. *Intl. Biodet. Biodegrad,* **62**(2): 204-209.

Mallick, B. K., Banerji, A., Shakil, N. A., and Sethunathan, N. N. 1999. Bacterial degradation of chlorpyrifos in pure culture and in soil. *Bull. Environ. Contam. Toxicol.,* **62:** 48-55.

Latha, M. R., Indirani, R, and Kamaraj, S. 2001. Bioremediation of polluted soils. *J. Environ. Res.,* **11**(1): 27 - 29.

Li, X., He, J. and Li., S. 2007. Isolation of a chlorpyrifos-degrading bacterium, Sphingomonas sp. strain Dsp-2, and cloning of the mpd gene. *Res. in Microbio.* **158:** 143-149.

Liu, B., McConnell, L.L., and Torrents, A. 2001. Hydrolysis of chlorpyrifos in natural waters of the Chesapeake Bay. *Chemosphere,* **44:** 1315-1323.

Nash, R. G., Beall, Jr., and E. A. Woolson. 1970. Plant uptake of chlorinated insecticides from soils. *Agron. J.,* **62:** 369-372.

Oliver, G. R., Bolles, H. G., and Shurdut, B. A. 2000. Chlorpyrifos: probabilistic assessment of exposure and risk. *Neurotoxicol,* **21:** 203-208.

Putnam, R. A., Nelson, J. O., and Clark, J. M. 2003. The persistence and degradation of chlorothalonil and chlorpyrifos in a cranberry bog. *J.Agric. Fd Chem.,* **51**(1): 170-176.

Racke, K.D., Laskowski, D.A. and Schultz, M. R. 1990. Resistance of chlorpyrifos to enhanced biodegradation in soil. *J. Agric. Fd Chem.,* **38:** 1430-1436.

Richins, R. D., Kaneva, I., Mulchandani, A. and Chen, W. 1997. Biodegradation of organophosphorus pesticides by surfacexpressed organophosphorus hydrolase. *Nature Biotechnol.,* **15:** 984-987.

Robertson, L. N., Chandler, K. J., Stickley, B. D. A., Cocco, R. F. and Ahmetagic, M. 1998. Enhanced microbial degradation implicated in rapid loss of chlorpyrifos from the controlled release formulation suSucon (R) Blue in soil. *Crop Prot.* **17:** 29-33.

Romyen, S., Luepromchai, E., Hawker, D. and Karnchanasest. 2007. Potential of agricultural by-product in reducing chlorpyrifos leaching through soil. *J. App. Sci.* **7** (18): 2686-2690.

Sardar, D. and Kole, R. K. 2005. Metabolism of chlorpyrifos in relation to its effect on the availability of some plant nutrients in soil. *Chemosphere.* **61** (9): 1273-1280.

Schnoor, J., Licht, L., McCutcheon, S., Wolfe, N. and Carreira, L. 1995. Phytoremediation of organic and nutrient contaminants. *Environ. Sci. Technol.* **29:** 318A-323A.

Serdar, C. M., Gibson, D. T., Munnecke, D. M., and Lancaster, H. M. 1982. Plasmid involvement in parathion hydrolysis by *Pseudomonas diminuta. App. and Environ. Microbiol,* **44:** 246-249.

Serrano, R., Lopez, F. J., Hernandez, F. and Pena, J. B. 1997. Bioconcentration of chlorpyrifos, chlorfenvinphos and methidathion in Mytilus galloprovincialis. *Bull. Environ. Contam. Toxicol.* **59:** 968-975.

Sethunathan, N. N., and Yoshida, T. 1973. A Flavobacterium that degrades diazinon and parathion. *Canadian J. Microbiol.* **19:** 873-875.

Singh, B.K., Walker, A., Alun, J., Morgan, W., and Wright, D. J. 2003. Effects of soil pH on the biodegradation of chlorpyrifos and isolation of a chlorpyrifos-degrading Bacterium. *App. Environ. Microbiol.* **69**(9): 5198-5206.

Singh, B., Phogat, R. K., Bhan, M. and Bhan, V. M. 1984. Beitr. *Trop. Landwirtsch. Veterinarmed.* **22:** 391.

Singh, B.K., Walker, A., Morgan, J.A. and Wright, D.J. 2004. Biodegradation of chlorpyrifos by Enterobacter strain B-14 and its use in bioremediation of contaminated soils. *App Environ. Microbiol,* **70**: 4855-4863.

Spitzer, E. 2000. In United States Environmental Protection Agencies Preliminary Risk Assessment for Chlorpyrifos Reregistration Eligibility Decision, Docket Control Number 0PP - 34203.

Velasco-Arjona, A., Manclus J. J., Montoya, A. and de Luque Castro, M. D. 1997. Robotic sample pretreatment-immunoassay determination of chlorpyrifos metabolite (TCP) in soil and fruit. *Talanta,* **45**: 371-377.

Vidali, M., 2001. Bioremediation. An Overview. *Pure Appl. Chem.,* **73** (7): 1163-1172.

Vischetti, C., Capri, E., Trevisan, M., and Casucci, C. 2004. Biomassbed: A biological system to reduce pesticide point contamination at farm level. *Chemosphere,* **55**: 823-828.

Walton, B. T., Hoylman, A. M., Perez, M. M., Anderson, T. A., Johnson, T. R., Guthrie, E.A. and Christman, R. F. 1994. Rhizosphere microbial community as a plant defense against toxic substances in soils. In: Bioremediation through rhizosphere Technology (Anderson T A and Coats J R, eds.). ACS Symposium Series No. 563. American Chemical Society, Washington, D C, 82-92.

Wang, L., Jiang, X., Yan, D., Wu, J., Bian, Y. and Wang. F. 2007. Behavior and fate of chlorpyrifos introduced into soil-crop systems by irrigation. *Chemosphere,* **66**(3): 391-396.

Wu, H. and Zhu, G. 2003. Study on the degradation of chlorpyrifos in sterilized and nonsterilized soil. *Nongyaoxue Xuebao,* **5**(4): 65-69.

Xu, G., Li, Y., Zheng, W., Peng, X., Li, W. and Yan, Y. 2007 Mineralization of chlorpyrifos by co-culture of *Serratia* and *Trichosporon spp. Biotechnol Lett.,* **29:** 1469-1473.

Xu, G., Zheng, W., Li, Y., Wang, S., Zhang, J. and Yan, Y. 2008. Biodegradation of chlorpyrifos and 3, 5, 6-trichloro-2-pyridinol by a newly isolated *Paracoccus sp.* strain TRP. *Intl Biodet. and Biodegrad.* **62** (1): 51-56.

Yang, C., Liu, N., Guo, X.M. and Qiao, C. 2006. Cloning ofmpd gene from a chlorpyrifos-degrading bacterium and use of this strain in bioremediation of contaminated soil. *FEMS Microbiology Letters.* **265:** 118-125.

Yang, L., Zhao, Y.-H., Zhang, B.-X., Yang, C.-H. and Zhang, X., 2005. Isolation and characterization of a chlorpyrifos and 3, 5, 6-trichloro-2-pyridinol degrading bacterium. *FEMS Microbiology Letters,* **251:** 67-73.

Zayed, SMAD., Farghaly, M., and El-Maghraby, S. 2003. Fate of 14C-chlorpyrifos in stored soybeans and its toxicological potential to mice. *Fd Chem. Toxicol,* **41:** 767-772.

Part - III

Chapter-10

Fungal Diseases and Fungicides in Horticultural Crops

Rajender Singh
Department of Plant Pathlogy, CCS Haryana Agricultural Univeristy, Hisar

Plant pathogens are known to cause yield reduction of almost 20 per cent in the principal food and cash crops worldwide (Oorke *et al.*, 1994). These losses may be more severe when highly susceptible cultivars are widely grown in a particular region of the globe. The increased world's population is posing a major global challenge to provide sufficient quatity of food with high quality fruits and vegetables. Nearly, 570 million US $ is now spent annually on research and development by the leading 15 agricultural Companies to combat with the important diseases of major food, fruits and vegetable crops. There are more than 150 active ingredients registered as fungicides worldwide (Mc Dougal, 1996). In the present era of intensive agriculture, fungicides play an important role in our fight against plant diseases and reduce the losses in yield and quality of farm produce. The global fungicide sale has been estimated to be $ 6.0 billion US. The cereal and vine industries of the Western European have become largest fungicide market followed by North America. However, the fungicide use is not intensive in Asia and the new world.

This article presents the historical development of fungicides, mode of their action, development of resistance to fungicides, fungicidal residue and chemical control of various diseases of horticultural and plantation crops.

Historical Development of Fungicides

Social impact of a plant disease was at its greatest, when the potato region in Europe witnessed threats of famine due to late blight of potato epidemics recurring over the years in 1845. In Ceylon (Sri Lanka), coffee rust washed out the coffee in 1878 and its cultivation was withdrawn. In Europe, downy mildew of grape posed a serious threat to Wine Industry. Bordeaux mixture (copper sulphate + lime) was formulated by Prof. Millardet in France in 1882 to control the downy mildew of grape (Millardet; 1885). This unique and versatile chemical preparation was later found effective to control the late blight of potato and several other plant diseases. Thus, it became the first generation fungicide, along with other inorgarnic chemicals with fungicidal properties. Forsyth (1802) advocated the use of lime sulphur for the control of powdery mildew of fruit trees. Since then it has been commonly used in the form of element sulphur to control this disease occurring on several other crops. It is commonly used as a dust or wettable sulphur powder form because of higher efficacy and environmental safety. Lime sulphur is also used when eradication of some pathogens is required.

The second generation fungicides include organic chemicals dates from 1934 with development of dithio-carbamates, organotins, chloronitrobenzene, phthalimides, chlorothalonil, quinones, captan and related compounds. These compounds are surface protectants like inorganic fungicides as they don't penetrate plant tissue. Thus, these are only effective if applied in advance of infection. Most of these fungicides are multicylic contact compounds of low biochemical specificity. With the development of fixed copper compounds like copper oxychloride, have the occasional phytotoxic effects due to excessive copper ions. Oil based paints of copper sulphate and its aqueous paste with lime are effectively used as tree wound dressings in fruit orchards. These fungicides are able to provide effective control of foliar and fruit diseases of horticultural and plantation crops. The use of captafol is now restricted to seed treatment only because of toxicological consideration (Thind and Chahal, 2000). Dithianon, a quinone fungicide is effective against foliar diseases of grapevine, pome, and stone fruits. Dodine, a guanidine fungicide is a protectant fungicide with some eradicant action and is recommended for control ot *Venturia* spp, *Taphrina deformans*, *Monilinia lauxca*, *Mycospheaerella* and *Cladosporium* spp. occurring on various fruit, vegetable and ornamental crops. Fluaginan, a phenylpyrimidine fungicide developed in mid 1990s is a protective fungicide used for the control of *Botrytis*, *Colletotrichum*,

Phytophthora and *Venturia* causing severe damage on several fruits and vegetables. Fenitrophan, a nitroparaffin compound developed in 1981, is a broad spectrum protectant and has been found very promising as foliar fungicide against scab and powdery mildew of apple, chinomethionat, a quinoxaline compound has specific activity against powdery mildew of fruits and vegetables. The third generation fungicides include several organic compounds that penetrate the plant tissues and control established infection(s). These fungicides are in the mainly systemic in nature, *viz.* 2-aminopyrimidines, benzimidazoles or benzimidazole generating compounds, carboxamides, phenylamides, fosetyl-Al, azoles and related compounds, and morpholines (De Ward *et al.*, 1993). These fungicides are very effective against most of pathogens of Ascomycetes, Deuteromycetes and some of Basidiomycetes. These are used as foliar sprays and post harvest application in fruits and vegetables. Thiabendazole is used to control various types of post harvest infections by fungi like *Penicillium* spp. and *Oospora* spp. in fruits and vegetables. Oxycarboxin and Bupirimates are used to control coffee and rust powdery mildew of Apple, respectively. Phenylamides are highly effective against fungal pathogens belonging to Oomycetes of fruits, vegetables and ornament crops (Thind and Chahal, 2002). Carbamates also check downy mildew fungi of potato and fruits. Alkylphosphonates and Fosety-Al or fosetyl have demonstrated their high activity against *Plasmopara viticola* in grapevine and *Phytophthora humuli* in citrus and pineapple, Triamorphos and organophosphorus fungicides developed in 1970 are active against powdery mildew of apples, pear and grapes. Hydroxypyrimidine is highly specific fungicide against powdery mildew of cucurbits and apple. Phenylamides are also highly effective against oomycetes including *Phytophthora, Pythium* and downy mildews. Carbamates [prothiocarb and propamocarb) are generally used to check downy mildew of fruits and potatoes. Cyanoacetamideoximes have exhibited promising effective against downy mildew of grapevine and late blight of potato. Triforine, the only piperazine based fungicide is mainly used as foliar sprays against powdery mildews and scabs. *Colletotrichum* spp., *Monilinia* spp. and several other pathogens of fruits and vegetables are also controlled by this fungicide (Schultz and Scheinpflug, 1988). Among pyridines, buthiobate is the only fungicide which is generally used for the control of powdery mildew of cucurbits and top fruits. Pyrifenox was introduced as foliar fungicide for *Cercospora, Septaria* and *Monilina* spp. in fruits and vegetables (Zobrist *et al.*, 1986).

Seven commercial fungicides belong to morpholines (tridemorph, dodemorph, aldimorph fenpropimorph, trimorphamide), piperidines (fenpropidine) and spiroketals (Spiroxamine) inhibit sterol biosynthesis of erysiphales (Bohrnen *et al.*, 1986; Balwdin and Corran, 1995).

Triazoles among the sterol biosynthesis inhibitors (SBIs) constitute one of the largest groups of highly effective fungicides. Nearly, more than 25 of these compounds are commercially available. Among the commonly used Triazoles developed by various agrochemical companies are triadimefon, triadimenol, bitertanol, propiconazole, flusilazol, penconazole, tebuconazole difenconazole, mylobutanil, cyproconazole, hexaconazole flutriafol expoxyconazole, triticonazole and metconazole. These fungicides are applied for the control of apple scab, grape downy mildew, powdery mildew of cucurbits, sigatoka of banana and coffee rust (Thind and Clerjeau, 1992).

Imazalil is used as post harvest dip treatment against *Penicillium* spp. in citrus and bananas. Triflumizole is used as a foliar fungicide for the control of several species of *Venturia* and *Monilinia* occurring on various fruits and vegetables. Triforine, the only piperazine based fungicide is mainly used as foliar sprays against powdery mildew and scab of important fruits and vegetables.

The fourth generation of fungicides consists of compounds that are non-fungitoxic in vitro but trigger defense mechanisms in plants. Non-fungicide disease control compounds are valuable since side effects on non-target organisms will be less severe than those produced by conventional fungicides.These fungicides are antipenetrants. Pyriquilon, chlobenthiazone, phthalide are used against Ascomycetes and Deuteromycetes.Phenylphrroles (Pyrrolnitdrin) is used against *Botrytis cinerea* (Gullino, 1998}. Anilinopyrimidines also known as pyrimidinamines are broad spectrum and have potential to control the diseases of a large variety of crops. Mepanipyrim and pyrimethanil are active against *Botrytis cinerea* on grapevine and *Venturia inequalis* on apples (Neumann *et al.*, 1992; Daniels *et al.*, 1994). Oxaolidinedione group is represented by famoxadone (DlP x -JE 874) is a broad spectrum fungicide with specific activity against *Plasmopara viticola, P. infestans* and *A. solani*. Phenoxyquinoline was introduced in 1996 as Quinoxyfen (DE 795) for the control of Uncinula necator of grapevine (Longhurst *et al.*, 1996). Spiroketalamines (Spiroxarnine) is a novel ergosterol biosynthesis inhibitor released in 1996 against powdery mildew of grapes (Dutzrnann *et al.*, 1996).

Validamycin is a secondary metabolite of *Streptomyces hygroscopicus* is used against black scurf of potato and sheath blight of rice. Polyoxin

D is used in Japan against black spot of Japanese pear (Gooday, 1995). Strobilurins analogue compounds have promising activity against major Asscomycetes, Basidiomycetes, Deuteromycete and Oomycete (*Venturia inequalis, V. necator, P. viticola, P.grisea, Alternaria* spp., *P. infestans* and *P. Ieucotrichas*).

Chronological Development of Fungicides for use in Plant Disease Control

Year	Fungicide
1802	Lime sulphur
1821	Robertson of England stated sulphur IS effective against peach mildew
1833	Kendrick (1833) of USA proposed lime sulphur preparation against grape mildew
1885	Bordeaux mixture
1887	Burgandy mixture
1906	Butler recommended Bordeaux mixture brushing for control of Koleroga of arecanut
1913	Organomercurials
1927	Kamat found Bordeaux paste effective against gummosis of citrus
1930-34	Fixed coppers, Dithiocarbamates, PCNB
1935-40	Chloronitrobenzene, O- phenyl phenol
1941-50	Biphenyl, Hexachlorobenzene, Dichlone, Glyodin, Cyclohexamide, Dinocap
1951-60	Captan, Organotins, Dodine, Anlazine, Dicolaran
1961-70	Chloroneb, Chlorothalonil, Polyoxin, Organophosphates, Hydroxypyrimidines, Oxathiins, Benzimidazoles, Morpholines, Thiophanates
1971-80	Dicarboximides, Sterol biosynthesis inhibitors, Phenylamides, Alkyl phosphonates, Carbamates, Isoxazoles, Tricyclazoles, Cyanoacetamide-oximes, Melanin biosynthesis inhibitors
1981-90	N-phenyl carbamates, Pencycuron
1991-2000	Strobilurins,Phenylpyrroles,Triazoles,Anilinopyrimidines, piroxamines, Probenazole, Benzothiadiazole, Phenyl phyridylamines, Qunolines

Mode of Action of Fungicides

Sulphur, copper, mercury and tin based compounds disrupt cell function in a non-specific manner and are also known as general cell toxicants. They disrupt several biochemical processes through their ability to bind with chemical groups such as thiol moities, common to many enzymes and form chelates with heavy metals with fungal cells. Captan, Captafol and Folpet preferably act on the enzymes with sulphydryl groups but may also attack amino groups. Chinomethionat and chlorothalonil apart from their binding to -SH groups also react with mercapto groups (Tillman *et al.*, 1973). The mode of action of dithianon is attributed to its strong affinity to -SH and thio dependent enzymes.

Crotonate compounds or nitrophenol derivatives such as dinocap, Fluazinam and tin compounds act as uncouplers of mitochondrial oxidative phosphorylation. Dodine disrupt permeability and nutrient uptake. Dicarboxamide inhibit spore germination and cause hyphal branching, swelling and lysis. Fenitropan, a nitroparaffin compound containing SH groups resulting in disruption of enzyme function. Aromatic hydrocarbon include hexachlorobenzene, quintozene, chloroneb, dichloran and phenyl phenol induce hyphal tip swelling and germ tube lysis. Carboxamides inhibit succinate dehydrogenase activity in fungal respiration (Buchenauer, 1990). Benzimidazoles and Thiophanates inhibit fungal mitosis by interfering with spindle formation at the level of tubulin biosynthesis (Daniels *et al.*, 1994). Organophosphorus especially pyrazophos inhibit conidial germination and appressoria formation. Bupirimate belonging to Hydroxypyrimidines inhibits adenosine deaminase, an enzyme which plays an important role in the purine salvage pathway during nucleic acid metabolism, thereby inhibiting germ tube elongation and appresorium formation. Acylalamines phenyl amides groups inhibit synthesis of ribosomal RNA via the RNA polymerase X - template complex (Davidse and Vander Berg-Valthius, 1989). It results in the disruption of protein synthesis. Carbamates influence amino acid metabolism, while Cyanoacetamideoximes inhibit nucleic acid and protein biosynthesis. Triazole compounds inhibit sterol and ergosterol biosynthesis. Phenylcarbamates disrupt mitosis through binding the tubulin protein (Fujirrrura *et al.*, 1990). Strobilurins and their analogues prevent electron transfer between cytochrome band C in complex HK 6f the mitochondrial electron transfer chain (Leroux, 1996). In general, tricyclazole blocks polyketide pathway that results into inhibition of melanin biosynthesis. Tricyclazole also prevents melanization of appressorial walls essential for development of infection hyphae and penetration of host epidermis. Phenoxyquinolines inhibit dihydro-orotatedehydrogenase in pyrimidine biosynthesis pathway (Gustafson, 1995). Famoxadone inhibits the function of ubiquinol cytochrome C oxidoreductase at complex HI. Antibiotic inhibit transfer of aminoacryl RNA to the ribosomes.

Resistance of Fungi to Fungicides

Dekker (1985) proposed that pathogenic fungi may become resistant to fungicides by the following mechanism:

1. A change at the site of inhibitor action which results in a decreased affinity to the fungicide.
2. Decreased uptake or accumulation of the fungicide in the fungus.
3. Detoxification of the fungicide before reaching the site of action or lack of conversion of a compound into the fungi toxic principle(s).
4. Compensation for the inhibitory effect e.g. by an increased production of an inhibited enzyme.
5. Circumvention of the blocked site by operation of an alternative pathway or mutation in fungi.

Fungicidal Resistance in Some Horticultural Crops

Fungicide	Year of observation	Pathogen/crop
Dodine	1969	*Venturia inaequalis* – apple (Gilpartick, 1982)
Benzimidazoles		Many target pathogens, viz., *Botrytis cinerea, Erysiphe* sp., *Venturia* spp., *Penicillium* spp. on various crops (Smith, 1988)
2- Aminopyrimidines	1971	*Sphaerotheca fuliginea* - cucumber (Brent, 1982)
Phenylamides	1980	*Phytophthra infestans*- potato; *Plasmopara viticola* - grape (Staub, 1984)
Dicarboximides	1982	*Botrytis cinerea* - grape (Lorenz, 1988), and strawberries (Maraite *et al.*, 1980)
Demethylation	1982	*Sphaerotheca fuliginea* - cucurbits; *Uncinula necator* inhibitors - grape (Steva *et al.*, 1988); *Venturia inaequalis* - apple (Thind *et al.*, 1986); *Penicillium digitatum* - oranges citrus (Eckert, 1982)
Aromatic hydrocarbon	1960	*Penicillium* - citrus and oranges (McDonald *et al.*, 1979)
Benomyl, MBL		*Botrytis cinerea* - grapes (Bhrenhardt *et al.*, 1973)
Carbendazin		*Gleocosporuim ampelophagum* grapes (Kumar and Thind, 1992)

Source: Brent (1995)

Fungicidal Resistance Risk

Based on resistance risk, the available fungicides may be grouped into following classes:

Resistance risk	Fungicide
High	Benzimidazoles, dicarboximides, phenylamide
Moderate	2-Aminopyrimidines,anilinopyrimidines,dimethomorph phenylpyrroles, aromatichydrocarbon: Triazoles, carboxamilides, cymoxanil-fentins, pyrimidine carbinols, strobilurins
Low	Dithiocarbamates, chlorothalonil, coppers phthalimid, fluazinan, probenazole, sulphur, quinoxyfen tricylazole, SAR compounds

Management of Fungicide Resistance

Important host resistance management strategies include the use of fungicides only when necessary at recommended dose, avoidance of multiple applications of fungicide(s) with same mode of action like triazoles, use of various combinations of products with different mode of action either in mixture or as alterations in a spray programme (Hew, 1998). Alternating fungicides with different modes of action is used to combat resistance development. Early detection of resistance in a field population through regular monitoring programme is important for taking timely measures to encounter resistance development (Thind and Mohan, 1996). Long term strategies to manage resistance development include exploitation of new target sites, development of compounds with negative cross resistance to known fungicides, synergism between fungicides, systemic acquired resistance, development of rapid diagnostic technology based on either immunology or DNA-probe resistance modelling and integrated disease control. The multisites, protectant fungicides are reliable and valuable future tools for controlling plant disease resistance. Thus, their protection is imperative to ensure their continued availability to the growers and hence more new fungicides of this generation need to be developed (Thind and Chahal, 2002). Systemic Acquired resistance molecules may be a valuable tool for anti-resistance strategies, it has been used against *Mycosphaerella* spp. in banana.

Fungicide Residue

Public concern over reports of fungicidal residue in food items has resulted in the imposition of stringent Maximum Residue Limits by regulatory authorities. Surveillance of pesticides in food is routinely carried out by many countries. Food and dairy administration of US demonstrated that pesticide residue level is generally well below those set by the Environmental Protection Agency (Anon, 1996). No residue were found in 64 per cent of domestic and 66 per cent of import surveillance samples, and less than 1 per cent of samples had residue level that were over tolerance. Fungicides that give low or non-detectable residue in the crop are actively sought in research programmes.Earlier, fungicides were selected for development based largely on their fungicidal performance and safety to the consumers and operators. However, there is now a greater awareness of the need to define the environmental profile of new fungicides early in the discovery process, before development (Knight *et al.*, 1997).The most important properties investigated include the persistence potential in soil and safety to wildlife, beneficial organisms and aquatic organisms (Lynch, 1995). Balancing fungicidal potency and improved performance with low environmental impact remains a challenge for fungicide research.

New fungicides demonstrate advantages over current products and benefits offered in safety or environmental profile are becoming increasingly important to the regulator, consumers and the operators. Food Quality Protection Act of 1996 (FQPA) requires EPA to fully consider all potential exposure to pesticides taking into account exposure in the diet, from drinking water, in homes, school and any other environmental source. The FQPA also mandates consideration of the effects of multiple chemical exposure, particularly of chemicals with same mode of fungicidal action, even though the mechanism of toxicity of a chemical agent may be distinct from its mode of fungicidal activity. According to new Dutch criteria, pesticide usage will not be registered. If

a) Application results in concentration in ground water 2 m below soil surface above 0.1 $\mu g\ l^{-1}$

b) Applications leads to concentrations in surface water 0.1 of the LC_{50} for fish, algae and crawfish

c) Compounds have 50 per cent degradation time longer than 60 days.

In general, acute toxicity to mammals from the available fungicides is low. The fungicides that rank in the three highest acute toxicity class (LD_{50} rat 1-50, 50-100 and 100-1000 mg kg^{-1}) are mostly conventional and belong to first and second generations, e.g. copper and tin derivatives (Anonymous, 1991). Most of the third generation fungicides have (LD_{50} values higher than 1,000 or even 10,000 mg kg^{-1}. Consequently, the Acceptable daily Intake (ADI) of most fungicides is relatively high, and is in excess of the level likely to be found in human diets (Rigsdale *et al.*, 1991). Thus, the threat to human health from consumption of food contaminated with fungicide residues is lower than for other classes of pesticides (Ames, 1989; Berry, 1990). Registration of new active ingredients requires, a huge investment of resources, of which toxicological and environmental studies represent a significant component. The cost of developing abroad spectrum, global fungicide is estimated at approximately $ 130 million. The time from discovery to development is now between 8-12 years.

Future research concentrates on development of compounds with lower persistence and mobility in the environment. In US, National Research Council establish tolerance limit of pesticide (maximum level of residues allowed for pesticide and its significant metabolites or breakdown products on food crops). Fungicides should not disturb or modify physiology of fruits and vegetables. They do not have any adverse effect on parasite/predators. Industries, government and educational institutions need to joint forces to foster a better understanding of risk management and assessment. Fungicide residues of both benomyl and thiabendazole have been detected in fresh pears following treatment of fruit with 500-1200 mg /l of these fungicides (Ben-Arie and Guelfat Reich, 1973). Pre-harvest sprays of benornyl, dichloran and other fungicides provided surface residues sufficient to protect the fruits from infection after harvest. Fosetyl-Al and related compounds sprayed on the foliage of apple trees and injected into trunks of citrus trees several weeks before harvest resulted in checking the surface infection of these fruits by zoospores of *Phytophthora* spp. (Barak *et al.*, 1984; Buchanan *et al.*, 1985). Effectiveness of carbendazim, benonyl, and mancozeb has also been recorded against the storage fungi of onion with 2.1-3.4 mg/kg bulb's residual toxicity (Anonymous, 1974). Estimated residue of carbendazim was 2.1 -3.4 mg/kg bulbs. This residual level of carbendazim in the treated bulb was found within safety limit (Raju and Nark, 2006) as the acceptable daily intake of carbendazim for human being is 0.125 mg/kg body weight. Residual levels in fruits after treatment with 100 mg l^{-1} fludioxonil at 20°C or with 25 mg l^{-1} at 48°C was recorded approximately 0.6-2 mg kg^{-1}, which were notably lower than the maximum residue limit (5 mg kg^{-1}) allowed in USA for stone fruits (Salvatore *et al.*, 2007). The residue of cyproconazole, hexaconazole, kresoxim-methyl, myclobutanil, penconazole, tetraconazole and triadimenol on grapes fruits after their application (a few tens of grams per hectare) could not be detected at harvest. Pyrimethanil residue was found to be constant upto harvest, whereas fluazinan, cyprodinil, mepantpyrim, azoxystrobin and fludioxonil showed different disappearance rates. After fermentation of grapevine to wine, pesticides residues were always less than those on grapes and even non detectable residues were found (Cabras and Angioni, 2000). Dicarboximide fungicide residue was found below the maximum residue limit of 5 mg/kg in all samples of Table Grapes by 21 days post application (Chaido *et. al.*, 2000). Carbendazim and Thiabendazole residues were extracted from 0.4 ppm in fresh mushrooms to 10 ppm in whole lemon homogenated of agriculture commodities (Lawloski and Porle, 1998).

Chemical Control of Horticultural Crops Through Fungicides

Name of crop	Name of disease	Name of chemical
Apple	Apple scab	Captan 600 g, Dodine 200 g, Ziram 600 gm, Carbendazim 100 g/Hexaconazole 60 ml,Mancozeb 60 g, Copper oxychloride 600 g spray in 200 litre of water (Gupta and Sharma, 2001)
	Powdery mildew	Sulphur (WP) fungicide, Tridemorph (0.1-1.0) spray (Gupta and Sharma, 2001)
	Canker	Dithane Z-78 (0.25%) and Captan (0.2%) (Gupta and Sharma, 2001)
	White root rot	Drenching of Carbendazim (0.1%) (Gupta and Sharma, 2001)
Pear	Leaf spots	Carbendazim (0.1%) spray followed by Dithane Z-78 spray (Shah *et al.*, 1985)
	Scab	Dodine 200 g (Gupta and Sharma, 2001) spray in 200 1 of water
	Short/Fruit blight	Carbendazim (0.1%) spray (Singh *et al.*, 1989)
	Powdery mildew	Karathane (0.1%) spray (Gupta and Sharma, 2001)
Peach	Leaf curl	Karathane followed by Bavistin (0.1 %) (Rana and Jain, 1992), Spray of Bavistin (0.05%) at demand or Bud swell stage (Thakur *et al.*, 1991)
Stone fruits (*Prunus* spp.)	Rust	Delan (0.05%) at bud swell as foliage (Thakur *et al.*, 1991)
	Shot hole	0.2% Captan spray (Gupta *et al.*, 1979)
	Powdery mildew	Delan (0.05%) spray (Thakur *et al.*, 1991)

Contd.

Mango	Powdery mildew	Three sprays of Microsul (0.2%) or Bavistin (Parkash and Raoof, 1985)
	Anthracnose	Bavistin (0.1%) spray at 15 day interval (Parkash and Mishra, 1988)
	Mango mal formation	Bavistin 200 ppm (Siddiqui *et al.*, 1987)
Citrus	Citrus canker	Bordeaux (2:2:50) mixture and streptomycin (150 mg/ ml spray)
	Powdery mildew	Sulphur, Carbendazim (0.1%) spray (Cheema *et al.*, 2001)
	Gummosis	Fosetyl-Al, Ridomil MZ (0.2%) and Alliete (0.2%) (Khan and Ravise, 1985;Gaur *et. al.* ,2011)
Grapes	Downy mildew	Copper oxychloride with Zineb (protectant), Metalaxyl, Fosetyl Al (eradicant)
	Powdery mildew	Triazole group (Chandershekhar and Thind, 2001)
	Anthracnose	Bavistin (0.1%) (Chandrashekhar and Thind, 2001)
Banana	Sigatoka	Carbendazim (0.1 %) + Tridemorph (0.1)
	Anthracnose	Prochloraz (0.15%) and Chlorothalonil (0.2%) (Rawal and Ullasa, 1989)
	Panama disease	Dipping of sucker in Bavistin (0.1%) (Rawal and Ullasa, 1989)
Guava	Wilt	Water soluble Quinolinol sulphate (0.1 %) injection (Mishra, 2001)
	Canker	Spray Bordeaux mixture or lime sulphur spray (Mishra, 2001)
	Anthracnose	Pre and post spray Thiabendazole (0.1 %) and Aureofungin 500 ppm (Mishra, 2001)
Papaya	Stem/collar/ fruit rot	Seed dressing with Captan or Bordeaux mixture (Rawal *et al.*, 1983)
	Anthracnose	Bavistin (0.1%) spray (Rawal *et al.*, 1983)
	Powdery mildew	Bavistin (0.1%) spray (Rawal *et al.*, 1981)
Litchi	Leaf spots	Zineb (0.1%) spray (Mishra and Pandey, 2001)
	Red rust	Ziram (0.25%) spray (Mishra and Pandey, 2001)
Nut fruits	Shot hole	Copper oxychloride (0.3%) (Putto and Razdan, 1988)
	Rust	Captan, Maneb spray (Bhardwaj, 1991)
	Powdery mildew	Delan (0.05%) spray (Thakur *et al.*, 1991)
Potato	Late blight	Ridomil-MZ and Dimethomorph (0.1%) (Thind and Mishra, 1997)

Contd.

	Early blight	Carbamate fungicide spray (Thind and Jhooty, 1982)
	Black scurf	Tuber treatment with Carbendazim (Khanna *et al.*, 1991)
Tomato	Early blight	Mancozeb or Copper oxychloride spray
	Late blight	Ridomil-Mz and Indofil M-45 (Shyam and Gupta, 1996)
	Buckeye lot	Ridomil MZ and Indofil M -45 (Shyam and Gupta, 1996)
	Fusarium wilt,	Benomyl (0.1%) soaking of seedling or
	Pythium	Captan and Thiram seed treatment (Tripathi and Grover, 1978)
Cucurbits	Downy mildew	Ridomil- MZ (Thind *et al.*, 1991)
	Powdery mildew	Benomyl and Triadimefon @ 0.1 % (Thind *et al.*, 1991)
	Anthracnose	Benomyl and Indofil M 45 Diathane M 45 alternate spray
Onion and garlic	Stemphyllium blight	Mancozeb @ 0.25% alongwith sticker Triton spray
	Purple blotch	Mancozeb (Srivastava *et al.*, 1995)
Pea	Powdery mildew	Bayleton followed by 3 spray of Karathane (Kapoor and Thakur, 1997)
	Wilt	Bavistin seed dressing
Cruciferous	Damping off	Thiram/ Carbendazim @ 2 g/kg seed
vegetables	Sclerotinia-rot	Benomyl seed treatment Bavistin (0.05%) spray (Kapoor, 2001)
	White rust	Metalaxyl and Mancozeb (Srivastava and Verma, 1989)
	Black leaf spot	Rovral and Ridomil MZ-72-50
Chilli and brinjal	Anthracnose	Dithane M -45 and Blitox spray (Thind and Jhooty, 1990)
	Root-rot, wilt	Seed treatment with Bavistin or Thiram
	Phytophthora blight	Seed dressing with Captan and Ridomil-MZ (Kaur *et al.*, 2001)
Carrot, beet and colocasia	Alternaria blight	Dithane M-45 (0.25%) or Blitox-50 (0.25%) foliar application
	Stemphyllium root rot	Seed treatment with Captan or Carbendazim
	Powdery mildew	Hexaconazole application 10 g 9%/ha
	Soft rot	Preharvest application of Carbendazim (Gupta, 2001)
Ginger	Rhizome rot	Soil dressing with Zineb or Mancozeb followed by rhizome treatment with Carbendazim (Srivastava, 1994)
	Yellow disease	Rhizome dipping in Carbendazim (0.1%)
	Phyllosticta leaf spot	Benomyl +Mancozeb (0.2%) spray (Dohroo *et al.*, 2001)

Post Harvest Diseases of Fruits

Name of crop	Name of disease	Name of chemical
Pear	Storage decay	Preharvest spray Benomyl/ Carbendazim (Sharma, 1990)
Apple	Storage rots	Preharvest spray Carbendazim (Kaul and Sharma, 1995)
Apple	Storage scab	Bitertanol dipping (Sharma and Kaul, 1995)
Peach, plum, cherry and apricot	Rhizopus rot and decay	Benomyl and Iprodione treatment (Sharma *et al.*, 1990)
Citrus, mango, pomegranate	Alternaria rot	Benomyl and Carbendazim (Majumadar and Singh, 1999)

Diseases of Plantation and Management

Name of crop	Name of disease	Name of chemical
Tea	Blister blight	Copper oxychloride and Triazole fungicides (Chandra Mouli and Agnihothri, 1999)
	Black rot	Spray of Dithianon at 280-375 g/ha (Mustafee, 1996)
	Red rust	Copper oxychloride (1:400) spray (Dutta and Barthakur, 1998)
Coffee	Leaf rust	Spray of 0.5% Bordeaux mixture(BM) or Copper fungicides (Mustafee, 2000)
	Black rot	Spray of 0.5% Bordeaux (BM) or Copper fungicides (Mustafee, 2000)
Rubber	Stem rot/ Bark conker/Abnormal leaf fall	Prophylactic spray of 1 % Bordeaux mixture
	Powdery mildew	Sulphur dusting at 10-12.5 kg/ha (Mustafee, 2000)
Coconut/ Arecanut	Bud rot/Fruit rot/ Nut fall	Two sprays of 1% Bordeaux mixture (Dasgupta, 1997)
	Stem bleeding	Carbendazim spray
Cashewnut	Damping off	Drenching with 0.2% Captan
	Dieback	Captan or Carbendazim 0.1 % (Mustafee, 2000)
Arecanut	Koleraga	0.5% Fosetyl
	Anaberoga	Drenching Captan (0.3%) or Carbendazim (0.3%) (Kumar and Nambiar, 1990)

Contd.

Turmeric	Rhizome rot /Leaf spot	Spray of Mancozeb (0.25%) (Sharma *et al.*, 1994)
Cumin	Wilt and Leaf blight	Spray of Mancozeb (0.2%) (Mehta and Solanki, 1990)
	Powdery mildew	Spraying Karathane or Sulphur dusting
Coriander	stem gall	Seed and soil treatment with Thiram (Mustafee, 2000)
	Powdery mildew	Sulphur application
Fenugreek	Collar rot	Captan soil drench (Mustafee, 2000)
	Powdery mildew	Sulphur application
Cardamom	Clump-rot	Bordeaux mixture (1%) or Copper oxychloride (0.2%) (Mustafee, 2000)
Black pepper	Foot rot/Quick wilt	Spray of Metalaxyl (0.4%) followed by Captafol (Mustafee, 2000)
	Basal wilt	MEMC (Mustafee, 2000)
	Anthracnose	Bavistin @ 0.5 kg/ha (Mustafee, 2000)

Chapter-11

Pesticide Residues in Fruits of India

J.K. Dubey and S.K. Patyal
Directorate of Extension Education
Department of Entomology and Apiculture
Dr Y.S. Parmar University of Horticulture and Forestry, Nauni, Solan (HP) 173 230

India is the world's second largest producer of fruits, of which mango, banana, citrus, guava, grape, pineapple and apple are the major ones. Apart from these, fruits like papaya, sapota, aonla, phalsa, jackfruit, ber, pomegranate in tropical and sub-tropical parts and peach, pear, almond, walnut, apricot and strawberry in the temperate parts are also grown in a sizeable area. Date palm and fig cultivation is also finding favour in some areas. The major fruit growing states are Maharashtra, Tamil Nadu, Karnataka, Andhra Pradesh, Bihar, Uttar Pradesh and Gujarat.

Production of fruits in India is 44.04 million tones from an area of 3.72 million hectares. One of the bottlenecks in fruit production is the losses incurred by insect pest and diseases which necessitated the use of pesticides to curb the pest problem. Use of pesticides in India is very low 0.350 - 0.5 kg/hectare as compared to 7.0 kg/hectare in USA, 2.5 kg/hectare in Europe, 12 kg/hectare in Japan and 6.6 kg/hectare in Korea. Despite of low pesticide consumption in India, there is a great problem of pesticide residues vis-a-vis other countries and these residues have entered into food products and underground water because of non-prescribed use of chemical pesticides, wrong advice and supply of pesticides to farmers by vested interests, non- observance of prescribed waiting period, pre-marketing pesticide treatments during storage and transport, use of sub-standard pesticides, effluents from pesticide manufacturing units, continued use of persistent pesticides for public health programme, lack of awareness and lack of aggressive educational programmes for farmers/consumers.

Pesticide residues lead to adverse effects on human health and environment. Food safety means security of food free from pesticide residues. Intake of fruits, containing pesticide residues cause chronic toxicity (toxicity due to repeated exposure to pesticide residues) in human beings. Acute poisoning (toxicity due to single dose) in human beings occurs rarely due to deliberate attempt or accidental incidence but never due to intake of fruits containing pesticide residues. Acute poisoning is not wide spread like chronic toxicity because pesticide contaminated fruits can be distributed throughout inter and intra country. Therefore, public is more concern about the chronic toxicity as it may result into mutagenic, teratogenic and carcinogenic effects which are not curable and resulted deformities are transmitted to the next progeny. Hazard is the probability that pesticide can cause toxicity in a proposed quantity and manner (Hazard = toxicity x exposure).

Although organic fruit production, is also emerging as a new concept of farming prohibiting the use of synthetic pesticides in conventional or widely adopted technology. However, Baker *et al;* (2002) observed low levels of pesticide residues in few organic fruit samples quoting explanation on account of mislabeling samples, violation of organic methods, and unavoidable environmental contamination from pesticides due to wind, contaminated water supplies or persistent pesticide residues in soil previously used to grow conventional crops. An example of the last is organochlorine insecticides (such as DDT) that were banned for agricultural use in the year 1993 but still persist in soil and contaminate both organic and conventional crop produce.

In conventional farming technology, numerous pesticides are used to spray on the tree and even after harvesting fruits are dipped by farmers illegally to combat post harvest losses due to pests. On the other side, consumers unknowingly eat the treated fruit if producers sell their treated produce immediately to the market. Persistence of pesticide is studied through supervised field trials. On the basis of persistence data and Maximum Residue Limit (MRL) a safe waiting period between the last application and harvest are recommended. This minimizes risk factor i.e. residues in the final produce are either totally absent or below the maximum residue limit (MRL). The MRL values are fixed in a manner that the total amount of residues reaching the consumer should be toxicologically acceptable or below the acceptable daily intake (ADI) and at the same time should be high to control the pests adequately.

1.0 SUPERVISED FIELD STUDIES

Supervised field studies are conducted by treating crop with pesticide in the field and edible samples are collected at different day's intervals for residue analysis to know the persistence behavior. On the basis of persistence data and MRL value, safe pre harvest interval or waiting periods (WP) are calculated. Waiting period is defined as time required by the initial deposits to dissipate below the MRL. Persistence of a particular pesticide in the field also depends upon RL_{50}. RL_{50} is the time required by the initial deposits to degrade to its half. Greater the value of RL_{50}, more will be its persistence. Safe waiting periods are suggested for the safety of consumers because pesticides are poisons and targeted to kill pests and diseases.

1.1 Apple

Apple fruit crop is attacked by number of insect pests and diseases like apple scab *Venturia inaequalis,* San Jose scale *Quadrespidiotus pernicious,* wooly apple aphid *Eriosoma lanigerum,* fruit scrapper *Archips pomivora,* defoliating beetles *Adoretlus* spp., *Holotrichia* spp. and tent caterpillar *Malacosoma indica.* Numbers of pesticides like chlorpyriphos, endosulfan, carbendazim, propineb, mancozeb etc. are applied to control these pests. After spraying/treatment, pesticide residues gets deposited on the fruits and dissipate slowly depending upon number of factors like physiochemical characteristics of pesticide, weather conditions, time after treatment etc. Pre harvest interval or waiting period between spray and harvest is required for safe consumption of fruits. Waiting period and RL_{50} values has been given in Table 1 for some of the important crops.

Table1: Safe waiting periods and residue half life of insecticides and fungicides

Crop	Pesticide/ Formulation	Dose (a.i./ha/conc.)	MRL (mg/kg)	Waiting period (days)	RL_{50} (days)	Reference
Apple	Carbendazim (Bavistin50 WP)	1500 g a. i./ha	5.0	1.7	2.0	Sharma *et al.*,1997
	Propineb (Antracol 70 WP)	2500 g a.i./ha	3.0	9.6	6.9	Sharma *et al.*, 2003
	Dichlorvos (Nuvan 76 EC)	500 ga.i./ha	0.1	8.3	2.2	Patyal *et al.*, 2001
	Lindane (Kanodane 20 EC)	500 ga. i./ha	3.0	0	5.3	Patyal *et al.*, 2001
	Chlorpyriphos (50 WP)	0.02%	0.5	0.0	2.3	Nath *et al.*, 1997
	Chlorpyriphos (20 EC)	0.02%	0.5	3.2	4.8	Nath *et al.*, 1997
	Endosulfan (35 WP)	0.05%	2.0	0	2.0	Nath *et al.*, 1997
	Endosulfan (50 WP)	0.05%	2.0	0	3.5	Nath *et al.*, 1997
	Endosulfan (Thiodan 35 EC)	500 g a. i./ha	2.0	2.5	2.84	Patyal *et al.*, 2004
Acid Lime /Citrus	Permethrin (Ambush 50 EC)	0.015%	0.5	7.5	2.6	Awasthi, 1989
	Cypermethrin (Cymbush 50 EC)	0.01%	2.0	Nil	2.0	Awasthi, 1989
	Fenvalerate (Fenval 20 EC)	0.01%	2.0	1.1	2.64	Awasthi, 1989
	Deltamethrin (Decis 2.8 EC)	0.0015%	0.05	Nil	2.9	Awasthi, 1989
	Monocrotophos (Nuvacron 36WSC)	0.05%	0.2	14.0	4.5	Awasthi, 1989
Grapes	Carbendazim (Bavistin 50 WP)	250 g a.i./ha 500 g a.i./ha	5.0	1.2 4.3	5.0 4.7	Mohapatra *et al.*, 1998
	Benomyl (Benlate 50 WP)	500 g a.i./ha	5.0	0	4.7	Banerjee *et al.*, 2001
		1000 g a.i./ha		3	5.1	Reddy *et al.*, 2007
	Penconazole (10EC)	250 g a. i./ha	0.5	2.5	4.9	Reddy *et al.*, 2007
	Hexaconazole (5EC)	500 g a.i./ha	0.1	5.1	2.7	Reddy *et al.*, 2007
	Myclobutanil (10 WP)	500 g.a.i./ha	0.1	2.5	2.6	Reddy *et al.*, 2007

Contd.

Mango	Permethrin (Ambush 50 EC)	0.02%	*NA	7.27	5.67	Awasthi, 1989
	Cypermethrin (Cymbush 50 EC)	0.01%	*NA	3.45	5.58	Awasthi, 1989
	Fenvalerate (Fenval 20 EC)	0.01%	*NA	11.27	5.33	Awasthi, 1989
	Deltamethrin (Decis 2.8 EC)	0.002%	*NA	0	3.81	Awasthi, 1989
	Bifenthrin (Talstar 10 EC)	0.5mL/L	0.5	6.0	11.30	Mohapatra *et al.*, 2007
Pomegr - nate	Endosulfan (35 EC)	500 g.a.i./ha	NA	NA	NA	Dubey and Nath,1994
	Cypermethrin (25 EC)	75 g.a.i.lha	NA	NA	NA	Dubey and Nath,1994
	Quinalphos (25 EC)	375 g.a.i./ha	NA	NA	NA	Dubey and Nath,1994

Not available

1.2 Ber

Insecticides are applied on ber to control ber beetle *(Adoretus* sp.*)*, fruit fly *(Carpomyia vesuviana)*, weevils *(Xanthochelus supercilosus* and *Myllocerus spp)* and mealy bug *(Drosichiella tamarindus)*. Considering the MRL of monocrotophos as 0.2 mg/kg and that of quinalphos as 0.25 mg/kg, safe waiting period of 6 days was recommended whereas, residues of dimethoate, phosphamidon and oxydemeton methyl were below detection limit (Agnihotri, 1999). Residues of lambda- cyhalothrin and beta- cyfluthrin persisted for 10 days on ber *(Zizyphus mauritiana* Lamark) fruits (Gyi *et al.*, 2003).

1.3 Banana

The major insect pests of banana are root stock/corn weevil *Cosmopolites sordidus*, stem weevil *Odoiporus longicollis* and fruit beetle *Nodostoma subcostatum*. The common diseases are wilt/Panama disease *Fusarium oxysporum* var. *cubense*, bacterial wilt *Pseudomonas solanacearum* and bunch top of banana (a viral disease). Application of phorate and carbofuran to roots/sucker at the time of planting and repeated after 120-150 days leaves no detectable residues of either phorate or carbofuran in whole fruit or pulp at harvest (120-150 days of last application) (Agnihotri, 1999).

1.4 Citrus

The important insect pests of acid lime are psylla, *Diaphorina citri,* whitefly *Aleurocanthus* sp. and *Aleurolobus* sp., lemon butterfly *Papilio* sp. and leaf miner *Phyllocnistis citrella.* The diseases are citrus canker *Xanthomonas compestris* var. *citri* and powdery mildew *Oidium tingitanium.* Waiting period of 4, 15 and 7 days have been suggested for monocrotophos, quinalphos and phosphamidon, respectively on acid lime (Agnihotri,1999). Awasthi (1989a) observed half life values for the disappearance of insecticides from fruits, 2.6 and 3.9 days for permethrin, 2.0 and 2.7 days for cypermethrin, 2.6 and 3.8 days for fenvalerate, 2.9 and 2.6 days for deltamethrin and 4.5 and 3.2 days for monocrotophos at recommended and double the recommended dose, respectively with waiting period of 8-13, 0-1, 1-4, 0-2 and 14-15 days, for respective , insecticides put. Mohapatara and Rekha (2005) reported that dicofol residues in the whole acid lime fruit persisted for 20 days and reached below detectable level in 30 days. In juice, dicofol residues were detected after 10 days, reached maximum of 0.22 μg/mL after 15 days and there after decreased. Based on MRL of 5.0 mg/kg, 1 and 2 days waiting periods were suggested as initial deposits were 3.11 and 5.60 mg/kg at recommended dose 0.05% and double the recommended dose 0.1 %. Kumari and Kathpal (2005) applied carbofuran in the soil of sweet orange *(Citrus sinensis* (L)) and detected residues of carbofuran and its metabolites in peel and pulp (0.015-0.066 mg/kg) but were below MRL of 0.1 mg/kg.

1.5 Grapes

Heavy losses of grapes *(Vitis vinifera* L.) are occurred due to the attack of insect pests and diseases like scale insects *Aspidiotus cydoniae,* leaf-roller *Sylepta lunalis,* mealy bugs *Planococcus citri,* powdery mildew *Uncinula necator* downy mildew *Plasmopara viticola* and anthracnose *Elsinoe ampelina* reduce the yield of grapes and required to be treated with pesticide which leave harmful residues. Grapes are mostly consumed as such and pesticide residues directly enter into the body of consumer. Application of carbofuran in the soil at pruning stage is safe as residues in fruits at harvest were below MRL of 0.1 mg/kg. Mancozeb residues were below detection limit at harvest. The repeated spray of fenvalerate and carbendazim should be avoided an grapes and it is desirable to stop spraying one month prior to harvest (Agnihotri, 1999).

Mohapatra *et al.* (1998) observed that carbendazim persisted beyond 15 days on grape berries with a half-life of 5 days and suggested a waiting period of 5 days. Banerjee *et al.* (2001) suggested waiting period of 3 days for benomyl in grapes however half-life of initial deposits were 5 days.

1.6 Guava

During fruiting period, guava is attacked by fruit flies *Bactrocera spp,* and fruit spot *Pestalotia psidii.* Malathion is used to control fruit flies. Malathion applied @ 0.25 and 0.50 mg/kg at fruiting time leaves no residues in fruits at harvest (Agnihotri, 1999).

1.7 Mango

Number of synthetic pesticides are used to control mango crop pests like mango hopper *Idioscopus clypealis,* psylid galls *Apsylla cistellata,* mango mealy bug *Drosicha mangiferae,* fruit fly *Bactrocera* sp., anthranose *Colletotrichum gloeosporides,* powdery mildew *Oidium mangiferae* and malformation *Fusarium moniliforme.* Deltamethrin do not require any waiting period but cypermethrin require 11 days waiting period. Although mango is eaten after removing the peel but residues on peel also get their way into the consumer by contact. The residues of dimethoate, quinalphos, oxydemeton methyl and carbaryl on mango fruits at harvest were below the detectable limits. However, the residues of mancozeb and lindane were detected in fruit samples particularly when applied at double the recommended dose, but these residues were within the permissible limit (Agnihotri, 1999). Repeated spray of bifenthrin @ 0.5 and 1.0 ml/L on mango from flowering to 1 month before harvest leads to 0.82 and 1.45 µg/g residues, respectively which persisted on peel for more than a month and rate of degradation was very low. Half-life of bifenthrin was 11 days. The amount of residues in pulp was very low and became non-detectable after 15 days. No residues were there where bifenthrin given only at flowering, lime size or at both stages (Mohapatra *et al.*, 2007).

1.8 Pear

Ethion is used for the control of pear mite *Eutetranychus orientalis* (Klein) which extracts cell sap from leaves and fruits during dry and hot months and renders the plant weak. Arora *et al* (2004) recorded 1.53 and 3.11 mg/kg initial deposits of ethion on pear at recommended and double the recommended dose which dissipated more than 50% in three days.

Initial deposits of ethion were below MRL of 2.0 mg/kg at recommended dose but if compared to its acceptable intake (ADI) value of 0.002 mg/kg (body weight/day), the situation is of serious concern. Accordingly, the daily consumption of ethion should not exceed 0.02 mg/day for child of 10 kg. It implies that a fruit having residues <0.10 mg/kg will be safe assuming a child consumes 200 g of it. So, Arora *et al* (2004) recommended a waiting period of 10 days for safe consumption of pear fruits following ethion application.

1.9 Pomegranate

Pomegranate is grown in tropical to temperate climate regions. It is attacked by insect pests like anar butterfly *Virachola isocrates,* anar white fly *Siphoninus finitimus* and grey weevil *Myllocerus* spp. Although most of the pesticides used are contact poisons and seed part of fruit is consumed therefore, there are least chances of pesticide residues in seed. However, contamination of hand during fruit opening can contaminate seeds. Therefore, one day waiting period has been suggested from consumer safety point of view for endosulfan, cypermethrin and quinalphos. Residues of systemic insecticides like monocrolophos, dimethoate and phosphamidon, were also below their detection limit (Agnihotri, 1999). All the pesticides schedules evaluated for their persistence have shown residues below the detectable limit and to be considered safe for human consumption.

2.0 MONITORING STUDIES

Pesticides residues not only threaten the human health but also cause imbalance in ecosystem. Monitoring of commodities revealed the current use pattern of pesticides and status of pesticide residues in fruits. Based upon the monitoring studies, prevention and control measures can be initiated by the government before the problem becomes too serious and wide spread to be controlled. Monitoring of food commodities for pesticide residues also check whether Good Agricultural Practices (GAP) has been followed or not. Higher residues are unacceptable to importer and act as non-tariff barrier. Pesticide residues in the marketable fruits should be within their prescribed MRL value for both domestic as well as export market. In India, there are two acts which regulate pesticides:

(1) The Insecticides Act, 1968 and Rules, 1971 under which guide lines are laid for the registration of safe pesticides,

(2) The Prevention of Food Adulteration Act, 1954 under which maximum residue limits for pesticides are laid and are applicable in India.

But other countries have their own MRL values and exporter has to obey their limits. Sanitary and Phytosanitary measures under the world trade organisation (WTO) have been setup as the basic rules for food safety.

The consumption of pesticide in India is low as compared to other countries even then there is widespread contamination of food commodities with pesticide residues due to non-judicious use of pesticides. Earlier survey carried out by Indian Council of Medical Research (ICMR) New Delhi, revealed that 51 % of our food commodities contained pesticide residues and out of these 20 % had pesticides residues above the MRL values, as compared to 21 % contamination with only 2% samples above the MRL on world-wide basis (Fig. 1) (Agnihotri, 1999).

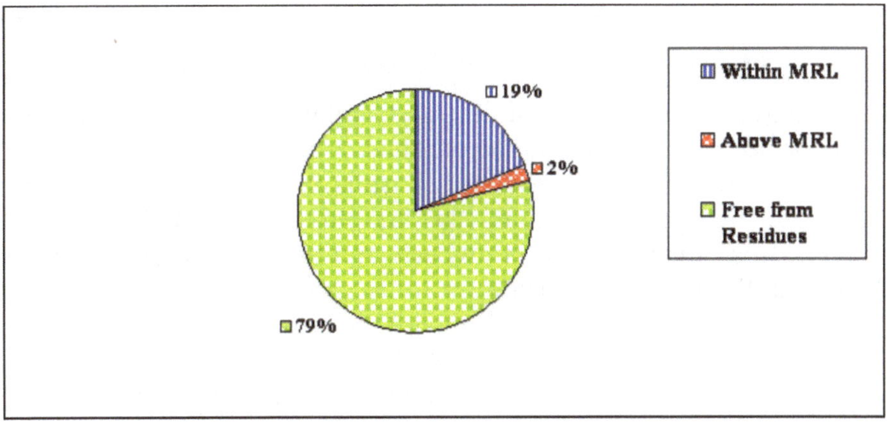

Fig.1: Pesticidal contamination of food commodities in world

Now the Indian scenario has started changing very rapidly as new pesticide molecules are being introduced every year whose application rate is very low and their persistence in the environment is also very less. The hard pesticides have been either banned or put under restricted use. Maximum pesticides in India are used on cotton and rice. Malwa area of Punjab is famous for cotton growing has been named as cancer belt of Punjab because pesticides have contaminated the whole environment only 13-14 % is used on fruits and vegetable even then half of fruits and vegetables were found contaminated with pesticides residues. Pesticide residues in 10% samples were above the MRL value (Fig.2).

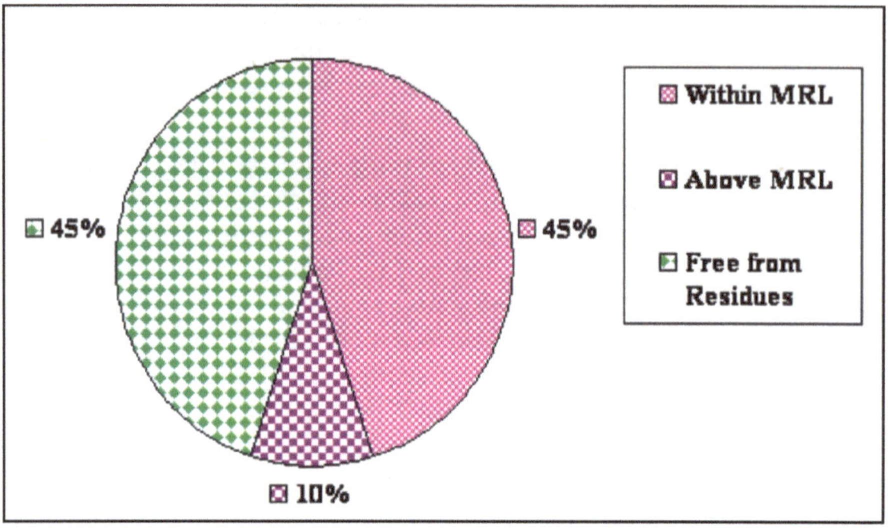

Fig. 2: Pesticidal contamination of fruits and vegetables in India

Residues of methyl parathion, endosulfan, chlorpyriphos, DDVP, dimethoate, fenitrothion, monocrotophos, cypermethrin, deltamethrin, and copper etc, were above the MRL in fruits and vegetables (Agnihotri, 1999). Monitoring of more than 1000 samples of fruits (1988-2003) like apple, grapes, mango, litchi, strawberry and ber from different parts of Himachal Pradesh for pesticides residues revealed 0.3, 0.9, 2.0, 3.3 and 10.1 % fruit sample contamination with the residues of copper, methyl benzimidazole carbamate (MBC), organophosphorus (OP), organocarbamate (OC) and ethylene-bis-dithiocarbamate (EBDC) pesticides, respectively. EBDC residues in nine apple fruit samples were above the MRL of 3.0 mg/kg. Residues of HCH, chlorpyriphos, methyl parathion, endosulfan, captafol, carbendazim, EBDC and copper were in the range of 0.01-0.191, 0.032-0.325, 0.010-0.159, 0.136, 0.060-0.222, 0.089-1.555, 0.05-013.448 and 0.024-0.050 mg/kg, respectively. One sample of strawberry contained EBDC and chlorpyriphos residues which were below MRL values. In Himachal Pradesh, one percent fruit samples contained pesticide residues above the MRL out of 16% contaminated samples (Fig.3) (Dubey and Nath,2005).Uday and Archana (1998) monitored Delhi fruit markets and found 0.11-0.48 mg/kg, 0.17 - 0.23mg/kg EBDC residues and 0.15-0.68 mg/kg, 0.10-0.21 mg/kg ethylene thio urea (ETU) residues in apple and grapes, respectively.

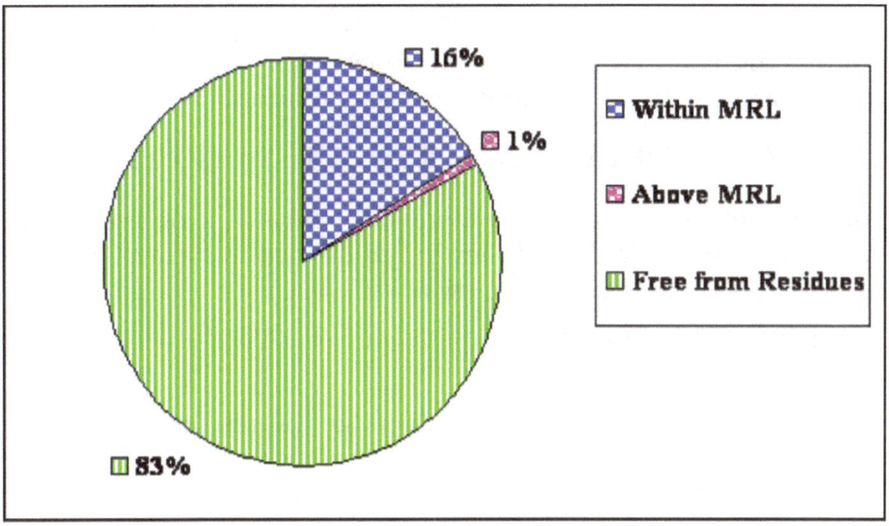

Fig.3: Pesticidal contamination of fruits in Himachal Pradesh (HP)

3.0 RESIDUES IN PROCESSED FRUIT PRODUCTS

Fruits are utilized at all stages of its development both in immature and mature stages. Raw fruits are used for making chutney, pickles and juices. The ripe fruits besides being used for desert are also utilized for preparing several products like squashes, syrups, nectars, jams and jellies. Therefore, pesticides in the fruit make its way into the processed product. During fruit processing, heating convert the pesticide into harmful product. Mancozeb is converted into ethylenethiourea (ETU) which is carcinogenic (FAO/WHO 1981). The Codex Committee on Pesticide Residues (CCPR) FAO/WHO (1989) fixed Temporary-MRLs (TMRL) as 0.02 mg/kg for apple and pear. Propineb is converted upon heating into carcinogen propylenethiourea (PTU). Kumar *et al* (2005) observed that when apple fruit containing 3.94 mg/kg propineb residues were processed into juice and then jam contained no residues whereas juice contained 0.07 mg/kg residues. Drying of treated apple fruit slices cause decline of propineb residues to 0.15 and 0.05 mg/kg residues in unpeeled and peeled apple slices, respectively. Same trend of decline in the residue was observed of endosulfan treated apple fruits (Patyal *et al;* 2004). The reduction in pesticide residue in dried slices could be due to UV radiation, decomposition, microbial degradation and enzymatic degradation. Nath (1996) detected no residues in wine prepared from 20 days old mancozeb treated apple fruits containing 3.00 - 3.10 mg/kg residues. Kumari *et al* (2007) found organochlorine residues in all fruit jam samples, synthetic pyrethroids and organophosphate pesticides in 25-75% jam samples, however, the residues did not exceeded the MRL values. Major contaminants were endosulfan, HCH, fenvalerate, methyl parathion and triazophos.

4.0 DECONTAMINATION OF RESIDUES IN FRUITS

Fruits are consumed as raw therefore, it is desirable to decontaminate the fruit from pesticide residues to reduce health hazard. Various pesticide decontamination processes like:

(a) Washing of treated fruit with water or dilute solutions of non toxic chemicals, and

(b) Peeling off the fruit skin, dislodge the residues to varying degrees depending on constitution of the fruit, chemical nature of the pesticide and environmental factors

Washing

General household fruit washing procedures is carried out with running or standing water at normal temperatures. However, addition of non toxic chemicals like detergents, sodium chloride and sodium hydroxide to the wash water improves the effectiveness of the washing procedure. The washing effect depend on the physico-chemical properties of the pesticides, such as water solubility, hydrolytic rate constant, volatility and octanol-water partition coefficient, actual physical location of the residues on the fruit and temperature. Washing coupled with gentle rubbing by hand under tap water dislodges pesticide residues and enhance the effectiveness of the technique .Systemic and lipophilic pesticide residues are not removed significantly by washing. The effectiveness of washing varies with the age of the residues. Freshly deposited residues are comparatively easy to remove from the fruit surface than field-weathered residues. It also varies with the kind of produce and variety. The reduction in pesticide concentration is greater when initial deposit is higher. Washing with detergents of high degree of detergency and a high pH is efficient in the removal of carbamate residues. Patyal *et al;* 2004) reported 49.00-74.19 % reduction of endosulfan residues by washing the treated apple fruits under running tap water. Reduction in endosulfan residues by washing was higher in 0 day sampled fruits than in 10 day's sampled fruits. Sodium chloride solution is used to decontaminate the pesticide residues from different fruits and vegetables. Washing of treated fruits in the solutions like 2% (w/v) NaOH, 1 % (w/v) sodium lauryl sulfate, 0.05%(v/v) HCl and 0.05% (w/v)sodium bicarbonate gave good reduction , 78.8-91.8,

66.7-79.3, 72.7-92.0 and 60.6-81.9% from propineb residues, respectively (Kumar *et al;* 2005) and 68.49-77.04, 61.20-70.03,65.16-75.97 and 56.09-71.15% reduction of endosulfan residues, respectively (Patyal *et al;* 2004).

Peeling

Non-systemic surface pesticide residues are concentrated in the peel and are easy to remove by peeling. Polar pesticides can be partly dislodged by simple washing but lipophilic pesticides like organochlorines and pyrethroids dissolve in the wax of pericarp and become difficult to remove by simple washing. Peel takes away along with it the major pesticide residue load and left little in the edible portions or pulp. This is especially important for fruits which are not eaten with their peels, such as bananas or citrus fruits. It is also important in fruits which are rarely eaten along with peel like mango and apple. In case of systemic pesticides, peeling may not be as effective may not be effective as in case of non systemic pesticides.

Peeling of fruits after washing under tap water is the best technique amongst decontamination processes (Nath, 1996; Patyal *et al;* 2004). Peeling alone is not so effective in freshly treated fruits because during peeling, loose pesticide on the fruit surface get stick to hand and knife and contaminate the pulp. Peeling fruits after washing provided maximum relief 75-93 % from endosulfan residues (Patyal *et al;* 2004) and 60-98 % relief from mancozeb residues (Nath, 1996) in apple. However, Awasthi (1993) reported complete removal complete of dimethoate, fenthion, fenvalerate and cypermethrin residues by peeling of mango fruit.

Pyrethroid residues did not move from the Acid lime fruit pericarp to the fruit juice, whereas monocrotophos residues were rapidly absorbed and were present in the fruit juice within 3 days of insecticide application. Hence, immobility of pyrethroids on fruit could be useful, as residues can be removed by peeling away the pericarp (Awasthi, 1989).

If GAP is followed to manage the pests complex, the problem of residues in produce/harvested crop can be minimized and the economy of the growers will increase based upon the monitoring studies. Consumers will not only worried about the safety but will be ready to remunerate price in the market.

References

Agnihotri, N.P. 1999. *Pesticides Safety Evaluation and Monitoring.* In: All India Coordinated Research Project on Pesticides Residues. Division of Agricultural Chemicals Indian Agricultural Research Institute, New Delhi, 173 p.

Arora, P.K. Battu, R.S. and Singh, Balwinder 2004. Bioefficacy of ethion against mite *Eutetranychus orientalis* (Klein) and its residues on pear fruits. *Pestic. Res. J.* **I6**(2): 33-35.

Awasthi, M.D. 1989. Persistence of synthetic pyrethroid residues on mango fruits. *Acta Horticul.* **231**: 612-616.

Awasthi, M.D. 1989a. Dissipation of synthetic pyrethroid and monocrotophos residues on acid lime *(Citrus aurantifolia)* fruits. *Indian J. Agric. Sci.* **59**(10): 660-664.

Awasthi, M.D. 1993. Decontamination of insecticides residues on mango by washing and peeling. *J. Food Sci. Technol.* **30**(2): 132-133.

Baker, B.P. Benbrook, C. Groth, IIIE and Benbrook KL 2002. Pesticide residues in conventional, integrated pest management (IPM) - grown and organic foods: Insights from three US data sets. *Fd. Addit. Contam.* **19**: 427-446.

Kumari, Beena and Kathpal, T.S. 2005. Residues of carbofuran and its metabolites in citrus (C. *sinensis* (L) Osbeek (Sweet orange) fruit. *Pestic. Res. J.* **I7**(1): 54--56.

Kumari Beena and Kathpal TS 2007. Analytical method for the estimation of multiresidues in fruit jam - A processed food. *Pestic. Res. J.* **19**(1): 131-135.

Banerjee K. Sawant, S.D. and Indu S. Sawant 2001. Persistence of benomyl in grapes. *Pestic. Res. J.* **13**(1): 106-108.

Dubey, J.K. and Nath, A. 1994, Persistence of some insecticides on pomegranate fruits. *Indian J Pt Prot.* **22** (2): 194-197.

Dubey, J.K. and Nath A 2005. Monitoring of apple for pesticide residue contamination -.A Survey Report. *Acta Horti.* **696** : 441-447.

FAO/WHO, 1981. Pesticide residues in food- 1980 evaluations. FAG Plant Production and Protection Paper 26 sup: 180-194.

FAO/WHO, 1989. Guide to Codex Maximum Limits for Pesticide Residues. Part 2:Codex Alimentarius Commision/ Pesticide Residues 2. Pesticide classification number 108.

GYi M M, Dikshit, A.K. and Lal, O.P. 2003. Residue studies and bio-efficacy of lambda-cyhalothrin and beta- cyfluthrin in ber *(Zizyphus manritiana* Lamark) fruits. *Pestic. Res. J.* **15**(1):26-27.

Nath, A. Patyal, S.K. and Sharma ID. 1997. Persistence and bioefficacy of endosulfan and chlorpyrifos on apple. *Pestic. Res. J.* **9**(1):*92-96,*

Reddy, C.N., Reddy, K.N., Reddy, T.N. and Rao, C.S. 2007. Dissipation of penconazole, hexacunazole and myclobutenilin grapes. Pestic. Res. J. **19**(2) : 246-247.

Mohapatara, Soudamini , Awasthi, M.D. Ahuja, A.K. and Sharma, Debi. 1998. Persistence and dissipation of carbendazim residue in/on grape berries. *Pestic. Res. J.* **10**(1): 95-97.

Mohapatara, Soudamini and A Rekha 2005. Persistence of dicofol residues in/on acid lime. *Pestic. Res. J.* **17**(1): 64-65.

Mohapatara Soudamini, Ahuja A.K. and Sharma Debi 2007 Persistence of bifenthrin residues on mango *(Mangifera indica)* fruit. *Pestic. Res. J.* **19**(1): 110-112.

Sharma, ID : Patyal S.K. Dubey, J.K. and Nath, A. 2003. Persistence of propineb (fungicide) in apple and tomato. *Pestic. Res. J.* **15**(1): 60-63.

Sharma, ID; Nath, A. and Patyal S.K. 1997. Persistence of carbendazim on apple *(Malus domestica)* in Himachal Pradesh. *Indian J of Agric. Sci.* **67**(12): 608-609.

Sharma, ID and Nath, A. 2005. Persistence of different pesticides in apple. *Acta Hort.* **696**: 437-440.

Uday, K. and Archana, K. 1998. Residues of EBDCs and ethylenethiourea (ETU) in grapes and apples. *Pestology* **22**(2):25-27.

Chapter-12

Pesticide Residues in Spices

Reena Chauhan, Shashi Madan* and Beena Kumari**
Department of Chemistry and Physics
*Department of Biochemistry
**Department of Entomology
CCS Haryana Agricultural University, Hisar - 125 004 (Harayana)

Spices are defined as "A strongly flavored or aromatic substance of vegetable origin, obtained from tropical plants, commonly used as a condiment". American Spice Trade Association (ASTA) defines spices as "Any dried plant product used primarily for seasoning purposes". The word "spice" came from the Latin word "species," meaning specific kind. The name reflects the fact that all plant parts have been cultivated for their aromatic fragrant, pungent or any other desirable properties including the seed (aniseed, caraway, coriander), leaf (cilantro, kari, bay, mint), berry (allspice, juniper, black pepper), bark (cinnamon), kernel (nutmeg), aril (mace), stem (chives), stalk (lemongrass), rhizome (ginger, turmeric, galangal), root (lovage, horse radish), flower (saffron), bulb (garlic, onion), fruit (star anise, cardamom, chile pepper) and flower bud (clove). Spices are distinguished from herbs, which are leafy, green plant parts used for flavoring purpose. Herbs, such as a basil or orengano, may be used fresh and are commonly chopped in to smaller pieces; spices however are dried and usually ground in to a powder (Anonymous, 2007).

In ancient times, spices were as precious as gold, and as significant as medicines, preservatives and perfumes. India is known as the home of spices and produces a wide variety of spices like black pepper, cardamom (small and large), ginger, garlic, turmeric, chilli and a large variety of tree and seed spices. These have nutritionally insignificant quantities as a food additive for the purpose of flavoring. They are good not only for our taste buds but also for our health. They have medicinal properties and also used as preservatives. Spices are well known appetizers, digestives and considered essential in culinary art all over the world. Some of them have antioxidant properties, while others possess strong antimicrobial and antibiotic activities. Spices increase the secretion of saliva rich in ptyalin, which facilitates starch digestion in the stomach, rendering the meals, which are rich in carbohydrates, more digestible. They inhibit thrombus formation and accelerate thromobolysis (Anonymous, 2007a).

Spices have profound influence on the course of human civilization. They permeate our live from birth to death. In every day life, spices succor us, cure us, relax us and excite us. Ancient peoples such as Egyptian, the Arab and the Roman made extensive uses of spices, not only to add flavour to food, beverages, but as medicines, disinfectants, incense, stimulants and even as aphrodisiacs agents. No wonder they were sought after in the same manner gold and precious metals (Chomchalow, 1996).

The basic effect of spices when used in cooking and confectionary can be for flavoring, deodorizing/masking, pungency and coloring (Table1). The major color components of spices are given in Table 2.

According to International Organization for Standardization (ISO), there are 109 spices belonging to 31 families are grown around the world, the majority of which are in Asia (Table 3). In India, about 52 spices are cultivated and classified in various ways. Based on plant part used, they are classified as (I) Rhizome and root spices, (II) Bark spices, (III) Leaf spices, (IV) Flower spices, (V) Fruit spices, (VI) Seed spices (Parthasarathy, 2008). The physico-chemical properties of spices depend upon the species, variety and climatic conditions of the area where the crop is raised (Table 4).

Table1: Use of spices

Basic Functions	Spices as major function	Spices as sub function
Flavouring	Parsley, cinnamon, allspice, dill, mint, tarragon, cumin, marjoram, star anise, basil, anise, mace, nutmeg, fennel, sesame, vanilla, fenugreek, cardamom, celery.	Garlic, onion, bay leaves, clove, thyme, rosemary, caraway, sage, savory, coriander, pepper, oregano, horse radish, Japanese pepper, saffron ginger, leek, mustard
Deodorizing/masking	Garlic, savory, bay leaves, clove, leek, thyme, rosemary, caraway, sage, oregano, onion, coriander.	
Pungency	Garlic, savory, bay leaves, clove, leek, thyme, rosemary, caraway, rage, oregano, onion, coriander, Japanese pepper, mustard, ginger, horse radish, Red pepper, pepper.	Parsley, pepper, allspice, mint, tarragon, cumin, star anise, mace, fennel, sesame, cardamom, mustard, cinnamon, vanilla, horse radish, Japanese pepper,
Colouring	Paprika, turmeric, saffron.	Nutmeg, ginger

Source: Ravindaran *et al.* (2006)

Table 2 : Colour components in spices

Color component	Tint	Spices
Carotenoid		
β-Carotene	Reddish orange	Red pepper, mustard, paprika, saffron
Cryptoxanthin	Red	Paprika, red pepper
Lutin	Dark red	Paprika, parsley
Zeaxanthin	Yellow	Paprika
Capsanthin	Dark red	Paprika, red pepper
Capsorbin	Purple red	Paprika, red pepper
Crocetin	Dark red	Saffron
Neoxanthin	Orange yellow	Parsley
Violaxanthin	Orange	Parsley, sweet pepper
Crocin	Yellowish orange	Saffron
Flavonoid	Yellow	Ginger
Curcumin	Orange yellow	Turmeric

Source: Ravindaran *et al.* (2002)

Table 3 : Plant species that are grown commercially as spices in Asia

Common name	Scientific name	Part used
Anise	*Pimpinella anisum*	Seed
Basil	*Ocimum basilicum*	Leaf
Black pepper	*Piper nigrum*	Berry
Caraway	*Carum carvi*	Seed
Cardamom (large)	*Amomum subulatum*	Capsule
Cassia	*Elettaria cardamomum*	Capsule
Celery	*Cinnamomum cassia*	Bark
Chili pepper	*Apium graveolens*	Leaf & seed
Cinnamon	*Capsicum annuum*	Fruit
Clove	*Cinnamomum verum*	Bark & leaf
Coriander	*Coriandrum sativum*	Flower bud
Cumin	*Cuminum cyminum*	Leaf & seed
Curry leaf	*Murraya koenigi*	Seed
Dill	*Anethum graveolens*	Leaf & seed
Fennel	*Foeniculum vulgare*	Leaf & seed
Fenugreek	*Triagonella foenumgraecum*	Seed
Gamboge	*Garcinia cambogiana*	Fruit
Garlic	*Allium sativum*	Bulb
Ginger	*Zingiber officinale*	Rhizome
Greater galanga	*Kaempferia galanga*	Rhizome
Long pepper	*Piper longum*	Leaf
Marjoram	*Marjorana hortensis*	Leaf & stem
Mustard	*Brassica nigra*	Seed
Nutmeg & Mace	*Myristica fragrans*	Seed & aril
Oregano	*Origanum vulgare*	Leaf
Parsley	*Petroselinum crispum*	Leaf
Rosemary	*Rosemarinus officinalis*	Flower/seed & leaf
Saffron	*Crocus sativus*	Flower parts
Sage	*Salvia officinalis*	Leaf & seed
Savory	*Satureja hortensis*	Leaf
Star anise	*Ilicium verum*	Fruit
Sweet flag	*Acorus calamus*	Rhizome
Tamarind	*Tamarindus indicus*	Fruit
Tarragon	*Artemisia dracunculus*	Leaf
Thyme	*Thymus vulgaris*	Leaf
Turmeric	*Curcuma domestica*	Rhizome
Vanilla	*Vanilla planifolia*	Pod

Source: Nazeem (1995)

Table 4 : Physico chemical characteristic of spices and condiments

Spice (Dry)	Moisture (%)	Volatile Oil (%)	Total ash (%)	Water Soluble ash (%)	Alkalinity of H_2O soln ash (%)	Ash in HCl soln (%)	Protein	Alcohol extract (%)
Ajwan	9.01	3.01	2.75	6.80	-	-	16.20	-
Aniseed	11.43	2.40	3.46	6.22	2.00	0.25	20.68	23.36
Cardamom	8.49	2.80	4.01	2.15	0.90	0.42	7.38	7.02
Celery Seed	7.52	NIL	6.50	3.49	0.70	0.35	18.62	21.90
Coriander seed	7.52	NIL	4.31	1.78	1.60	0.19	15.93	22.16
Cinnamon	9.49	0.70	5.41	2.80	1.40	0.88	25.57	8.18
Clove	7.72	13.20	5.26	3.17	3.00	0.24	2.19	0.24
Garlic powder	6.10	0.04	3.46	3.40	0.76	NIL	12.75	12.22
Ginger	7.23	1.60	3.57	2.03	2.20	0.56	0.70	5.44
Mustard Powder	8.29	NIL	3.78	0.62	0.36	0.27	27.75	11.05

There is no other country in the world that produces as many kinds of spices as India. India is the largest producer, consumer, exporter of spices and spice products.As per Spices Board, although 52 spices are grown in the country but more than 90 per cent of the spices produced in the country are used for domestic consumption and the rest only is being exported as raw and value added forms (Nybe *et al.,* 2007).

The total production of spices is 4.3 million metric tonnes and the area covered is 2.56 million hectares (Anonymous, 2011). Because of the varying climates in India, from tropical to sub-tropical, 45°C to 0°C temperate, almost all spices are grown in this country. In almost all of the 28 states and seven union territories of India, at least one spice is grown in abundance. Maharashtra is one of the important spices growing states. The area under spices crops is about 1.62 lakh ha and production is 13.46 lakh MT. The major crops are chilies, ginger, turmeric, black pepper, cardamom, etc. and also cultivation of seed spices like coriander, fenugreek, etc. to some extent. Kerala is the leading producer of both pepper and cardamom. India is the largest chilly producer, with 40-50 per cent of the output coming from Andhra Pradesh and meets one third of the country's need (Singhal, 2003). About 80 per cent of world supply of coriander is produced in India of which, Rajasthan alone accounts for 70 per cent. In cumin again, India holds an enviable position, with about 90 per cent of the production coming from Rajasthan and Gujarat. In India, turmeric is grown over 1.65 lakh ha with an annual production of 5.6 lakh tones. The annual growth rate during 2005-06 was 3.7 and 9.1 per cent for area and production, respectively. In India, Andhra Pradesh tops the list in area and production followed by Orrisa and Tamil Nadu (Peter, 2006). India exported 23,000 tonnes of turmeric during 1996-97 to 67 countries (Peter, 1999) and after a span of ten years i.e during 2006-07, the export increased to 51,500 tonne. It is exported as dry, fresh, powder, oleoresin and oil.

Spices constitute an important group of agricultural commodities which are virtually indispensable in the culinary art. In India, spices are important commercial crops from both domestic consumption and export point of view. At present, India produces around 2.75 million tones of different spices valued at approximately 4.2 billion US $, and holds the premier position in the world spices market. There is comparison of export from India during year 2009-10 to 2010-11 (Table 5).

Infestation by insect pests is a major factor responsible for the low productivity of spice crops in India. Spice crops such as black pepper (*Piper nigrum* L.), cardamom (*Elettaria cardamomum* Maton), ginger (*Zingiber officinale* Roscoe) and turmeric (*Curcuma longa* L.) are infested by many species of insect pest, some of which cause severe crop losess. The chilli crop is also ravaged by a wide array of insect pests including 51 species of insect and 2 species of mites, belongs to 27 families under 9 orders in both and nursery as well as main field (Reddy *et al.*, 1983, Reddy *et al.*, 1984) resulting in yield losses. The combine infestation of mite and thrips may cause losses upto ~34 per cent (Ahmad, 1987). Thrips alone causes 60.5-74.3 per cent yield losses (Patel, 1998). Unit saved is unit produced; hence farmers forcibly go for management of these pests. The indiscriminate and repeated use of conventional insecticides accumulates the toxic pesticide residues on an agricultural produce and pose serious threat to health of the consumers. India is dominating the world market in chilli and other spices as spice product. The trend in export of chilli and other spices are very encouraging but on the other hand with a fall in quality of certain spices, mainly chillies, due to the presence of pesticides residues, seem to have adversely affected the export of Indian spices. India is known as the "Land of Spices" and our country earns a considerable amount of foreign exchange from the export of spices. The world consumption of spices is growing steadily year by year. Expansion of our export of spices to increase or even to retain our share of world market is imperative. India exports spices mostly to developed countries like US, UK, Germany and other European Countries, Japan and Canada. These countries have their own stringent food laws and regulations. On the ground of pesticide residues that were present in our spices, European countries start rejecting our consignments mainly of chillies. Under the rapid food alert system in European Union (EU), each country was strictly following the residue levels of pesticides in spices. In 1996, 75 containers were rejected by the US importers (Nair, 2003). During 2001, Germany and Greece rejected red chilli and chilli powder due to pesticide residues (Rao *et al.*, 2005). Many farm gate chilli samples showed the presence of insecticides residues (Singh *et al.*, 1999). Since then, the Spices Board and the forum had been working together to find a solution to the issue.

Table5: Estimated Export from India during April-February 2010-11 compared with April-February 2009-10

Items	April-February 2010-11				April-February 2009-10			
	Qty. (Tonnes)	Value (Rs. Lakhs)	Value (US $)	Rate (Rs/Kg)	Qty. (Tonnes)	Value (Rs. Lakhs)	Value (US $)	Rate (Rs/Kg)
Pepper	16,600	33,462.25	73.42	201.58	18,425	29,299.93	61.60	159.02
Cardamom (S)	865	10,041.25	22.03	1160.84	1,765	14,362.15	30.19	813.59
Cardamom (L)	700	3,885.40	8.53	555.06	795	1,343.30	2.82	168.97
Chilli	218,500	137,951.50	302.70	63.14	180,750	115,657.27	243.15	63.99
Ginger	12,000	7,728.75	16.96	64.41	5,100	4,151.25	8.73	81.40
Turmeric	42,500	60,884.55	133.59	143.26	46,575	34,286.88	72.08	73.62
Coriander	37,500	14,998.23	32.91	40.00	41,150	20,207.11	42.48	49.11
Cumin	27,500	33,372.75	73.23	121.36	44,800	49,448.13	103.96	110.38
Celery	3,250	2,236.89	4.91	68.83	4,575	2,430.95	5.11	53.14
Fennel	6,950	6,300.75	13.83	90.66	6,050	4,848.58	10.19	80.14
Fenugreek	17,000	6,168.10	13.53	36.28	19,650	6,511.10	13.69	3.14
Other Seeds (1)	11,750	5,145.55	11.29	43.79	14,750	5,670.34	11.92	38.44
Garlic	16,700	6,654.30	14.60	39.85	10,050	2,799.97	5.89	27.86
Nutmeg & Mace	1,550	6,976.82	15.31	450.12	3,210	8,891.82	18.69	277.00
Other Spices (2)	21,500	14,915.30	32.73	69.37	18,800	13,274.93	27.91	70.61
Curry Powders / Paste	13,250	17,875.50	39.22	134.91	13,100	17,329.79	36.43	132.29
Mint Products (3)	16,250	154,174.00	338.29	948.76	17,725	110,144.62	231.57	621.41
Spice Oils & Oleoresins	6,800	80,302.45	176.20	1180.92	6,225	65,168.35	137.01	1046.88
Total	**471,165**	**603,074.33**	**1,323.28**		**453,495**	**505,826.46**	**1,063.44**	

1. Include Mustard, Aniseed, Ajwanseed, Dill seed, Poppy seed etc.
2. IncludeTamarind,Asafetida,Cassia,Saffronetc.
3. Include Mint oils, Menthol and Menthol crystals

Source: Estimate based on DLE from customs, report from RO'S etc.

As compared to other foodstuffs, spices are very concentrated and full of complex aromatic compounds. This means that analyzing them is technically challenging.

Analytical Methods

The samples were analyzed by the American Spice Trade Association (ASTA) and India by the multi-residue procedures summarised below in Table 6.

Table 6 : Multi-residue methods used in the monitoring programmes of ASTA and India

Pesticides	Analytical method Summary	Analytical method Summary
Organophosphorus(OP)	U.S. Food and Drug Administration Pesticide Analytical Manual (FDAPAM) Vol.1, 3rd Edition, 1997, E4and C5)	Extraction: acetone-water Clean-up: FlorisilDetection: capillary GC with NP detector (PAM DG 17)
Organochlorine(OC)	Official Methods of AOACInternational, Vol. 1, 17th Edition,2000. Method 970.52	Extraction: acetonitrile-water Clean-up: sulfuric acid Detection: capillary GC with ECDdetector (PAM DG 13)
Pyrethroid (PY)	Official Methods of AOAC International, Vol. 1, 17th Edition, 2000. Method 970.52	Extraction: acetonitrile-water Clean-up: Florisil (PAM C5) Detection: capillary GC with ECD detector (PAM DG 10)

Supervised Trials

Supervised trials are those under which complete study of particular pesticide like its persistence, degradation, half- life period and safe waiting period etc. are calculated. These are primarily conducted to generate residue data required for registration of pesticides, setting up of maximum residue limits (MRLs) and finding safe waiting periods for their application. To obtain the necessary data for these purposes, the crops are grown under different agro-climatic conditions and treated with pesticides using *good agricultural practices (GAP).* Under *good agricultural practices,* the pesticide treatments are in accordance with the registered or approved usage pattern, taking into account the minimum quantities necessary to achieve an adequate control. The crop commodities are then analyzed for the remaining pesticides residues. The data pertaining to half-life and safe waiting periods of commonly used pesticides on different spice crops are given in Table 7.

Table 7 : Safe waiting periods and half-life values of various insecticides on different spices

Spices	Variety	Location	References	Insecticides	Dose (kg/ha or g a.i./ha) or percent	Half -Life Period (days)	Safe Waiting period	MRL (mg/kg)	Processing Factor
Cardamom	TDK 1	Tamil Nadu	Rajabaskar and Regupathy, 2008	Diafenthiuron	600	1.7 (Green Cardamom)	7	-	-
					1200	3 (Green Cardamom)	11	-	-
					600	3 (cured Cardamom)	4	-	-
					1200	4 (cured Cardamom)	10	-	-
Chilli	Pusa jawala	Hyderabad	Reddy *et al.*, 2007a	Fipronil	0.01	16.8	5	0.1	-
				Profenofos	0.1	41	19	0.05	-
	Pusa jawala	Hyderabad	Reddy *et al.*, 2007	Endosulfan	350	9.2	8.2	2.0	-
				Dicofo l	450	4.1	6.2	1.0	-
				Dimethanoate	300	1.0	4.7	2.0	-
				λ-cyhalothrin	50	4.2	6.4	0.2	-
	Pusa jawala	Gurgaon	Sanyal *et al.*, 2008	Acetamiprid	20 40	2.24 4.84	1	- -	- -

Contd.

	Pusa jawala	Mohanpura	Roy *et. al.*, 1997	Fluvalinate 20AF	20	1-1.7	-	-	-
	Sada Bahar, Vaikali Mukta-keshi, Brinjal Dotis, Son-6, JC-2	Punjab, Rahuri, Kalyani, Anand, Jabalpur, Vellyani and Jorhat	Agnihotri, 1999	Quinalphos 25EC	500	-	2-7		-
					1000	-	2-7		-
	BRS-51, Arka Keshar, Morbi, Surya, Gulabi	Punjab, Bangalore, Anand, Vellayani Hyderabad		Lindane20EC	0.35	-	1-5	3.0	-
					0.70	-	1-6		-
	BRS-51, Arka Keshar, Morbi-4, Surya	Punjab, Bangalore, Anand, Vellayani,		Triazpohos 40EC	0.35	-	3-7	0.1	
					0.70	-	3-9		-
	Pusa purple round	Assam	Duara *et al.*, 2003	Cypermethrin 10EC	22.5	3.30	-	-	-
					45	3.39	-	-	-
					75	3.16	-	-	-
				Fenvalerate20EC	30	2.98	-	-	-
					60	3.26	-	-	-
					112.5	3.76	-	-	-
				Endosulfan 35EC	525	5.29	-	-	-

Contd.

Chilli	IIHR-NP-46	Bangalore	Awasthi, 1994	Permethrin 25EC	0.015	-	-	1.0	-
					0.030	-	-		-
				Cypermethrin 25EC	0.01	-	-	0.5	-
					0.02	-	-		-
				Fenvalerate 20EC	0.01	-	-	1.0	-
					0.02	-	-		-
				Deltamethrin 2.8EC	0.0015	-	-	0.2	-
					0.0030	-	-		-
				Monocrotophos	0.05	-	-	1.0	-
					0.10	-	-		-
Chilli	Agni Rekha, Jwala, Peprika	Rahuri, Bangalore Hyderabad	Agnihotri, 1999	Lindane	150	1-2	1.0-2.8	0.5	-
					300	1-4	2.2-2.5		-
	Jwala Mukhi	Vellayani		Lindane	250	2	1.2		-
					500	3	3.3		-
	Local	Coimbatore		Lindane	350	2	5.2		-
					700	4	7.0		-
	Agni Rekha, Jwala, Peprika	Rahuri, Bangalore , Hyderabad,		Quinalphos	150	1-10	1.6-3.3	0.25	-
					300	2-14	1.7-3.6		-
	Arka Jwala	Bangalore	Awasthi, 1995	Fenvalerate	100	5.80	-	2	-
				Fenvalerate	100	2.50	-		-
	Local, Jwala Mukhi	Coimbatore, Vellayani		Quinalphos	250	6-7	1.9-2.0 2.2-2.6		-
					500	9-10			-

Contd.

	Agni Rekha, Local	Rahuri, Coimbatore	Agnihotri, 1999	Dicofol	250 500	1 1-2	1.8-3.3 2.0-3.1	5.0	- -
				Dicofol	175	1	8.2		-
					300	1	14.6		-
	Agni Rekha, Local	Rahuri, Coimbatore		Mancozeb	750 1500	1-5 1-6	3.7-6.9 3.9-6.4	0.5	- -
	Jwala Mukhi	Vellayani		Mancozeb	1000 2000	2 4	1.3 2.8		- -
	Hisar Shakti	Hisar, Haryana	Anonymous, 2008	Cypermethrin Dicofol Ethion	0.15 0.03 0.05	- - -	- - -	- - -	4.52 3.62 3.84
Chilli	-	Kalyani, West Bengal	Papia *et al.*, 2010	Chlorfenapyr	75 100	2.93 2.96	- -	- -	- -
Chilli		B.C.K.V., Mohanpur, Nadia, West Bengal Mahatma Phule Krishi Vidyapeeth Maharashtra	Mukhopadhyay *et al.*, 2011	Difenoconazole	62.5 125	2.15 2.32	- -	- -	- -
		PAU, Ludhiana	Sahoo *et al.*, 2011	Trifloxystrobin 25%	250 500	1.81 1.58	- -	- -	- -
				Tebuconazole 50%	250 500	1.37 1.41	- -	- -	- -

Contd.

Chilli	-	Jalgoan city, Maharashtra	Waghulde *et al.*, 2011	Chlorpyriphos	32 g a.i ha^{-1}				
				Endosulphan	14 g a.i ha^{-1}	3.22 4.99	15.75 18.08	- -	- -
				Dimethoate	6 g a.i ha^{-1}	4.7	9.38	-	-
				Malathion	31.68 g a.i ha^{-1}	5.49	12.01	-	-
	-	Vellayani,	Varghese	Dimethoate	300 g a.i ha^{-1}	1.94	13.63	0.5	-
		Kerala	*et al.*, 2011	Ethion	375 g a.i ha^{-1}	3.43	28	1.0	-
				Oxydemeton -methyl	500 g a.i ha-1	4	7	0.01	-
		PAU, Ludhiana	Pandher *et al*; 2012	Deltamethrin 10 EC	17.5 35	0.36-1.99 0.38-2.06		0.05	
Coraindar		Udaipur, Rajasthan	Shyam *et al.*, 2011	Imidacloprid	160	2.37	4.97 (Leaves)	-	-
					2.19 (Seed)	3.32	-	-	
Fenugreek		PAU, Ludhiana	Singh *et al.*, 2006	Malathion 50EC	625 1250	- -	- -	3 -	- -
Garlic		Wufeng, Taiwan	Lin *et al.*, 2007	Pendimethalin 34% EC	1.19 2.38	13-17 -	- -	- -	- -
		CSK HPKV, Palampur	Sharma *et al.*, 2011	Oxyfluorfen (23.5 EC)	125 250 500	23.2-30.12		0.05	
Mustard	T-59	ARS, Durgapur, Jaipur	Gupta *et al.*,2001	Lindane 25 EC	500 1000			0.05	

Contd.

	Pusa Kalyani	IARI, New Delhi	Kumar *et al.*,2000	Alphamethrin 10EC	40 80	5.38 5.45	1	0.05	
	Pusa Bold	IARI, New Delhi	Gopal *et al.*, 2002	β-Cyfluthrin	12.5 25	4.2 5.0			
				Imidacloprid	20 40	4.3 5.0		0.05	
	Pusa Bold	IARI, New Delhi	Mukherjee *et al.*, 2003	Iprodione 50WP	500	6.02 (Leaves) 5.00 (Pod)		0.5	
Turmeric	-	Anand, Gujarat	Shah *et al.*, 2009	Chlorthalin Chlorpyriphos Endosulfan	0.45 0.12 0.21	- - -	- - -	- - -	2.33 0.64 4.67

Refernces

Anonymous, 2008. Annual Progress Report and Proceedings 2008. All India Network Project (AINP) on Pesticide Residues. Project coordinating cell, Division of Agricultural Chemicals, India. Agricultural Research Institute, New Delhi, p. 13.

Anonymous, 2010. Annual Progress Report and Proceedings 2008-09. AICRP on Pesticide Residues. Project Coordinating Cell, Division of Agricultural Chemicals, Indian Agricultural Research Institute, New Delhi, p. 82.

Anonymous, 2011.www.evergreenexports.net/spices-production-india.

Anonymous, 2007(a).http://indianfood.indianetzone.com/spices/(online accessed on 9-3-2007)

Anonymous, 2007.http://www.refrence.com/browse/wiki/spice(online accessed on 9-3-2007)

Ahmad, K. Mohamed, M.G. and Murthy, N.S.R. 1987.Yield losses due to various pests in hot pepper. *Capsicum News Letters:* 83-84.

Agnihotri, N.P. 1999. Pesticide: Safety Evaluation and Monitoring. All India Coordinated Research Project on Pesticide Residues, Division of Agricultural Chemicals, Indian Agricultural Research Institute, New Delhi pp 67--93.

Awasthi, M.D. 1994. Studies on dissipation and persistence of pyrethroid residues on chilli fruits for safety constants. *Pestic. Res. J.*, **6**(1): 80-83.

Awasthi, M.D. 1995. Persistence pattern of fenvalerate residues on chilli fruits following applications of emulsifiable concentrate and dust formulations. *Pestic. Res. J.*, **7**(2): 162-164.

Chomchlow, N.1996. Spice production in Asia - An overview unpublished paper presented at IBC'S Asia spice market'96 confrence Singapore, 27-28 May, 1996. (www.journal.av.edu/au-techno/2001/oct2001/article6.pdf) on accessed on 8-3-2007.

Chowdhury, P. and Sukul, P.1997.Residue behaviour of fluvalinate in chilli (*Capsicum annum L.*) under Indian climatic condition. *Bull. Environ.Contam. Toxicol.*, **59**:723-727.

Duara, B. Baruah A.A.L.H., Deka, S.C. and Barman N. 2003.Residues of cypermethrin and tenvalerate on brinjal. *Pestic. Res.J.*, **15**(1): 43-46.

Gopal, M. Mukherjee, I. Chander S.2002. Behaviour of B- cyfluthrin and Imidacloprid in Mustard Crop:Alternative Insecticide for Aphid Control. *Bull. Environ.Contam. Toxicol.*, **68**:406-411.

Gupta, A. Parihar, N.S. and Bhatnagar, A. 2001. Lindane, chlorpyriphos, and quinalphos residues in mustard seed and oil. *Bull. Environ.Contam. Toxicol.***67**:122-127.

Kumar, R. Dikshit, A.K. and Prasad, S.K. 2000.Persistence and safety evaluation of alphamethrin on mustard (*Brassica campestris* Linn.). *Bull. Environ.Contam. Toxicol.*, **65**:200-206.

Lin, HT. Chen, S.W. Shen C.J. and Chu, C. 2007.Dissipation of pendimethalin in the garlic (*Allium sativum L.*) under subtropical condition. *Bull. Environ.Contam. Toxicol.* **79**(1):84-86.

Mukhopadhyay, S. Das, S. Bhattacharyya, A. and Pal, S. 2011. Dissipation study of difenoconazole in/on chili fruit and soil in India. *Bull. Environ.Contam. Toxicol.*, 10.1007/s00128-011-0275-2

Mukherjee I, Gopal, M. Chatterjee, S.C. 2003. Persistence and effectiveness of iprodione against alternaria blight in mustard. *Bull. Environ.Contam. Toxicol.* **70**:586-591.

Nair, G.K. 2003. Fix minimum residue level for pesticides: Spice exporters.

Nazeeem, P.A. 1995. The Spices of India. *The Herb, Spice, and Medicinal Plant Digest* **13**(1):1-5.

NRCS, 1993, Spices Biotechnology at National Research Centre for Spices, NRCS, Calicut, Kerala, India.

Nybe, E.V. Mini Raj, N. and Peter K.V. 2007. Spices. (ed. K.V.Peter) Horticulture Science Series-5. New India Publishing Agency.

Official methods of AOAC International, Vol. 1, 17th Edition. Method 970.52.

Pandher, S. Sahoo, S.K. Battu, R.S. Singh, B. Saiyad, M.S. Patel, A.R. Shah, P.G. Reddy, C. Narendra, Reddy J.D., Reddy N.K., Rao Ch. Sreenivasan, Banerjee T, Banerjee D, Hudait R.K., Banerjee, H. Tripathy, V and Sharma, K.K. 2012. Persistence and dissipation kinetics of deltamethrin on chili in different agro-climatic zones of India. *Bull. Environ.Contam. Toxicol.***88**(5):764-768.

Papia, D. Das, S.P. Sarkar, P.K. and Bhattacharyya Anjan. 2010. Degradation dynamics of chlorfenapyr residue in chili, cabbage and soil. *Bull. Environ.Contam. Toxicol.* 10.1007/s00128-010-9994-z

Parthasarathy, V.P. Kanndiannan, K. and Srinivasan. 2008. Organic Spices. (eds. Parthasarathy VP, Kanndiannan K, Srinivasan). Pub. NIPA, New Delhi

Patel, V.N. and Gupta H.C.L. 1998. Estimation of losses and management of thrips infesting chillies. p: 99. In : National Seminar Ento. in 21st century. Biodiversity, Sustainability, environmental safety and human health held at Udaipur.

Pesticide Analytical Manual, US Food and Drug Administration, Vol.1, 3rd Edition. 1997. E4 and C5.

Peter, K.V. Nybe, E.V. and Raj, N.M. 2006. Available technology to raise yield. *The Hindu: Survey of Indian Agriculture*, pp. 82-86.

Peter, K.V. 1999. Handbook of herbs and spices. pp. 297-300.

Rajabaskar D and Regupathy A. 2008.Persistence of diafenthiuron in cardamom. *Pestic. Res. J.*, **20**(2): 247-249.

Ravindaran, P.N. 2006. Spices: Defination, Classification, History, Properties, uses & role in Indian life. In. (eds. P.N. Ravindarn, K. Nirmal Babu, K.N.Shiva, Jhony, A.K.) Advances in spices research-History Achievements of Spices Research in India since Independence, Agrobios (India, Jodhpur).

Ravindaran, P.N. Jhony, A.K. and Nirmal Babu, K. 2002. Spices in our daily life. *Satabdi Smaranika* 2. Arya Sala, Kottakkal.

Reddy, K.N. Satyanarayana S. and Reddy K.D. 2007. Persistence of some insecticides in chillies. *Pestic. Res. J.* **19**(2): 234-236.

Reddy, D.K. Reddy, N.K. and Mahalingappa, P.B. 2007a. Dissipation of fipronil and profenofos residues in chillies (*Capsicum annum* L.) *Pestic. Res. J.*, **19** (1): 106-107.

Reddy, D.N.R. and Puttaswamy.1983. Pest infesting chilli in the transplanted crop . *Mysore J. of Agri.*, 17: 246- 251.

Reddy D.N.R. and Puttaswamy.1984. Pest infesting chilli (*Capsicum annum* L.) in nursery. *Mysore J. of Agri.*, **18:** 122- 125.

Rao, C.S. Bour, T.B. and Reddy, K.N. 2005. Pesticide residues : Impact on Indian agriculture exports in WTO era. pp. 54-57. In: Proc. sym. on "Pesticide Residues and Their Risk Assessment", January 20-21:

Sanyal, D. Chakma, D. and Alam, S. 2008 .Persistence of a neonicotinoid insecticide, acetamiprid on chili (*Capsicum annum* L.). *Bull. Environ.Contam. Toxicol.*, **81**(4): 365-368.

Singh, B. Gupta, A. Bhatnagar, A. and Parihar, N.S. 1999. Monitoring of pesticide residues in farm gate samples of chilli. *Pestic Res. J.*, **11**(2):207-209.

Singhal, Vikas. 2003. Indian economics data center-New Delhi *Indian J. of Agri.*, 565 - 570 pp.

Singh, B. Battu, R.S. and Dhaliwal, J.S. 2006. Dissipation of malathion on fenugreek (*Trigonell foenum-gracaeum* L.). *Bull. Environ.Contam. Toxicol.*, **77**: 521-523.

Sahoo, S.K. Jyot, G. Battu, R.S. and Singh, Balwinder 2011. Dissipation kinetics of trifloxystrobin and tebuconazole. *Bull. Environ.Contam. Toxicol.*, **88**(3):368-371.

Shah, P.G. Diwan, Kalpana Raj, M.F. and Patel, A.R. 2009. Effect of processing of turmeric on Chlorthalonil, chlorpyriphos and endosulphan. *Pestic. Res. J.*, **2**(1) : 86-88.

Shyam, M.R. Gupta, H.C.L and Sharma, R.P. 2011. Evaluation of insecticides against coriander aphid and estimation of residues. *Annals. Pl. Protec. Sci.*, **19**(1):101-105.

Varghese, S.T. Mathew, B.T. George, T. Beevi, N.S. and Xavier, G. 2011. Dissipation study of dimethote, ethion and oxydemeton methyl in chilli. *Pestic Res. J.*, **23**(1):68-73.

Waghulde, P.N. Khatik, M.K. Patil, V.T. and Patil, P.R. 2011. Persistence and dissipation of pesticides in chilly and okra at north maharashtra region. *Pestic Res. J.*, **23**(1):23-26.

Chapter-13

Pesticide Residues in Vegetables

Reena Chauhan, Anil Duhan* and Beena Kumari**
Department of Chemistry and Physics
*Department of Agronomy
**Department of Entomology
CCS Haryana Agricultural University, Hisar - 125 004 (Haryana)

The noun vegetable means an edible plant or part of a plant, but usually excludes seeds and most sweet fruits. This typically means the leaf, stem, or root of a plant. Vegetables are eaten in a variety of ways, as part of main meals and as snacks. The nutritional content of vegetables varies considerably, though generally they contain little protein or fat (Woodruff,1995 and Whitaker, 2001) and varying proportions of vitamins such as Vitamin A, Vitamin K and Vitamin B6, provitamins, dietary minerals and carbohydrates. They contain a great variety of other phytochemicals, some of which have been claimed to have antioxidant, antibacterial, antifungal, antiviral and anticarcinogenic properties (Gruda, 2005; Steinmetz and Potter,1996). Some vegetables also contain fiber, important for gastrointestinal function. Some contain important nutrients necessary for healthy hair and skin as well. However, vegetables often also contain toxins and antinutrients such as á-solanine, á-chaconine,(Science direct) enzyme inhibitors (of cholinesterase, protease, amylase, etc.), cyanide and cyanide precursors, oxalic acid, and more (Bad Bug Bock > BBB – *Clostridium botulinum*). Depending on the concentration, such compounds may reduce the edibility, nutritional value, and health benefits of dietary vegetables. Cooking and/or other processing may be necessary to eliminate or reduce them. Diets containing recommended amounts of fruits and vegetables may help to lower the risk of heart diseases and two type of diabetes. These diets may also protect against some cancers and decrease bone loss. The potassium provided by vegetables may help to prevent the formation of kidney stones.

India has taken a bold step towards self sufficiency in food. However, self sufficiency in the true sense can be achieved only when each individual in the country is assured of balanced diet. Varied agro-climatic conditions in India make it possible to grow a wide variety of vegetable crops all the year round in one part of the country or another. In vegetables production, India is next only to China with an annual production of 87.53 million tonnes from 5.86 million hectares having a share of 14.4 per cent to the world production (Anonymous, 2011a).The year wise production and area of vegetable all over the India is given in Table 1.

Vegetable crops are attacked by a number of insect-pests and diseases as they are grown intensively, sometimes even two or more crops are taken in a season where prevailing weather conditions permit and suitable varieties are available. For better yield and quality, insecticides are repeatedly applied during the entire period of growth and sometimes even at fruiting stage. India is the third largest consumer of pesticides in the world and highest among the South Asian countries. Pesticides have become most essential inputs in modern agriculture for ensuring food security particularly in the developing countries where population far exceeds the agricultural growth but the indiscriminate use of pesticides to protect the crop particularly at fruiting stage and non-adoption of safe waiting period leads to accumulation of pesticide residues in edible portion of crop, which may be hazardous to humans beings. The presence of residues above the permissible limit is also a major bottleneck in the acceptance of food commodities by importing countries in context to World Trade Organization (WTO) challenge. So, there is an urgent need to find eco-friendly and efficient alternative for the control of pests.

Table1: Area, Production and Productivity of Vegetables In India

Year	Area (in 000' ha)	Production (In 000' Mt)	Productivity (in Mt/ha)
1991-92	5593	58532	10.5
2001-02	6156	88622	14.4
2002-03	6092	84815	13.9
2003-04	6082	88334	14.5
2004-05	6744	101246	15.0
2005-06	7213	111399	15.4
2006-07	7581	114993	15.2
2007-08	7848	128449	16.4
2008-09	7981	129077	16.2
2009-10	7985	133738	16.7
2010-11	8495	146555	17.3

Source: Indian Horticulture Database, 2011b

Pesticide

Pesticide means any substance intended for preventing, destroying, attracting, repelling, or controlling all pests including unwanted species of plant, or animals during the production, storage, transportation, distribution and processing of foodstuffs, agricultural commodities or animals feed or which may be administered to the animals for the ectoparasites. Pesticides include such chemicals as insecticides, herbicides, fungicides, acaricides and nematicides.

According to food and agricultural organization (FAO) 1986, the pesticides have been defined as any substance or mixture of substances intended for preventing or controlling any pest, and they include any substance or mixture of substances intended for use as plant growth regulator (PGR), defoliant, or desiccant. The term excludes fertilizers and antibiotics or other chemicals administered to animals to stimulate their growth or to modify their reproductive behaviour.

Toxicological Consideration

Any chemical substance may evoke one or both of two toxic effects, first, the acute effect, is the one more readily comprehensible to the nonprofessional, and normally occurs shortly after contact with a single dose of poison, and second, the chronic effect occurs when an organism is exposed to repeated small and non-lethal doses of a potentially harmful substance. A frequently used measure of acute toxicity is the 50% lethal dose or LD_{50}. It is the amount of poison that kills half the organisms and is expressed in terms of milligrams poison per kilogram body weight of experimental animals (mg/kg), thus, the dose that kills half the organism is a more sensitive index of toxicity than any other dose, and this is why, the LD_{50} is usually adopted as a standard for comparing the relative toxicity of the substances.

An empirical scale of assessment of levels of toxicity was introduced by Du Bois and Geiling (1959). On the scale they proposed, a substance with an LD_{50}between 1 and 50 mg/kg is extremely toxic, between 50 and 500 mg/kg is highly toxic and 500 to 5000 mg/kg as moderately toxic and above 5000 is slightly toxic.

Maximum Residue Limit

To regulate the pesticides residues to a safe level, a concept was introduced by Joint FAO/WHO Expert Committee on Food Additives (1955) and Codex Alimentarius Commission was established in 1964. Maximum residue limit (MRL) is the maximum concentration for a pesticide residue on crop or food commodity resulting from the use of pesticides according to good agricultural practice (GAP). The concentration is expressed in milligram of pesticide residues per kilogram of the commodity (mg kg^{-1}/µg g^{-1}/ppm).

The problem of pesticide residues in vegetables is particularly more serious in India because these are invariably harvested at short intervals, and many vegetables are consumed in raw form as salad. This section is divided into two heads, i.e., *Monitoring studies* and *supervised trials.* This chapter attempts to deal with the extent of contamination of vegetables and spices with pesticides residues as revealed by surveillance and compliance studies carried out in India. Finally, some kitchen processes have been discussed for reducing the residues in vegetables, along with some suggestions to keep the residues within the safe limits.

Pesticide Residues in Vegetables

Monitoring studies

For monitoring studies in the general survey, all the samples are monitored for residues for all the applied pesticides. The data pertaining to residues of pesticides in different vegetables are given in Table 2. Seasonal vegetables consisting of green leafy vegetables, tomato, potato, cauliflower, okra, cabbage etc. have been analyzed for pesticidal contamination in almost all states of India. As is evident from the data, the estimations have been made for organochlorine insecticides such as DDT and its analogues, HCH, aldrin, dieldrin, heptachlor, endosulfan and lindane before 1988 but analyses afterwards have been done for organochlorines (OC), organophosphates (OP), synthetic pyrethroids (SP) and carbamates. A perusal of the data indicates that organochlorine insecticides have dominated contamination scene in all the four phases of analysis, i.e., 1970-1980, 1981-1987, 1988-1998, 1999-2006 and then onwards. The incidence of contamination has slightly increased in the last one and then onwards half decade. Further, it becomes clear from the data that level of contamination has clearly gone down from 1970 to 2006 which could possibly be attributed to the shift of pesticide use from organochlorine to other groups of insecticides such as OPs, SPs and carbamates. On an average, 55.1 % contamination of seasonal vegetables was recorded up to 1998 with 9.5% samples containing residues above maximum residues limit (MRL).

During 2000-06, the residues of different pesticides exceeded their maximum residue limits in about 5-9% samples (Anonymous, 2001; 02; 03-04). The data given in Table 3 reveal that the major vegetables, like tomato, okra, cabbage, brinjal, capsicum, potato and cauliflower (2930 samples), showed contamination with OC, OP, SP, and carbamate insecticides during 1988-98. Among these, 57% samples were found contaminated with about 11 % samples having residues above maximum residue level (MRL). On comparing our data with other countries, it has been observed that samples containing the residues above MRL values are found maximum in India (25%) followed by USA whole country (4.8%), California (1.6%), and European Union, i.e., 1.4% (Shaw *et al.,* 1999). Probably due to regular surveillance studies and excellent implementation of regulatory measures, these countries have been able to minimize pesticidal contamination with their food commodities. During 2001-04, the situation has improved in India and samples having residues above MRL values have decreased, i.e., 5.3-9.0% samples as compared to 11 % during 1988-98. From 2007-2009, out of 85 analysed samples of vegetables all over India, only 2.3 per cent showed residues above MRL values (Anonymous, 2010)This clearly indicates the adoption of good agricultural practices and judicious use of pesticides. Baig *et al.* (2009) reported that on the whole, 35 samples out of 108 seasonal vegetables were found to be contaminated with selected type of (organophosphate) pesticide-residues. From the results, it was clear that although all selected vegetables were found to be contaminated with the pesticide-residues, yet okra was seen to be affected the most from the organophosphate pesticides. Out of 35 pesticide-contaminated samples, 9 were contained residue more than their MRLs. In all, 320 samples of fruits were collected for pesticides analysis. Organochlorines were detected in all the samples. There were no detectable residues of organophosphates, carbamate, and pyrethroids in the surveyed samples. The total percentage pesticide residues found in pawpaw, tomato, and imported apple samples were in the order of 59%, 27%, and 20%, respectively reported by Bempah and Donkor, 2011.

Table 2 : Residues of insecticides in seasonal vegetables

Location	Year	Vegetables	No. of samples analyzed	No. of samples contaminated	(%)	Incidence Insecticide(s) detected	Residues (ppm)	Reference(s)
Karnataka, Andhra Pradesh, Punjab, Maharashtra, Delhi	1970-80	Seasonal vegetables both leafy and root tubers	2109	1080	51.21	DDT	Tr-169	Kathpal and Kumari 1993; 1998; 2000
						HCH	Tr-60	
						Aldrin	Tr-0.13	
						Dieldrin	0.02-1.40	
						Heptachlor	Tr-2.0	
						Enctosulfan	Tr-1.0	
						Endrin	Tr-2.5	
						Lindane	0.2-7.60	
Maharashtra	1981-87	Seasonal vegetables	583	297	50.11	DDT	Tr-10.20	Khandekar *et al.*, 1982
						HCH	Tr-5.42	
						Aldrin	0.49-0.60	Jadhav; 1986
						Dieldrin	0.31-2.10	Banerji; 1989
						Heptachlor	0.11-0.80	
All Over India	1988-98	Seasonal vegetables	4111	2265	55.11	OC	0.10-0.82	Agnihotri; 1999
						OP	0.04-0.68	
						SP	0.04--0.78	
						Carbamates	0.120-3.5	
						Fungicides	0.02-0.200	
All Over India	2001-04	Seasonal vegetables	2054	1242	60.00	OC		Kathpal and Kumari, 1998
						OP		Anonymous, 2001; 2002; 2004
						SP		
Haryana	2002-04	Seasonal vegetables	224	132	58.92	OC	0.00I-0.910	Kumari *et al.*, 2002; 2003; 2004
						OP	0.00 1-1.284	
						SP	0.002-0.275	
						Carbamates	0.003-0.277	
Total			**9081**	**15016**	**55.23**			

OC: Organochlorines; OP: Organopnosphates; SP: Synthetic pyrethroids
MRL values (PFA, 1954 for vegetables): DDT- 3.5 pm; Lindane- 3.0 ppm; Aldrin and dieldrin- 0.1 ppm;
Heptachlor- 0.05 ppm

Table 3 : Residues of insecticides in some major vegetables all over India (1988-99)

Vegetable	Number of samples analyzed	Number of samples contaminated	Incidence (%)	Above MRL
Tomato	598	271	45.3	5.4
Okra	468	281	60.0	15.8
Cabbage	302	189	62.6	7.0
Brinjal	843	493	58.5	9.8
Capsicum	124	39	31.5	12.0
Potato	219	137	62.6	9.6
Cauliflower	376	249	66.2	16.8
Total	2930	1659	56.62	10.9

Pesticides detected: OC, OP, SP, carbamates and fungicides

Supervised Trials

Supervised trials are those under which complete study of particular pesticide like its persistence, degradation, half- life period and safe waiting period etc. are calculated. These are primarily conducted to generate residue data required for registration of pesticides, setting up of maximum residue limits (MRLs) and finding safe waiting periods for their application. To obtain the necessary data for these purposes, the crops are grown under different agro-climatic conditions and treated with pesticides using *good agricultural practices (GAP)*. Under *good agricultural practices,* the pesticide treatments are in accordance with the registered or approved usage pattern, taking into account the minimum quantities necessary to achieve an adequate control. The crop commodities are then analyzed for the remaining pesticides residues. The data pertaining to half-life and safe waiting periods of commonly used pesticides on different vegetable crops are given in Table 4.

Table 4 : Safe waiting periods and half-life values of various insecticides on different vegetables

Vegetables	Variety	Location	References	Insecticides	Dose (kg/ha or g a.i./ha)	Half -Life Period (days)	Safe Waiting period	MRL (mg/kg)
Cowpea	Variety 1552	Delhi	Kumar and Agnihotri, 1991	Deltamethrin	12.5 25.0 50.0	1.8 1.9 2.4	- - -	
	VS 271	Hisar	Kumari *et al.*, 1996	Endosulfan	350 700	4.1 4.0	3 3	2.0
				Lindane	250 500	4.8 5.1	3 3	
Chickpea	Pusa 256	IARI, New Delhi	Mukherjee *et al.*, 2010	Bifenthrin	25 50	4.30 2.89	1	0.02
Pigeon Pea	Pusa 855				25 50	3.34 3.34	1	
Cowpea		IARI, New Delhi	Dikshit, 2002	Deltamethrin	2 3	12.84 13.20		58.10 60.32
Greengram					2 4	13.12 13.74		58.00 61.26
Lentil					2 3	14.78 15.23		60.20 68.43
Soyabean	PK-327	Delhi	Pal and Handa, 1994	Endosulfan		2.61 (Leaves) 3.78 (Pods)	9.31 (Leaves) 3.10 (Pods)	2

Contd.

Cauliflower	Snowball	Bangalore, Kalyani, Solan, Rahuri and Ludhiana	Agnihotri, 1999	Endosulfan35EC	0.50 1.0	1.0-2.0 2.0-3.0	1-5 3-8	2.0
	Snowball	Bangalore, Kalyani, Solan, Pusa, Bihar and Rahuri		Monocrotophos 36 WSC	0.35 0.70	1.0-6.0 1.0-2.0	1-15 2-16	1.0
	Snowball	Bangalore, Kalyani, Solan, Pusa, Bihar and Rahuri		Fenvalerate 20EC	0.075 0.150	1.0-6.0 2.0-7.0	1-4 14	2.0
	Snowball	Ludhiana		Fenvalerate 20EC	0.050	-	3	2.0
				Cyprmethrin 25EC	0.050	-	3	1.0
				Deltamethrin 2.8EC	0.012	-	1	0.05
	Pusa Drumhead	West Bengal	Pandit *et al.*, 1996	a-cypermethrin 10EC	30 45 60	2.1 2.6 1.9	3.5 4.6 5.2	0.5
	Snowball -16	Hisar, Haryana	Deivendran *et al.*, 2005	Endosulfan 35EC	350	1.81	1	2
				Dichlorvos	115	2.088	1	0.5

Contd.

Cauliflower	NS-60N	IIHR Bangalore	Sharma and Awasthi,2002	λ-cyhalothrin 2.5 EC	15	2.02	4.5	0.2
					30	2.24	5.0	
				λ-cyhalothrin5 EC	15	2.18	4.2	
					30	2.39	5.2	
Cabbage	Pusa Rubey	Jaipur	Agnihotri, 1999	Monocrotophos 36WSC	0.40		15	1.0
					0.80	-	15	
				Dimethoate30EC	0.30	-	6	1.0
					0.60	-	7	
	-	Anand		Endosulfan 35EC	0.35	-	7	2.0
	Tropics	Ludhiana		Fenvalerate 20EC	0.050	-	3	1.0
				Cyprmethrin 25EC	0.050	-	3	0.5
				Deltamethrin 2.8EC	0.012	-	1	0.05
	Capitata.L	Ludhiana	Singh *et al.*, 1992	Cyprmethrin 25EC	50	-		2.0
					100	-	-	
				Fenvalerate 20EC	50	-	-	2.0
					100	-	-	
				Deltamethrin 2.8EC	12	-	-	0.5
					24	-	-	
	Golden acre	Hyderabad	Malathi*et al.*, 1999	Endosulfan	0.07%	3.76-3.96	-	2.0
	Doli-5	Gujrat	Patel *et al.*, 1999	Chlorpyriphos	0.04%	1.04	10.47	0.01
					0.08%	0.33	13.94	
	Capitata.L	Ludhiana	Agnihotri, 1999	Fenvalerate 20EC	0.050	-	3	
				Cyprmethrin 25EC	0.050	-	3	
				Deltamethrin 2.8EC	0.012	-	5	

Contd.

	Rupali	Rahuri		Mancozeb75WP	0.40	-	1	3.0
	Golden acre	Solan		Carbaryl 50WP	0.75 1.50	- -	4 5	10.0
				Endosulfan 35EC	0.50 1.00	- -	6 6	2.0
	-	Hisar	Jaglan *et al.*, 1995	Cypermethrin proprietary	0.01	2.5	6.10	
				Decamethrin proprietary	0.001	2.48	8.89	
				Fenvalerate proprietary	0.01	3.14	8.89	
				Fenpropathrin proprietary	0.01	4.26	6.86	
				Cypermethrin+ Xlene+Triton X-100	0.01	1.97	5.17	
				Decamethrin+ Toluene+ Triton X-100	0.001	2.47	6.66	
				Fenvalerate+ C-IX+ Tween-80	0.01	3.68	10.19	
				Fenpropathrin+ Aromax+ Swascofix DP-50	0.01	3.85	5.98	
Cabbage		PAU, Ludhiana	Urvashi *et al.*, 2012	Indoxacarb (Avant[R] 14.8 EC)	52.2 104.4	2.88 1.92		3.0
	Indian	ARF,	Aktar	Quinalphos 20AF	500	1.27	5.28	0.01

Contd.

	Rare ball	Baruipur, kolkata	*et al.*, 2010		1000	1.38	6.7	
Okra	Pusa Sawani	West Bengal	Biswas *et al.*, 1991	Fenvalerate	75 150	2.59 3.01	- 2.8	2.0
				MonocrotoPhos	350 700	1.10 1.10	4.73 6.26	0.2
	Arka Anamika	Hyderabad	Masood Khan *et al.*, 1999	Cypermethriin	0.012%	2.25	5.91	0.2
	PB-57	Gujrat	Raj *et al.*, 1999	Triazophos	350 700	1.7 1.7	- -	-
	Arka Anamika	Chennai	Iiango and Devraj, 2003	Imidacloprid 70WS	100ml 200ml	2.61 2.95	3.51 5.73	
Okra	Pusa Sawani and Pusa Vadhu	Bangalore, Rahuri , Kalyani, Pusa, Jaipur, Bhubanas-war,	Agnihotri, 1999	Monocrotophos 36 WSC	0.35 0.70	1-2 2-6	4-11 5-12	1.0
	Pusa Sawani and Pusa Vadhu	Bangalore, Rahuri, Kalyani, Pusa, Jaipur, Bhubana-swar and Solan		Endosulfan 35EC	0.50	2-4	2-3	2.0
					1.00	2-4		

Contd.

Arka Anamika	Hyderabad	Masood Khan *et al.*, 1999	Dichlorvos	0.2%	1.86	2.34	0.50
			Dimethoate	0.06%	3.17	2.19	2.0
Pusa Sawani and Pusa Vadhu	Bangalore, Rahuri , Kalyani, Pusa, and Solan	Agnihotri, 1999	Fenvalerate 20EC	0.075 0.15	1-3 2-3	2-3 2-3	2.0
Pusa Sawani	Jaipur		Carbaryl 50WP	1.0 2.0	- -	12 12	10.0
			Phosphomidon 100EC	0.21 0.42	- -	12 12	2.0
Arka Ananiska, Pusa Sawani, Parbani Kranthi, PB-57	Bangalore, Jorhat, Hyderabad, Anand, Pusa Ludhiana and Coimbatore	Agnihotri, 1999	Triazophos 40EC	0.35 0.70	3-5 5-9	- -	0.1
	Hisar, Haryana	Nath *et al.*,2005	Polytrin C 44EC (P40%+C4%)	1L/ha	1.35(P) 4.11 (C)	-	
			Spark36EC (T35%+D1%)	1L/ha	2.55(T) 7.60 (D)	-	
Varsha Uphar	Hisar, Haryana	Deen *et al.*, 2009	Cypermethrin	60	3.3	4.7	0.2
			λ-cyhalothrin	15	5.2	8.6	0.2
			Endosulphan	300	3.8	8.3	2.0

Contd.

Okra	A4	IARI, New Delhi	Gupta *et al.*, 2009	Bifenthrin	25 30	1.32 1.58	1 1	- -
				Fipronil	50 100	0.65 1.12	3 3	- -
				Indoxacarb	70 140	0.58 1.02	1 1	- -
		GAU, Gujarat	Shah *et al.*, 1999	Triazophos(Spark)	0.036	2.12		
				Profenofos (Polytrin c)	0.044	1.35		
				Deltamethrin (Spark)	0.036	5.09		
				Deltamethrin (Decidan)	0.0014	2.83		
				Cypermethrin (Polytrin c)	0.044	3.95		
				Endosulphan (Decidan)	0.0014	1.81		
	Varsha Uphar	HAU, Hisar	Samriti *et al.*, 2011	Chlorpyriphos (Radar 20 EC)	200 400	3.15 3.46		0.2
Brinjal	Manjar Gota, Arka Shirish, Pusa Purple, Mukta -keshi, Punjab	Rahuri, Bangalore, New Delhi, Kalyani, Pusa, Ludhiana	Agnihotri, 1999	Endosulfan 35EC	0.5 1.0	2-4 2-5	0-5 1-7	2.0
				Fenvalerate 20EC	0.75 1.50	1-2 1-3	1-2 1-3	2.0
				Monocrotophos 36WSC	0.35 0.70	1-7 1-7	5-10 6-15	1.0

Contd.

Bahar,							
Doli-5	Guajrat	Raj *et al.*, 1991	Endosulfan	0.07%	1.9	0.3	2.0
Pusa Purple Long	Jorhat	Barooah and Yein, 1996	Quinalphos (Ekalux 20AF)	0.5 1.0	2.2 2.1	4.9 6.1	-
Morbi-4	Gujarat	Raj *et al.*,1999	Triazophos	350	1.4	-	-
				700	2.9	-	
Doli-5	Gujarat	Patel *et al.*, 1999	Chlorpyriphos	0.04%	0.4	0.46	0.2
				0.08%	0.4	1.46	
Pusa Chister	Bhubne -shwar	Agnihotri, 1999	Carbaryl 50WP	0.75 1.50	- -	10 3	10.0
			Malathion 50EC	0.50 1.00	- -	4 4	
Chamblika,	Punjab		Permethrin 50EC	0.05	-	2	1.0
			Cypermethrin 25EC	0.05	-	2	0.5
			Deltamethrin 2.8EC	0.012	-	2	0.5
			Fenvalerate 20EC	0.05	-	2	2.0
	HAU, Hisar	Chauhan and Beena, 2011	Endosulfan (Thiodan 35EC)	4.5 9	3.30 3.41	-	2
Punjab Jamuni Gola	PAU, Ludhiana	Kang *et al.*, 2008	propargite (Omite 57EC)	570 1140	3.07 3.54	1	2
Nishat Hybrid	PAU, Ludhiana	Jyot *et al.*,2005	Ethion (Fosmite 50EC)	375 750	4.92 2.95	4	1
Punjab Jamuni	PAU, Ludhiana	Mandal *et al.*,2010	β-cyfluthrin (purity	18 36	1.74 1.39	-	-

Contd.

	Gola			99.0%)				
				Imidacloprid (purity 98.6%)	42 84	2.31 2.18	-	0.2
	Pant rituraj	Pantnagar	Chandra *et al.*, 2009	Benfuracarb (40% EC)	0.25 0.50	3.54 3.90		
	BR 112	HAU, Hisar	Kaur *et al.*, 2011	Cypermethrin (Cymbush 25 EC)	43.75 87.50	1.16 1.18		0.2
				Decamethrin (Decis 2.8 EC)	11.20 22.40	1.33 1.42		0.05
		Durgapura, Jaipur	Pathan *et al.*,2012	Quinalphos (25 EC)	375 750	2 3	6 9	0.01
		Kalyani, West Bengal	Banerjee *et al.*,2012	β-cyfluthrin (purity 98.6%)	200 400	1.56 2.41		
				imidacloprid (purity 98.8%)	200 400	3.30 2.64		
		AAU, Anand	Chawla *et al.*,2011	Flubendiamde (Fame 480 SC)	90 180	2.68 2.55		
		Pau, Ludhiana	Takkara *et al.*,2012	Flubendiamde (Fame 480 SC)	90 180	0.62 0.54		0.01
Chillies	Sada Bahar, Vaikali Mukta -keshi, Brinjal Dotis, Son-6,	Punjab, Rahuri, Kalyani, Anand, Jabalpur, Vellyani & Jorhat	Agnihotri, 1999	Quinalphos 25EC	500 1000	- -	2-7 2-7	

Contd.

	JC-2							
	BRS-51, Arka Keshar, Morbi, Surya, Gulabi	Punjab, Bangalore, Anand, Vellayani, Hyderabad		Lindane20EC	0.35 0.70	- -	1-5 1-6	3.0
	BRS-51, Arka Keshar, Morbi-4, Surya	Punjab, Bangalore, Anand, Vellayani,		Triazpohos 40EC	0.35 0.70	- -	3-7 3-9	0.1
	Pusa purple round	Assam	Duara *et al.*, 2003	Cypermethrin 10EC	22.5 45 75	3.30 3.39 3.16		
				Fenvalerate20EC	30 60 112.5	2.98 3.26 3.76		
				Endosulfan 35EC	525	5.29		
Chillies	IIHR-NP-46	Bangalore	Awasthi, 1994	Permethrin 25EC	0.015 0.030			1.0
				Cypermethrin 25EC	0.01 0.02			0.5
				Fenvalerate 20EC	0.01 0.02			1.0
				Deltamethrin 2.8EC	0.0015 0.0030			0.2
				Monocrotophos	0.05			1.0

Contd.

			36WSC	0.10			
Agni Rekha, Jwala, Peprika	Rahuri, Bangalore Hyderabad	Agnihotri, 1999	Lindane 20EC	150 300	1-2 1-4	1.0-2.8 2.2-2.5	0.5
Jwala Mukhi	Vellayani		Lindane 20EC	250 500	2 3	1.2 3.3	
Local	Coimbatore		Lindane 20EC	350 700	2 4	5.2 7.0	
Agni Rekha, Jwala, Peprika	Rahuri, Bangalore, Hyderabad		Quinalphos 25EC	150 300	1-10 2-14	1.6-3.3 1.7-3.6	0.25
Arka Jwala	Bangalore	Awasthi, 1995	Fenvalerate 20EC	100	5.80	-	2
			Fenvalerate Dust	100	2.50	-	
Local, Jwala Mukhi	Coimbatore, Vellayani		Quinalphos 25EC	250 500	6-7 9-10	1.9-2.0 2.2-2.6	
Agni Rekha, Local	Rahuri, Coimbatore	Agnihotri, 1999	Dicofol 18EC	250 500	1 1-2	1.8-3.3 2.0-3.1	5.0
			Dicofol 18EC	175	1	8.2	
				300	1	14.6	
Agni Rekha, Local	Rahuri, Coimbatore		Mancozeb 80WP	750 1500	1-5 1-6	3.7-6.9 3.9-6.4	0.5
Jwala	Vellayani		Mancozeb 80WP	1000	2	1.3	

Contd.

	Mukhi				2000	4	2.8	
		PAU, Ludiana	Sharma *et al.*,2007	Spiromesifen 240 SC	96 192	2.2 2.3		
		ANGRAU, Hyderabad, Solan			96 192 96 192	2.4 2.5 2.2 2.5		
Chillies		Kalyani, West Bengal	Papia *et al.*, 2010	Chlorfenapyr	75 100	2.93 2.96		
		B.C.K.V., Mohanpur, Nadia, West Bengal Mahatma Phule Krishi Vidyapeeth Maharashtra	Mukhopadhyay *et al.*, 2011	Difenoconazole	62.5 125	2.15 2.32		
		PAU, Ludhiana	Sahoo *et al.*, 2011	Trifloxystrobin 25%	250 500	1.81 1.58		
				Tebuconazole 50%	250 500	1.37 1.41		
		Jalgoan city, Maharashtra	Waghulde *et al.*, 2011	Chlorpyriphos Endosulphan Dimethoate	32 14 6	3.22 4.99 4.7	15.75 18.08 9.38	
				Malathion	31.68	5.49	12.01	
		Vellayani, Kerala	Varghese *et al.*, 2011	Dimethoate Ethion	300 375	1.94 3.43	13.63 28	
				Oxydemeton-				

Contd.

				methyl	500	4	7	
		PAU, Ludhiana	Pandher *et al*;2012	Deltamethrin 10 EC	17.5 35	0.36-1.99 0.38-2.06		0.05
Tomato	Pusa Rubey	Gujarat	Raj *et al.*, 1991	Endosulfan	0.07	3.6	7.7	0.2
	-	Coimbtore	Jayakumar *et al.*, 1995	Mancozeb	0.25 0.50	3.9(Leaves) 4.2(Leaves)	- -	3.0
						1.1 (Fruit)	1.5	
						2.8(Fruit)	1.5	
	Pusa Rubey	Jaipur	Agnihotri, 1999	Monocrotophos 36WSC	0.40 0.80	- -	15 15	1.0
				Dimethoate 30EC	0.30 0.60	- -	6 7	1.0
	-	Anand		Endosulfan 35EC	0.35	-	7	2.0
	Rupali	Rahuri		Mancozeb 75WP	0.40	-	1	
	Punjab Tropics	Ludhiana		Fenvalerate 20EC Cypermethrin 25Ec Deltamethrin 2.8EC	0.05 0.05 0.40	- - -	3 3 1	1.0 0.5 0.05
	CV Arka Saurab (S-IV)	Bangalore	Hanumantharaju *et al.*, 2002	Metalaxyl	0.2%-0.4 7.82-8.75	3.13-4.14 (F) (Fruits)	15.83-16.83 (F43.08-50.61 (Fruits)	2.0
				Mancozeb	0.2-0.4%	13.56-13.87 (F) 5.23-6.95 (Fruits)	43.08-50.61 (F) 10.04-15.93 (Fruits)	3.0

Contd.

	IIHR – Selection-4	Bangalore	Mohapatra *et al.*, 1998	Carbofuran (Furadan 3G)	1	10	-	
				Metalaxyl (Ridomil MZ 72 WP)	1.5	12	-	
				Carbofuran (Combination)		14	-	
				Metalaxyl (Combination)		12	-	
	-	Solan	Sharma *et al.*, 2003	Propineb	2	2.35-3.27	2.79-3.40	3.0
					4	2.29-2.49	5.07-5.15	
	Pusa early dwarf	IARI, New Delhi	Gupta *et al.*,2010	Cypermethrin (Rocket 44EC)	40	2.0	1	-
					80	3.6	1	-
				Cypermethrin (Action 505EC)	40	2.5	1	-
					80	4.8	1	-
				Pofenophos (Rocket 44EC)	400	2.2	1	-
					800	5.4	1	-
				Chlorpyriphos (Action 505EC)	400	2.9	1	-
					800	3.3	1	-
	HS-102	Hisar	Chauhan *et al.*,2011	λ-cyhalothrin (Karate2.5 EC)	15	2.07	1	-
					30	1.88	1	-
			Chauhan *et al.*,2012	Bifenthrin (Brigade 2.5EC)	25	2.02	1	-
					30	2.32	1	-
French bean	Arkal Komal	Bangalore	Agnihotri, 1999	Monocrotophos 36WSC	0.35	-	10	1.0
					0.70	-	12	
				Quinalphos 25EC	0.35	-	15	0.25

Contd.

	Arkal Komal	Bangalore	Ahuja and Awasthi, 2002	Hexaconazole (Contaf 5EC)	525 1050	2.59 4.21	6.5 13.1	0.05
Pea	Bennevillia	Jaipur	Agnihotri, 1999	Carbaryl 50WP	1.0 2.0	- -	12 15	5.0
				Phosphamidon 10EC	0.21 0.42	- -	12 15	0.2
Bitter Gourd	-	Vellayani	Agnihotri, 1999	Carbofuran 3G	0.50	-	28	0.2
	Var. Preethi	Trivandrum	Naseema Beevi *et al.*, 1996	Mancozeb (Indofil M- 45)	0.2% 0.4%	2.3 2.5	1.6 4.0	3
Mushroom	White button	Solan	Agnihotri, 1999	Carbofuran 3G	0.15			
				Endosulfan 35EC	0.05	-	3	
					0.10	-		
		Bangalore		Carbofuran 3G	20.0			
					40.0			
				Lindane 1.3D	0.10			
					0.20			
		Solan, Bangalore, Hyderabad		DDVP 76EC	0.05 0.10	- -	1	0.5

F: Foliage; - Not mentioned

Reference

Anonymous. 2011a. Indian Horticulture Database (Chief ed. Bijay Kumar). www.nhb.gov.in.

Anonymous. 2011b.www. agricoop.nic.in/hort/hortrevo5.htm

Anonymous.2010. Monitoring of pesticide residues at National Level. Department of Agriculture and Cooperation, Ministry of Agriculture, Govt. of India Division of Agricultural Chemicals, Indian Agricultural Research Institute, New Delhi.

Anonymous. 2001. *Annual Progress Report and Proceedings 2001. AICRP on Pesticide Residues.* Project Coordinating Cell, pp. 1-66.

Anonymous, 2002. *Annual Progress Report and Proceedings AICRP on Pesticide Residues.* Project Coordinating Cell. Division of Agricultural Chemicals, Indian Agricultural Research Institute, New Delhi, pp. 1-69.

Anonymous, 2004. *Annual Progress Report and Proceedings AICRP on Pesticide Residues.* Project Coordinating Cell. Division of Agricultural Chemicals, Indian Agricultural Research Institute, New Delhi, pp. 1-43.

Agnihotri, N.P. 1999. *Pesticide: Safety Evaluation and Monitoring.* All India Coordinated Research Project on Pesticide Residues, Division of Agricultural Chemicals, Indian Agricultural Research Institute, New Delhi pp 67--93.

Ahuja, A.K. and Awasthi, M.D. 2002. Persistence and dissipation of hexaconazole residues in/on French beans. *Pestic.Res.,* **14**(2): 299-301.

Awasthi, M.D. 1994. Studies on dissipation and persistence of pyrethroid residues on chilli fruits for safety constants. *Pestic. Res. J.***6**(1): 80-83.

Awasthi, M.D. 1995. Persistence pattern of fenvalerate residues on chilli fruits following applications of emulsifiable concentrate and dust formulations. *Pestic. Res. J.,* **7**(2): 162-164.

Aktar, W.Md., Sengupta, D. Purkait, S. and Chowdhury, A.2010.Rish assessment and decontamination of quinalphos under different culinary processes in/on cabbage. *Environ. Monitor. Asses.,* **163:**369-377.

Baig, S.A. Skhtera, N.A. Ashlaq, M. and Asi., M.R. 2009. Determination of the organohosphorus pestide in vegetables by high-performance Liquid Chromatography. *American - Eurasian J. Agric. Environ. Sci.,* **6**(5) : 513-519.

Banerjee, T. Banerjee, D. Roy, S. Banerjee, H. and Pal, S.2012. A Comparative study on the persistence of imidacloprid and β-cyfluthrin in vegetables. *Bull. Environ. Contam. Toxicol.,* **89(1):**193-196.

Banerji, S.A. 1989. Pesticide residues in food items available in Bombay market. pp. 304-309. In: *Pesticide Residues in Food* .Consumer Education and Research Centre, Ahmedabad.

Barooah, A.K. and Vein, B.R. 1996. Residues of quinalphos in/on brinjal *(Solanum melongena)* fruits. *Pestic.Res. J.,* **8**(1): 43-48.

Bempah, C.K. and Donkor, A.K.2011. Pesticide residues in fruits at the market level in Accra Metropolis, Ghanna, a preliminary study. *Environ. Monitor. Asses.,* **175**:551-561.

Biswas, S.K. Sukul, P. Chakraborty, A. Bhattacharyya, A. Das, A.K. and Pal S.1991. Studies on residues of fenvalerate and monocrotophos in okra *(Abelmoschus eseulentus* Moench.). *Pestic. Res.J.,* **3** (2): 119-122.

Chandra, Raju , Srivastava, Anjana and Srivastava, C.Prakash.2009.Fate of benfuracarb insecticide in mollisols and brinjal crop. *Bull. Environ. Contam. Toxicol.*, **83(3)**:348-351.

Chawla, Suchi, Patel, A.R. Patel, H.K. Shah, P.G. 2011. Dissipation of flubendiamide in/on brinjal (*Solanum melogena*)fruits.*Environ. Monitor. Asses.*, **183**:1-4.

Chauhan, Reena and Kumari, Beena. 2011. Reduction of endosulfan residues in brinjal fruits during processing. *Sci. Rev. Chem. Commun.*, **1(1)**:42-48.

Chauhan, Reena, Samriti and Kumari, Beena. 2012. Dissipation of λ-cyhalothrin on tomato (*Lycopersicon esculentum* Mill.) and effect of processing on removal of residues .*Bull. Environ. Contam. Toxicol.*, **88**(3):352-357.

Chauhan, Reena, Samriti and Kumari, Beena. 2012. Dissipation and decontamination behavior of bifenthrin residues in tomato. *Bull. Environ. Contam. Toxicol.*, **89**(1):181-186.

Duara, B. Baruah, A.A.L.H. Deka, S.C. and Barman, N. 2003.Residues of cypermethrin and fenvalerate on brinjal. *Pestic. Res.J.*, **15** (1): 43-46.

Deen, M.K. Kumari, Beena and Sharma, S.S. 2009. Dissipation and decontamination of residues of three pesticides in okra fruits. *Pestic. Res. J.*, **21** (1): 80-82.

Deivendran, A. Kumari, Beena and Yadav, G.S. 2005. Dissipation of Endosulfan and dichlorvos residues in/on cauliflower curds. *Environ. Monitor. Asses.*, **116**(1-3): 307-313.

Dikshit, A.K. 2002. Stability of deltamethrin in pulses during storage and the effect of processing. *Pestic. Res. J.*, **14** (1):40-46.

Du-Bois, K.P. and Geiling, E.M.K. 1959. Textbook of Toxicology. Oxford Univ. Press, Oxford.

Gupta, S. Sharma, R.K. Gupta, R.K. Sinha, S.R. Singh, Rai and Gajbhiye, V.T. 2009. Persistence of new insecticides and their efficacy against insect pests of okra. *Bull. Environ. Contam. Toxicol.*, **82:** 243-247.

Gupta, S. Gajbhiye, V.T. Sharma, R.K. and Gupta, R.K. 2010. Dissipation of cypermethrin, chlorpyriphos and profenfos in tomato fruits and soil following application of pre-mix formulations. *Environ. Monit. Assesss.*, **174 (1-4):**337-345.

Gruda, N .2005. Impact of Environmental Factors on Product Quality of Greenhouse Vegetables for Fresh Consumption. *Crit. Rev. Plant Sci.* **24**(3): 227–247.

Hanumantharaju, T.H. Awasthi, M.D. and Siddaramappa, R. 2002. Persistence of fungicides in tomato following application of metalaxyl-mancojeb combined formulation. *Pestic. Res. J.*, **14**(2): 268-275.

Iiango, K. and Devraj, H. 2003. Residue level of imidacloprid in okra. *Pestic. Res. J.*, **15**(1): 47-49.

Jadhav, G.D. 1986. *Final Technical Report.* Department of Environment (19/58/79-Env), Govt. of India, New Delhi. .

Jaglan, R.S. Sircar, Poo and Dureja, P. 1995. Persistence of some synthetic pyrethroids in/on cabbage. *Petsic. Res.J.*, **7**(2): 105-109.

Jayakumar, R. Habeebulla, B. and Regupathy, A. 1995. Evaluation of mancozeb in tomato. *Pes tic. Res. J.*, **7**(1): 87-88.

Jyot, G. Sahoo, S.K. Battu, R.S. Kang, B.K. Singh, Balwinder 2005. Dissipation of ethion on brinjal (*Solanum melongena* L.) under subtropical conditions at Ludhiana, Punjab India. *Bull. Environ. Contam. Toxicol.*, **75(6)**:1094–1097

Kang, B.K. Jyot, Gagan Sharma, R.K. Sahoo, S.K. Battu, R.S. and Singh, Balwinder .2009.Dissipation Kinetics of Propargite in Brinjal Fruits Under Subtropical Conditions of Punjab, India . *Bull. Environ. Contam. Toxicol.*, **82(2)**:248-250.

Khandekar, S.S. Noronha, A.R.C and Banerji, S.A. 1982. Organochlorine pesticide residues in vegetables from Bombay markets: A three-year assessment. *Environ. Pollut.*, **48**: 127-134.

Kathpal, T.S. and Kumari, Beena. 1993. Pesticide contamination of non-fatty foods in developing countries. pp. 169-171. In: *Pesticides: Their Ecological Impact in Developing Countries* (eds. Dhaliwal, G.S. and Singh, R). Commonwealth Publishers, New Delhi.

Kathpal, T.S. and Kumari, Beena. 2000. Pesticide contamination of non-fatty food commodities. pp. 191-216. In:*Pesticides and Environment.* (eds. G.S. Dhaliwal and Balwinder Singh) Published by Commonwealth Publishers.

Kaur, Prabhjot Yadav, G.S. Chauhan, Reena and Kumari, Beena. 2011. Persistence of cypermethrin and decamethrin residues in/on brinjal fruits. *Bull. Environ. Contam. Toxicol.*, **87(6)**: 693-698.

Kumar, Y. and Agnihotri, N.P., 1991. Persistence and translocation of deltamethrin in/ on cowpea. *Pestic. Res.J,* **3**(2): 159-162.

Kumari, Beena, Kumar, R. Malik, M.S. Naresh, J.S. and Kathpal, T.S. 1996. Dissipation of endosulfan and lindane on sunflower seeds and cowpea pods. *Pestic. Res. J.*, **8**(1): 49-35.

Kumari, Beena, Kumar, R. Madan, V.K. Singh, R. Singh, lagdeep and Kathpal, T.S. 2003. Magnitude of pesticidal contamination in winter vegetables from Hisar, Haryana. *Environ. Monito. Assess.*, **87**:311-318.

Kumari, Beena, Madan, V.K. Kumar, R. and Kathpal, T.S. 2000. Monitoring of seasonal vegetables for pesticide residues. *Environ. Monito. Assess.*, **74**: 263-270.

Kumari, Beena, Madan, V.K. Singh, Jagdeep Singh, Shashi and Kathpal, T.S. 2004. Monitoring of pesticidal contamination of farmgate vegetables from Hisar. *Environ. Monito. Assess.*, **90**: 65-77.

Malathi, S. Sriramulu, M and Ramesh Babu, T. 1999. Dissipation of endosultan on cabbage *Brassica oleraeea* Var. Capitata. *Pestic. Res. J.*, **11** (2): 215-217.

Mandal Kousik, Chahil, G.S. Sahoo, S.K. Battu, R.S. and Singh, Balwinder. 2010. Dissipation kinetics of β-cyfluthrin and imidacloprid in brinjal and soil under subtropical conditions of Punjab, India. *Bull. Environ. Contam. Toxicol.*, **84 (2)**:225-229.

Masood Khan, M.A. Jagdishwar, D. and Venkateswara Rao, S. 1999. Dissipation of dichlorvos and dimethoate residues in okra. *Pestic. Res. J.*, **11** (2): 204-206.

Mohapatra, Soudamini, Awasthi, M.D. and Sharma, Debi . 1998. Effect of combination treatment of carbofuran and metalaxyl on their residue persistence in soil and tomato fruit. *Pestic. Res.J.*, **10**(1): 49-53.

Mukherjee, I. Singh, R. and Govil, J.N. 2010. Risk assessment of a synthetic pyrethroid on pulses. *Bull. Environ. Contamn .and Toxicol.*, **84** (3): 294-300.

Mukhopadhyay, S. Das, S. Bhattacharyya, A. and Pal, S. 2011. Dissipation study of difenoconazole in/on chili fruit and soil in India .*Bull. Environ.Contam. Toxicol.*, 10.1007/s00128-011-0275-2.

Naseema Beevi, S. Gokulapalan, P. Regunath, P and Visalakshi, A. 1996. Dissipation of mancozeb residues in bittergourd *(Momoridca charantia* L.) fruits. *Pestie: Res. J.*, **8**(I): 199-201.

Nath, P. Kumari, Beena, Yadav, P.R. and Kathpal, T.S. 2005. Persistence and dissipation of ready mix formulations of insecticides in/on okra fruits. *Environ. Monitor. Assess.*, **107**: 173-179.

Pal, R. and Handa, S.K. 1994. Persistence of endosulfan on soyabean. *Pestic. Res. J.*, **6**(1): 87-91.

Pandit, G.K. Bhattacharya, A. Bose, A.X. Bandopadhyaya, D. Das, A.K. and Adityachaudhary, N. 1996. Persistence of α-cypermethrin in cabbage and monocrotophos in three soils of Bengal. *Pestic. Res.* J., **8**(2): 132--138.

Pandher, S. Sahoo, S.K. Battu, R.S. Singh, B. Saiyad, M.S. Patel, A.R. Shah, P.G. Reddy, C. Narendra, Reddy, J.D. Reddy, N.K. Rao, Ch. Sreenivasan, Banerjee, T, Banerjee D. Hudait, R.K. Banerjee, H. Tripathy, V and Sharma, K.K. 2012. Persistence and Dissipation Kinetics of Deltamethrin on Chili in Different Agro-Climatic Zones of India. . *Bull. Environ.Contam. Toxicol.*, **88**(5):764-768.

Papia, D. Das S.P. Sarkar, P.K. and Bhattacharyya, Anjan. 2010. Degradation dynamics of chlorfenapyr residue in chili, cabbage and soil. *Bull. Environ.Contam. Toxicol.*, 10.1007/s00128-010-9994-z.

Patel, B.A. Shah, P.G. Raj, M.F. Patel, R.K. Patel, I.A. and Talati, J.G . 1999. Chlorpyriphos residues in/on cabbage and brinjal. *Pestic. Res. J.*, **11** (2): 194- 196.

Pathan, A.R.K. Parihar, N.S. Sharma, B.N. 2012. Dissipation Study of Quinalphos (25 EC) in/on Brinjal and Soil. *Bull. Environ. Contam. Toxicol.*, **88(6)**:894-896.

Raj, M.F. Patel, B.K. Shah, P.G. and Barevadia, T.N. 1999. Pendimethlin, fluchloralin and oxadiazon residue inion onion. *Pestie. Res. J.*, **11**(1): 68-70.

Raj, M.F. Shah, P.G. Patel, B.K. and Patel, J.R. 1991. Endosulfan residues in/ion tomato and brinjal fruits. *Pestie. Res. J.*, **3**(2): 135-138.

Sajjid, A.B. Akhtera, A.N. Ashfaq, M and Asi, M.R. 2009. Determination of the organophosphorus pesticide in vegetables by high performance liquid chromatography. *American-Eurasian J. Agric. Environ. Sci.*, **6**(5): 513-519.

Sahoo, S.K. Jyot, G. Battu, R.S. and Singh, Balwinder. 2011. Dissipation kinetics of trifloxystrobin and tebuconazole. *Bull. Environ.Contam. Toxicol.*, **88**(3):368-371.

Samriti, Chauhan, Reena and Kumari, Beena, 2011. Persistence and effect of processing on reduction of chlorpyriphos residues in okra fruits. *Bull. Environ. Contam. Toxicol.*, **87(2)**: 198–201.

ScienceDirect - Food Chemistry : Balance between nutrients and anti-nutrients in nine Italian potato cultivars.

Sharma, D. and Awasthi, M.D. 2002. Persistance of lambda cyhalothrin residues in cauliflower manuscript in *Indian Ins. Horti. Res.*, 195-198.

Sharma, I.D. Patyal, S.K. Dubey, J.K. and Nath, A. 2003. Persistence of propineb (fungicide) in apple and tomato. *Pestic. Res. J.*, **15**(1): 60-63.

Shah, B.H. Shah, P.G. Jhala, R.C. and Vyas, H.N. 1999. Studies on dissipation of some ready-mix insecticide combinations in/on okra fruits. *Pestology.*, **23** (6): 3–9.

Sharma, K.K. Cherukuri, S.R. Dubey, J.K. Patyal, S.K. Parihar, N.S. Battu, R.S. Sharma V. Gupta, P. Kumar, A. Kalpana, Jaya, M. Singh, B. Sharma, I.D Nath, A. Gour T.B. 2007. Persistence and dissipation kinetics of spiromesifen. *Environ. Monitor. Asses.*, **132:**25-31.

Shaw, I. 1999. Pesticides in food. Pesticide Chemistry and Bioscience. pp. 421-428. In: *The Food Environ. Chall.* (eds. Brooks, G.T. and Roberts, T.R.). Royal Soc. Of Chem., London,

Singh, B. Singh, P.P. Battu, R.S. and Kalra, R.L. 1992. Residues of some synthetic pyrethroid insecticides on cabbage. *Pestic. Res.J.,* **4**(2): 137-141.

Steinmetz, K.A. and Potter, J.D. 1996. Vegetables, fruit and cancer prevention: a review. *J. Am. Diet Assoc.,* **96** (10): 1027–39. DOI:10.1016/S0002-8223(96)00273-8. PMID 8841165.

Takkar, R. Sahoo, S.K. Singh, G. Battu, R.S. Singh, B. 2012. Dissipation pattern of flubendiamide in/on brinjal (Solanum melogena) fruits. *Environ. Monitor. Asses.,* **184(8):**5077-5083.

Urvashi, Jyot Gagan, Sahoo, S.K. Kaur, Sarabjit, Battu, R.S. and Singh Balwinder. 2012. Estimation of indoxacarb residues by QuEChERS technique and its degradation pattern in cabbage. *Bull. Environ. Contam. Toxicol.,* **88 (3):**372-376.

Varghese, S.T. Mathew, B.T. George, T. Beevi, N.S. and Xavier, G.2011.Dissipation study of dimethote, ethion and oxydemeton methyl in chilli. *Pestic Res. J.,* **23**(1):68-73.

Waghulde, P.N. Khatik, M.K. Patil, V.T. and Patil, P.R. 2011. Persistence and dissipation of pesticides in chilly and okra at north maharashtra region. *Pestic Res. J.,* **23**(1):23-26.

Whitaker Julian, M. 2001. *Reversing Diabetes.* New York: Warner Books. ISBN 0-446-67658-6. OCLC 45058465.

Woodruff Sandra, L. 1995. *Secrets of Fat-Free Cooking : Over 150 Fat-Free and Low-Fat Recipes from Breakfast to Dinner-Appetizers to Desserts.* Garden City Park, N.Y: Avery Publishing Group. ISBN 0-89529-668-3. OCLC 33142807.

Part - IV

Chapter-14

Preparation of Standard Solutions

Vinod Kumar Unvi
Central Laboratory
Chaudhary Charan Singh Harayana Agricultural University, Hisar - 125004

Need of Analysis Technique

There are various methods to determine the concentration of a substance present in the matrix of an organic matter, these methods are very expensive, require expertise, need skilled man power, expensive chemicals and sophisticated equipments. But the volumetric analysis requires very less equipment and its level of accuracy is fairly high. It is very beneficial in the preparation of normal and molar solutions which is the back bone of all the research. Scientific venture is useless without the preparation of accurate standard solutions which is to be used in the estimation of various substances by sophisticated techniques like colorimetry, spectrophotometry, atomic absorption spectrophotometer (AAS), inductively coupled plasma emission spectrophotometer (ICP) *etc.* So the emphasis has been laid on the preparation of quick and accurate preparation of the standards. Volumetric analysis is the simplest technique by which, we can determine the concentration of a substance in solution form by titration of unknown solution against a known concentration solution. One solution is called the standard (known concentration) and other is called unknown solution. In this, reaction takes place either by

1) As precipitation (ppt)
2) Change in colour

3) Evolution of gas

Normal Solution

Gram equivalent weight of solute dissolved in one litre of solution is called normal solution e.g. 40 gram of NaOH is present in 1 litre of water.

Sub Normal Solution

A solution containing a fraction of gram equivalent weight of the solute dissolved per litre e.g. A solution of NaOH containing 20 g of NaOH dissolved per litre is a sub normal solution(½ of g eq. wt.) .It is written as N /2 or 0.05 N solution.

Normality

Strength of a solution measured in gram equivalents per litre is called normality.

Mathematically

$$N = \frac{\text{Strength of g/ litre}}{\text{Equivalent weight}}$$

Normality Equation

It is the expression of the result of a titration in the form of an equation.

Suppose 8 ml of NaOH neutralises 10 ml of N/10 HCl

Then the normality equation for the data is

8 mol of 1N NaOH = 10ml of HCl

The sign = indicates "equivalent to" or "completely react with."

Normality Formula

Substances generally react in the ratio of equivalents (law of equivalents). Therefore equal volumes of equinormal solution exactly neutralises each other .It is represented by

$$N_1 V_1 = N_2 V_2$$

Here N_1 and V_1 is the normality and volume of one solution and $N_2 V_2$ is the normality and volume of the other solution.

Equivalent Weight (eq. wt.) of an Element

The number of parts by weight of that **element or compound** which combines with or displaces directly or indirectly1.008 parts by weight of hydrogen or 8 parts by weight of oxygen or 35.45 parts by weight of chlorine.

$$\text{Eq.wt.} = \frac{\text{Atomic weight of acid}}{\text{Valency}}$$

Equivalent Weight of a Compound

It may be defined as the number of parts by weight of that compound which combines with or displaces directly or indirectly 1.008 parts by weight of hydrogen or 8 parts by wt. of oxygen or 35.5 parts by weight of chlorine.

Equivalent Weight of an Acid

Equivalent weight of an acid is the number of parts by weight of it that contains 1.008 parts by weight of replaceable hydrogen.

Alternatively, it is the quantity by weight that supplies one mole of H^+.

Mathematically

$$\text{Eq. wt. of acid} = \frac{\text{Molecular weight of the acic}}{\text{No. of replaceable H atom}}$$

$$\frac{\text{Molecular weight of the acid}}{\text{Basicity of the acid}}$$

Basicity

Basicity of an acid is the number of replaceable hydrogen atoms present in molecule of the acid.

$$\text{Thus eq.wt. of HCl} = \frac{36.45}{1} = 36.45$$

$$\text{Eq.wt. of } H_2SO_4 = \frac{98}{2} = 49$$

Equivalent Weight of a Base

Equivalent weight of a base is the parts by weight of it that completely neutralises one gram equivalent of an acid.

Alternatively, it is the quantity by weight that reacts with one mole of H^+

Mathematically

$$\text{Eq. wt.} = \frac{\text{Molecular weight of the base}}{\text{No. of replaceable OH ions present}}$$

$$\text{Eq. wt.} = \frac{\text{Molecular weight of the base}}{\text{Acidity of the base}}$$

Acidity of an alkali is the number of replaceable OH ions present in one molecule of the base

Thus

Eq .wt of NaOH = 40 /1 = 40

Eq .wt of Ba $(OH)_2$ = 315 / 2 =157.5

Equivalent. wt. of Na_2CO_3

$$= \frac{\text{Molecular weight of the salt}}{\text{Total No. of positive or negative valences the radical}}$$

= 106/2 =53

Equivalent Weight of Oxidising and Reducing Agents

Equivalent weight of an oxidising agent is that weight of the compound which can supply 8 parts by weight of oxygen or can react with 1.008 parts by weight of hydrogen or can be oxidised by 8 parts by weight of oxygen.

Thus Eq . wt. of oxidising agent

$$= \frac{\text{Molecular weight of the compound}}{2 \times \text{No. of oxygen atoms used by the molecule}}$$

Thus Eq . wt. of a reducing agent

$$= \frac{\text{Molecular weight of the compound}}{2 \times \text{No. of oxygen atoms used by the molecule}}$$

Eq. wt.of $KMnO_4$ in acidic medium

$2KMnO_4 + 3H_2SO_4 \rightarrow 2(39 +55+64) =316$

$K_2SO_4 + 2\ MnSO_4 + 3\ H_2O + O \rightarrow 5\times 16 = 80$

$$\text{Eq. wt. of } KMnO_4 \text{ in acidic medium} = \frac{316 \times 8}{80} = 31.6$$

Eq. wt. of $KMnO_4$ in alkaline medium

$$2KMnO_4 + 2KOH \rightarrow 2\,K_2MnO_4 + H_2O + O$$

$$\text{Eq. wt.} = 316 \times 8 = \frac{158}{16}$$

Eq. wt. of $KMnO_4$ in neutral medium

$$2KMnO_4 + H_2O \rightarrow 2KOH + 2\,MnO_2 + 3\,O$$

$$\text{Eq. wt.} = \frac{316 \times 8}{48} = 52.67$$

Calculation for 1 Normal Solution of Acid

N_1V_1 = N_2V_2

Unknown Known

N_1 = Normality of the acid (written on the bottle)

N_2 = Normality of the acid to be prepared i.e. 1N

V_1 = Volume of the solution required.

V_2 = Volume of the solution to be prepared i.e. 1000 ml

For example

1N HCl

$$N_1V_1 = N_2V_2$$

$$V_1 = \frac{1 \times 1000}{12.1} = 82.645 \text{ ml}$$

i.e. 82.645 ml HCl is to be dissolved in 1 litre distilled water for the preparation of 1N HCl.

For I M H_2SO_4 Acid

$$N_1V_1 = N_2V_2$$

$$V_1 = \frac{1 \times 1000}{\text{Molarity}} = \frac{1 \times 1000}{18}$$

= 55.5 ml H_2SO_4 is to be dissolved in 1 litre distilled water for the preparation of 1M H_2SO_4

Brief Formula

For I M Acid

$$V_1 \text{ (ml Acid)} = \frac{1 \times 1000}{\text{Molarity}}$$

For I N Acid

$$V_1 \text{ (ml Acid)} = \frac{1 \times 1000}{\text{Normality}}$$

Equivalent weight of an acid

$$\text{Thus eq. wt. of HCl} = \frac{36.45}{1} = 36.45$$

$$\text{Eq. wt. of } H_2SO_4 = \frac{98}{2} = 49$$

*** Normality and molarity are generally written on the bottle**

Calculation for 1 Normal Solution of Alkali

Calculation for molar solution of an alkali i.e. sodium hydroxide (NaOH)

Molecular weight of an sodium hydroxide = 23+1+16 = 40

Dissolve 40 gram of sodium hydroxide in 1 litre distilled water

Calculation for 1 normal solution of sodium hydroxide (NaOH)

Again dissolve 40 gram of sodium hydroxide as molecular and equivalent weight of sodium hydroxide is the same.

Calculation for preparation of 1000 ppm solution

Steps involved in preparation of 1000 ppm solution of NaCl and KCl

NaCl

Mol.wt.	=	58.5		
Na	=	23		
Cl	=	35.5		
	=	23+35.5	=	58.5

23 mg is present in = 58.5mg

$$1 = \frac{58.5}{23}$$

1000 mg will present

$$= \text{Eq. wt. of } H_2SO_4 = \frac{58.5 \times 1000}{23} = 2543.9 \text{ mg}$$

Hence dissolve 2543.9 mg in one litre distilled water or 2.543 g / Litre

KCl

Molecular weight	=	74.5 mg
K	=	39
Cl	=	35.5
	=	39+35.5 = 74.5
39 mg is present in	=	74.5mg
1	=	$\frac{74.5}{39}$
1000 mg will present	=	$\frac{74.5 \times 1000}{39} = 1910.2 \text{ mg}$

or 1.9102 g of salt in one litre distilled water. Per cent purity assay should be extra i.e. 1.909g

Steps for the Dilution of 1000 ppm Stock Solution

Prepare **1000 ppm** solution and treat it as stock solution. Further dilution sequence is as follows:

Steps

Prepare 1000ppm solution

↓ Pipette out 10 ml and make volume 100 ml

Concentration of the solution will be 100ppm

↓ Now pipette out

0.1ml → Make the volume to100ml → Concentration will be 0.1ppm
1ml → Make the volume to100ml → Concentration will be 1ppm
2ml → Make the volume to100ml → Concentration. will be 2ppm
3ml → Make the volume to100ml → Concentration will be 3ppm
4ml → Make the volume to100ml → Concentration will be 4ppm
5ml → Make the volume to100ml → Concentration will be 5ppm
10ml → Make the volume to100ml → Concentration will be 10ppm

Steps for the Dilution of I Molar Stock Solution

Prepare **I Molar** standard solution after seeing the amount required from the Table1 for the acids and Table 2 for the bases or by calculating as is specified above. Further dilution sequence is as follows:

Steps

Prepare 1M solution

↓ Pipette out 10 ml and make volume 100 ml

Concentration of the solution will be 0.1M or M/10

↓ Now pipette out 10ml and make the volume100ml

Concentration will be 0.01M or M/ 100

↓ Now pipette out 10ml and make the volume 100ml

Concentration will be 0.001M or M/ 1000

Steps for the Dilution of IN Stock Solution

Preparation of 1N Stock Standard Solution

Prepare **I Normal** standard solution with the help of Table 1for the acids and Table 2 for the bases or by calculating as is specified above. Further dilution sequence is as follows

Prepare 1N solution

↓ Pipette out 10 ml and make volume 100 ml

Concentration of the solution will be 0.1N or N/10

↓ Now pipette out 10ml and make the volume100ml

Concentration will be 0.01N or N/ 100

↓ Now pipette out 10ml and make the volume 100ml

Concentration will be 0.001N or N/ 1000

Table 1: Composition of concentrated reagent grade acids (ml required to prepare 1normal, molar and equivalent solution are specified)

Chemical	Formula weight	Strength of concentrated reagent	Assay limits (% w/w)	Molarity of concentrated reagent	ml of acid required to prepare1L of I Molar solution	Normality of concentrated reagent	ml of acid required to prepare1L of I Normal solution
Acetic Acid glacial (CH_3COOH)	60.053	99.8	99.7-99.9	17.4	57.5	17.4	57.5
Hydrochloric acid (HCl)	36.461	37.2	36.5-38.0	12.1	82.5	12.1	82.5
Hydrfluoric acid (HF)	20.006	49.0	48.0-51.0	28.9	34.5	28.9	34.5
Nitric acid (HNO_3)	63.013	70.4	69.0-71.0	15.9	63.0	15.9	63.0
Phosphoric acid (H_3PO_4)	97.995	85.5	85.0-87.0	14.8	67.5	44.4	22.5
Sulfuric acid (H_2SO_4)	98.079	96.0	95.0-98.0	18.0	55.5	36.0	28.0

Table 2 : Weight in grams required to prepare normal and molar solutions of various bases and salts

Sr No	Name	Formula	Molecular Weight	Normal concentration	Molar concentration
1.	Ammonium acetate	CH_3COONH_4	77.08	77.08	77.08
2.	Ammonium borate	$NH_4HB_4O_7.3H_2O$	228.33	228.33	228.33
3.	Ammonium chloride	NH_4Cl	53.49	53.49	53.49
4.	Ammonium citrate	$(NH_4)_3C_6H_5O_7$	243.22	81.07	243.22
5.	Ammonium fluoride	NH_4F	37.04	37.04	37.04
6	Ammonium hydrogen phosphate	$(NH_4)_2HPO_4$	132.06	132.06	132.06
7.	Ammonium hydroxide	NH_4OH	35.05	35.05	35.05
8.	Ammonium lactate	$CH_3CHOH.COONH_4$	107.11	107.11	107.11
9.	Ammonium per sulphate	$(NH_4)_2 S_2O_8$	228.20	114.10	228.20
10	Ammonium hydrogen orthophosphate	$(NH_4)_2HPO_4$	132.06	132.06	132.06
11.	Ammonium phosphate monbasic	$NH_4H_2PO_4$	115.03	115.03	115.03
12.	Ammonium phosphate dibasic	$NH_4H_2PO_4$	115.03	115.03	115.03
13.	Ammonium dihydrogen ortho phosphate	$NH_4H_2PO_4$	115.03	115.03	115.03
14.	Ammonium thiocyanate	NH_2SCN	76.12	76.12	76.12
15.	Barium chloride	$BaCl_2.2H_2O$	244.31	122.16	244.31
16	Beryllium acetate	$(CH_3COO)_2$ Be	127.10	63.55	127.10
17.	Bis-(2-ethyl hexyl) hydrogen phosphate	$(C_6H_{17})_2HPO_4$	322.43	322.43	322.43
18.	Calcium chloride Fused	$CaCl_2$	110.99	55.50	110.99
19.	Calcium chloride dihydrate	$CaCl_{2..}2H_2O$	147.02	73.51	147.02
20.	Calcium hydroxide	$Ca(OH)_2$	74.09	37.05	74.09
21.	Cobalt sulphate	$CO SO_4$	155.00	77.50	155.00
22.	Copper sulphate	$CO SO_4. 7H_2O$	281.10	140.55	281.10
23.	Copper sulphate	$Cu SO_4.5 H_2O$	249.68	124.84	249.68
24.	E.D.T.A.(Ehytlene diaminetetra acetic acid)	$CH_2.N(CH_2COOH)_2$ or $C_{10}H_{16}N_2O_8$	292.25	73.06	292.25
25.	Hydrazine	$N_2H_4 H_2O$	50.06	25.03	50.06
26.	Iodine	I_2	253.81	126.90	253.81
27.	Iodine chloride	ICl	162.36	162.36	162.36
28.	Lead nitrate	$Pb (NO_3)_2$	331.20	165.20	331.20
29.	Magnesium chloride	$Mg Cl_2. 6 H_2O$	203.31	101.66	203.31
30.	Nitric acid	HNO_3	63.01	63.01	63.01

Contd.

31.	N-phenyl benzohydroxamic acid	C_6H_5 CON (OH)C_6H_5	213.24	213.24	213.24
32.	Potassium bromate	$KBrO_3$	167.01	74.56	167.01
33.	Potassium chloride	KCl	74.56	74.56	74.56
34.	Potassium dichromate	$K_2Cr_2O_7$	294.19	49.03	294.19
35.	Potassium hydrogen phthalate	C_6H_4 COOH. COOK	204.22	102.11	204.22
36.	Potassium hydroxide	KOH	56.11	56.11	56.11
37.	Potassium iodate	KIO_3	214.00	214.00	214.00
38.	Potassium iodide	KI	166.01	166.01	166.01
39.	Potassium 2-naphthal -3-6 disulphonic acid	$C_{10}H_5OH(SO_3K)_2$	380.49	190.25	380.49
40.	Potassium nitrate	KNO_3	101.11	101.11	101.11
41.	Potassium permanganate solution	$KMnO_4$	158.04	-	158.04
42.	Potassium sulphate	K_2SO_4	174.27	87.14	174.27
43.	Sodium acetate.	CH_3COONa	82.03	82.03	82.03
44.	Sodium carbonate	Na_2CO_3	105.99	53.00	105.99
45.	Sodium chloride	NaCl	58.44	58.44	58.44
46.	Sodium citrate	$Na_3C_6H_5O_7.2H_2O$	294.10	98.03	294.10
47.	Sodium hydroxide	NaOH	40.00	40.00	40.00
48.	Sodium 2-naphthol-3: 6 -disulphonic acid	$C_{10}H_5OH(SO_3Na)_2$	348.27	174.14	348.27
49.	Sodium nitrate	$NaNO_3$	85.00	85.00	85.00
50.	Sodium nitrite	$NaNO_2$	69.00	69.00	69.00
51.	Sodium thiosulphate	$Na_2S_2O_3.5H_2O$	248.18	248.18	248.18
52.	Tin	Sn_4	474.76	118.69	474.76
53.	Titanous sulphate	$Ti_2(SO_4)_3$	383.82	191.91	383.82
54.	Tributyl phosphate	$(C_4H_9)_3PO_4$	266.32	266.32	266.32

Table 3: Preferential choice of indicators generally used in acid base titrations

Indicator alkaline solution	pH range	Colour in acid solution	Alkaline solution	Used in titration of a solution of
Methyl orange		Pink	Yellow	Strong acid against a strong or weak base
Methyl red	4.4-6.3	Red	Yellow	Strong acid against a weak base
Litmus	6.0-8.0	Red	Blue	Strong acid against a strong base
Phenolphthalein	8.2-10.00	Colourless	Pink	Strong or weak acid against a strong base

Table 4 : Elements and their atomic number

Serial No	Element	Symbol	Atomic Number	Atomic Weight	Oxidation Number
1	Hydrogen	H	1	1	+1
2	Helium	He	2	4	0
3	Lithium	Li	3	6.94	+1
4	Beryllium	Be	4	9.01	+2
5	Boron	B	5	10.81	+3
6	Carbon	C	6	12	+2, ,±4
7	Nitrogen	N	7	14	±1, ±2, ±3, +4, +5
8	Oxygen	O	8	16	-2
9	Fluorine	F	9	19	-1
10	Neon	Ne	10	20.17	0
11	Sodium	Na	11	23	+1
12	Magnesium	Mg	12	24	+ 2
13	Aluminium	Al	13	27	+3
14	Silicon	Si	14	28	+2, ±4
15	Phosphorous	P	15	31	±3,+5
16	Sulphur	S	16	32	+4,+6,-2
17	Chlorine	Cl	17	35	±1, +5,+7
18	Argon	Ar	18	40	0
19	Potassium	K	19	39.09	+1
20	Calcium	Ca	20	40.078	+2
21	Scandium	Sc	21	44.95.	+3
22	Titanium	Ti	22	47.86	+2,+3,+4
23	Vanadium	V	23	50.94	+2,+3,+4,+5
24	Chromium	Cr	24	51.99	+2,+3,+6
25	Manganese	Mn	25	54.93	+2,+3,+4,+7
26	Iron	Fe	26	56	+2,+3
27	Cobalt	Co	27	58.93	+2,+3
28	Nickel	Ni	28	58.69	+2,+3
29	Copper	Cu	29	63.54	+1,+2
30	Zinc	Zn	30	65.39g	+2
31	Gallium	Ga	31	69.72	+3
32	Germanium	Ge	32	72.61	+2,+4
33	Arsenic	As	33	74.92	±3,+5
34	Selenium	Se	34	78.96	+4,+6,-2

Contd.

35	Bromine	Br	35	79.90	±1,+5
36	Krypton	Kr	36	83.80	0
37	Rubidium	Rb	37	85.46	+1
38	Strontium	Sr	38	87.62	+2
39	Yttrium	Y	39	88.90	+3
40	Zirconium	Zr	40	91.22	+4
41	Niobium	Nb	41	92.90	+3,+5
42	Molybdenum	Mo	42	95.94	+6
43	Technetium	Tc	43	97.90	+4,+6,+7
44	Ruthenium	Ru	44	101.07	+3
45	Rhodium	Rh	45	102.90	+3
46	Palladium	Pd	46	106.42	+2,+4
47	Silver	Ag	47	107.86	+2
48	Cadmium	Cd	48	112.41	+2
49	Indium	In	49	114.81	+3
50	Tin	Sn	50	118.71	+2,+4,-2
51	Antimony	Sb	51	121.76	±3,+5
52	Tellurium	Te	52	127.6	+4,+6,-2
53	Iodine	I	53	126.90	±1,+5,+7,
54	Xenon	Xe	54	131.29	0
55	Cesium	Cs	55	132.90	+1
56	Barium	Ba	56	137.32	+2
57	Lanthanum	La	57	138.90	+3
58	Cerium	Ce	58	140.11	+3,+4
59	Praseodymium	Pr	59	140.90	+3
60	Noodymium	Nd	60	144.24	+3
61	Promethium	Pm	61	144.91	+3
62	Samarium	Sm	62	150.36	+2,+3
63	Europium	Eu	63	151.96	+2,+3
64	Gadolinium	Gd	64	157.25	+3
65	Terbium	Tb	65	158.92	+3
66	Dysprosium	Dy	66	162.50	+3
67	Holomium	Ho	67	164.93	+3
68	Erbium	Er	68	167.26	+3
69	Thulium	Tm	69	168.93	+3
70	Ytterbium	Yb	70	173.04	+2, +3
71	Lutetium	Lu	71	174.96	+3

Contd.

72	Hafnium	Hf	72	178.49	+4
73	Tantalum	Ta	73	180.94	+5
74	Tungsten	W	74	183.84	+6
75	Rhenium	Re	75	186.20	+4,+6,+7
76	Osmium	Os	76	190.2	+3,+4
77	Iridium	Ir	77	192.217	+3,+4
78	Platinum	Pt	78	195.08	+2,+4
79	Gold	Au	79	176.96	+1,+3
80	Mercury	Hg	80	200.59	+1,+2
81	Thallium	TI	81	201.38	+1,+3
82	Lead	Pb	82	207.2	+2,+4
83	Bismuth	Bi	83	208.98	±3,+5
84	Polonium	Po	84	208.98	+2,+4
85	Astatine	At	85	209.98	±1,+5,+7
86	Radon	Rn	86	222	0
87	Francium	Fr	87	223	+1
88	Radium	Ra	88	226.02	+2
89	Actinium	Ac	89	227.02	+3
90	Thorium	Th	90	232.03	+4
91	Protactinium	Pa	91	231.03	+4,+5
92	Uranium	U	92	238.02	+3,+4,+5,+6
93	Neptunium	Np	93	237.04	+3,+4,+5,+6
94	Plutonium	Pu	94	244.06	+3,+4,+5,+6
95	Americium	Am	95	243.06	+3,+4,+5,+6
96	Curium	Cm	96	247.07	+3
97	Berkelium	Bk	97	247.07	+3,+4
98	Californium	Cf	98	251.07	+3
99	Einsteinium	Es	99	252.08	+3
100	Fermium	Fm	100	257.09	+3
101	Mendelevium	Md	101	258.10	+2,+3
102	Nobelium	No	102	259.10	+2,+3
103	Lawrencium	Lr	103	262.11	+3
104	Rutherfordium	Rf	104	261.11	+4
105	Dubnium	Db	105	262.11	
106	Seaborgium	Sg	106	263.811	
107	Bohrium	Bh	107	262.12	
108	Hassium	Hs	108		

109	Meitnerium	Mt	109	266
110	Ununilium	Uun	110	269
111	Unununium	Uuu	111	272
112	Ununbium	Uub	112	277
113	Uut		113	
114	Ununquadium	Uuq	114	285
115		Uup	115	
116	Ununhexium	Uuh	116	289
117	Uus		117	
118	Ununoctium	Uuo	118	293

Chapter-15

Gas Liquid Chromatography : An Analytical tool for Pesticide Residue Estimation of Horticultural Crops

Beena Kumari
Department of Entomology, CCSHAU, Hisar

Chromatography was first employed by Ramsay (1905) to separate mixtures of gases and vapours. The first scientist to recognize chromatography as an efficient method of separation was the Russian botanist Tswett (1906), who demonstrated that, when a plant extract was carried by petroleum ether through a column consisting of a glass tube packed with calcium carbonate powder, a number of dyes were separated, as shown in Figure 1. He named this analysis method "Chromatographie" after "chroma" and "graphos", which are Greek words meaning "color" and "to draw," respectively.

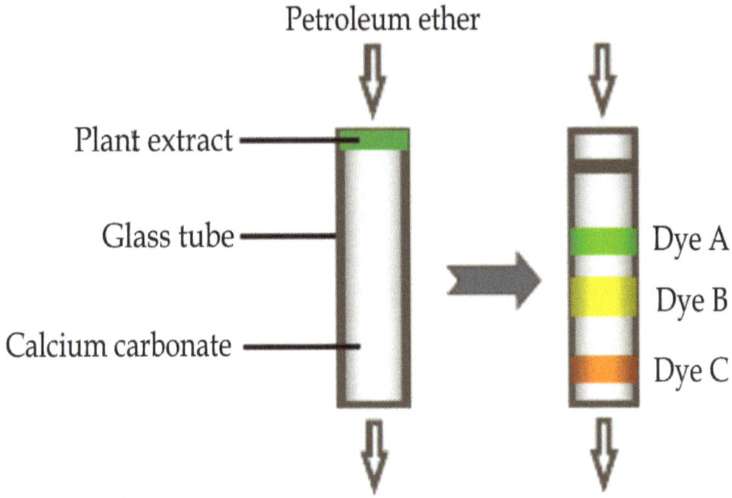

Figure 1. Diagram showing Tswett's experiment

Principle of Chromatography

Chromatography is a technique by which a mixture sample is separated into components. Although originally intended to separate and recover (isolate and purify) the components of a sample, today, complete chromatography systems are often used to both separate and quantify sample components.

Although color has little to do with modern chromatography, the name has persisted and, despite its irrelevance, is still used for all separation techniques that employs the essential requisites for a chromatographic separation, viz. a mobile phase and a stationary phase.

Gas chromatography (GC), is a common type of chromatography used in analytical chemistry for separating and analyzing compounds that can be vaporized without decomposition. Typical uses of GC include testing the purity of a particular substance, or separating the different components of a mixture (the relative amounts of such components can also be determined). In some situations, GC may help in identifying a compound. In preparative chromatography, GC can be used to prepare pure compounds from a mixture (Pavia *et al.,* 2006).

GC was invented by James and Martin (1952) and is a separation technique in which the mobile phase is a gas (usually helium or nitrogen) and the stationary phase is a liquid. In the original columns used by James and Martin, the liquid stationary phase was adsorbed on the surface of an inert support such as Celite (a diatomateous earth) or calcined Celite (a form of brick dust). The support was usually deactivated before use by acid treatment and subsequent reaction with hexamethyldisilazane. The technique was extensively used for the separation of a wide range of volatile substances. Thus, chromatography has been defined as a separation process that is achieved by distributing the components of a mixture between two phases, a stationary phase and a mobile phase. Those components held preferentially in the stationary phase are retained longer in the system than those that are distributed selectively in the mobile phase. As a consequence, solutes are eluted from the system as local concentrations in the mobile phase in the order of their increasing distribution coefficients with respect to the stationary phase.By separating the sample into individual components, it is easier to identify (qualitate) and measure the amount (quantitate) of the various sample components.

The basis for gas chromatographic separation is the distribution of a sample between two phases. One of these phases is a stationary bed of large surface area, and the other phase is a gas which percolates through the stationary bed. There are numerous chromatographic techniques and corresponding instruments. GC is one of these techniques. It is technique for separating volatile substances by percolating over a stationary phase. Chromatography is probably the most powerful and an versatile technique available to the chemist. It is estimated that GC can analyse 10 - 20% of the known compounds, and of the total pesticides used today, about 80% can be analysed using GC. An analyst can separate gases, and volatile substances by gas chromatography (GC), involatile / nonvolatile chemicals and materials of extremely high molecular weight (including biopolymers) by liquid chromatography (LC) and if necessary very inexpensively by thin layer chromatography (TLC). All the three techniques, GC, LC and TLC have common features that classify them as chromatography systems. Several references are useful for understanding the gas liquid chromatogrphy: McNair and Bonelli(1969), Agnihotri (1980),Ravinderanath (1989), Sant (1997), Grob (1995), Dean (1998), Handa *et al.,* (1999), Sharma (2007). Significant technological advances in the area of electronics, computerization and columns have yielded lower and lower detectable limits and more accurate identification of substances through improved resolution because of capillary columns and qualitative analysis techniques.

1. WHAT IT DOES?

Gas chromatographic analysis provides the chemist with a chromatogram and report of the composition of a sample containing a mixture of components. Generally, GC systems are configured and set up for analyzing primarily routine, quality control type samples or for primarily non routine research and development (R&D) analyses. GC systems used for routine analyses can be set up with the latest automation enhancements so that minimum operator assistance is required. One of the major reasons for the widespread use of gas chromatography is the ability to configure the instrument for one's specific needs and budget.

1.1 Advantages of GC over the other Analytical Techniques

- Separation can be achieved within minutes.
- Better resolution of closely related compounds.
- Qualitative and quantitative analysis.
- Very sensitive (up to pico and femtogram level).
- Simple to operate and understand the mechanism.
- Data interpretation is very simple.
- Performs dynamic separation and identification of all types of volatile organic compounds and several inorganic permanent gases.
- Performs quantitative and qualitative determination of compounds in mixtures.
- May be destructive or nondestructive, depending on the type of detector used.
- Can determine molecular conformation (structural isomers) and stereochemistry (geometrical isomers) with appropriate detector.
- Can be configured for analysis of specific compounds using a wide choice of options.
- Can be automated for analyses of solid, liquid, and gas samples and may include automated steps in sample preparation.

1.2 Instrument

The basic components of instrument for gas chromatography are illustrated in Figure 2. The basic parts of a gas chromatograph are:

i) Carrier Gas

ii) Injection port
iii) Oven/Column
iv) Detector
v) Recorder/Database unit.

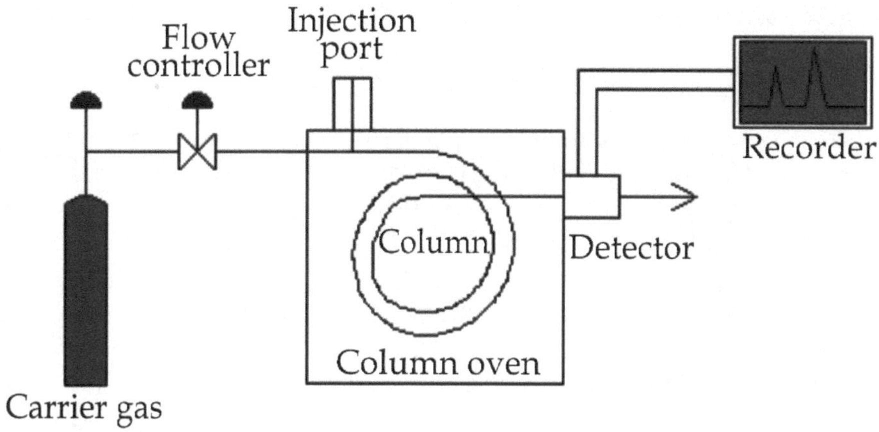

Figure 2 : Schematic diagram of a gas chromatograph

1.2.1 Carrier Gas

Main function of carrier gas is either to provide a flame or help in burning. Mostly hydrogen and oxygen is used for this purpose. The function of carrier gas is to carry vapours of analyte from injection port to detector through column. The carrier gas used in GC is generally chemically inert. Commonly used gases include nitrogen, helium, argon, and carbon dioxide. The choice of carrier gas is often dependant upon the type of detector which is used. Most commonly used carrier gas is nitrogen. However, hydrogen, helium and argon have also been used.

1.2.1.1 Characteristics of a Carrier Gas

- It should be inert to avoid interaction with sample or stationery phase.
- It should be available in purest form. Carrier gas purity should be better than 99.99% and 99.999% (referred to as five 9s or 99999 grade).
- The gas should be further purified by passing through a series of traps (filters) to remove contaminants (molecular sieve for moisture, charcoal for hydrocarbons and oxygen trap to remove O_2).

- The purifiers must be changed regularly and frequently to prevent saturation; change in colour indicates a requirement for replacement
- It should be moisture free.
- It should be suitable for the detector being applied in a particular analysis.

1.2.1.2 Carrier Gas Regulators

Normally two stage regulators with stainless steel (ss) diaphragms are used in GC. Although maximum pressure of 80 psi is acceptable but maximum pressure of 200 psi offers flexibility. When carrier gas enters chromatograph, it is usually controlled by one of the three methods use of needle valve, mass flow controller and pressure controller. Most chromatographs available presently have gas flow-rate controller built in the instrument. Flow rate of carrier gas is adjusted before gas chromatographic analysis. Small variations in the carrier gas flow rate affect column performance and retention times. For reproducible separations, a constant flow rate is maintained.

1.2.2 Injection Port

Injection port is a device to introduce the sample into the carrier gas stream and the substance to be analysed is injected in solution prepared in organic solvents like hexane or ethyl acetate. Its efficiency is reflected in the overall efficiency of the separation procedure and the accuracy and precision of the qualitative and quantitative results. For optimum column efficiency, the sample should not be too large. Because of ease and convenience, sample is injected with microlitre syringe. Syringe injection method is the most popular method because of the ease and convenience. Special syringes capable of delivering sample volumes in the range of 0.1 - 5000 μl are available. Most commonly used syringes are 1 and 10 μl fixed needle. Samples can be injected either manually or by auto injection. Good reproducibility is achieved with auto injectors than with manual injection. For packed columns, sample size ranges from tenths of a microlitre up to 20 microliters. Capillary columns, on the other hand, need much less sample, typically around 1 μL. Normally for capillary columns, split/splitless injection sytems is used (Figure 3).

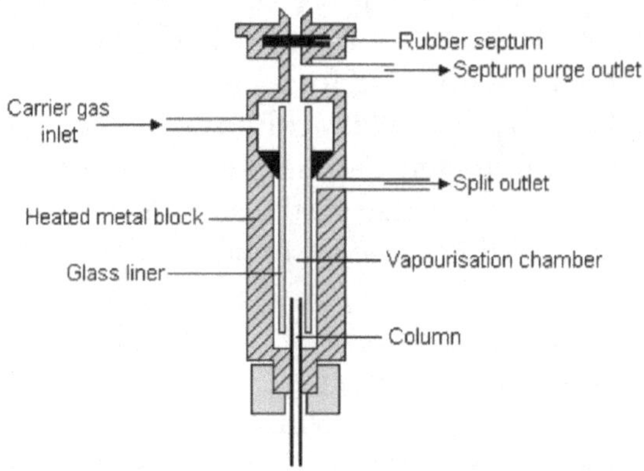

Figure 3 : Split/splitless injector (Source: http://www.shu.ac.uk)

The injector can be used in one of two modes, split or splitless. A split injection port is designed to allow only a fraction of the injected volume of sample in to the capillary column. The injector contains a heated chamber containing a glass liner into which the sample is injected through the septum. The carrier gas enters the chamber and can leave by three routes (when the injector is in split mode). The sample vapourises to form a mixture of carrier gas, vapourised solvent and vapourised solutes. A proportion of this mixture passes onto the column, but most of the mixture exits through the split outlet. Split ratio is calculated by using the following formula:

$$\text{Split ratio} = \frac{\text{Split vent flow rate} + \text{Column flow rate}}{\text{Column flow rate}}$$

1.2.3 *Oven/Column*

Chromatographic column is responsible for separation of component in the sample mixture and is called as heart of column. The shape of the column may be straight, bent or coiled. Columns may be made of metal (copper, aluminum and steel), glass and fused silica glass. The length of the column varies from 3-10 feet in case of glass and wide bore whereas it may vary 10-100m in case of capillary column. Efficiency of the column is inversely proportional to diameter of the column. Diameter of the packed column generally ranges from 3-4 mm and it is 320µm for capillary and megabore columns. Film thickness of the stationary phase in capillary column is 0.25µm and 2.65µm is of megabore column. There are two general types of column, packed and capillary (also known as open tubular). Packed columns contain a finely divided, inert, solid support material (commonly based on diatomaceous earth) coated with liquid stationary phase. Most packed columns are 1.5 - 10 m in length and have an internal diameter of 2 - 4mm. Packed or porous layer open tubular (PLOT) columns are used for analysis of low molecular weight gases (e.g. CO_2, CH_4, H_2). Capillary columns have an internal diameter of a few tenths of a millimeter. Capillary columns are made from fused silica and synthetic quartz, coated on the outside with polyamide, which make them rugged, flexible and easy to handle. They can be one of two types; wall-coated open tubular (WCOT) or support-coated open tubular (SCOT). Wall-coated columns consist of a capillary tube whose walls are coated with liquid stationary phase. In support-coated columns, the inner wall of the capillary is lined with a thin layer of support material such as diatomaceous earth, onto which the stationary phase has been adsorbed. SCOT columns are generally less efficient than WCOT columns. Both types of capillary column are more efficient than packed columns. In 1979, a new type of WCOT column (Figure 4) was devised - the fused silica open tubular (FSOT) column. Length of capillary columns generally ranges from 10 to 100 m, however, 15 or 30 m columns are commonly used for pesticide residue analysis. These have much thinner walls than the glass capillary columns, and are given strength by the polyimide coating. These columns are flexible and can be wound into coils. They have the advantages of physical strength, flexibility and low reactivity.

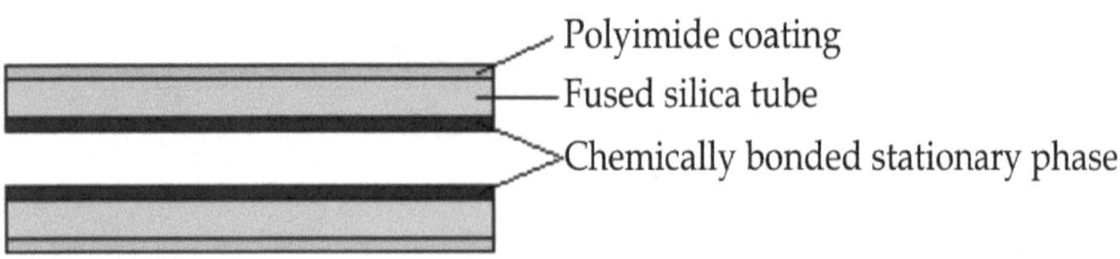

Figure 4 : Cross section of fused silica open tubular column (Source: http://www.shu.ac.uk)

1.2.3.1 Solid Support

Purpose of the solid support is to provide large uniform inert surface area for the distribution of liquid phase. It is important to select appropriate stationary phase of columns in optimizing gas chromatographic separation. The stationary phase of column system is chosen after considering polar characteristics of the analytes, their volatility range and column temperature programme. Two main solid supports are Chromosorb- P and Chromosorb-W, the later being more inert and good for polar compounds. Chromosorb is the registered trade mark for solid support material for GC. It is important to select appropriate stationary phase of columns in optimizing gas chromatographic separation. The stationary phase of column system is chosen after considering polar characteristics of the analytes, their volatility range and column temperature programme.

1.2.3.2 Desired Solid Support Properties

- Inert to avoid adsorption
- Thermally stable
- High crushing strength
- Mechanically strong
- Large surface area(1-20m^2/g)
- Smaller particle size to increase column efficiency
- Uniform packing

1.2.4 Liquid Phases

Liquid phases provide differential solubility to components of a substance, which help in their separation. Liquid phases are basically polymeric high melting point silicone greases. About 200 liquid phases are available but only six are most commonly used. Their names are as: SE-30 or OV-101 (dimethyl silicone), OV-17 (50% phenyl methyl silicone), carbowax 20M (polyethylene glycol), DEGS (poly diethylene glycol succinate), Silar-10C (cyanopropyl silicone) and OV-210 (trifluropropyl methyl silicone).Mostly liquid phase is used singly but some times mixed stationary phases are used to achieve selectivity and resolution of the mixture which are difficult to resolve in a single phase.A wide range of stationary phase is available for WCOT capillary columns. One example is a 100 % dimethyl polysiloxane polymer that is chemically bonded onto the interior wall of the column and provides an example of a nonpolar stationary phase. The 5% phenyl polymer composition (middle structure) is one of the most widely used separation phase. DB-5 is first choice of analyst for method development. The stationary phase of DB-5, (nonpolar column) is 5%-phenyl/95% methylpolysiloxane. DB-l has most non polar siloxane stationary phase, surface bonded with 100% dimethyl polysiloxane. DB-5 is similar to DB-l excepting that 5% of the methyl group is substituted with phenyl group, which is more polar. For medium-polar stationary phase (50%-phenyl-methyl polysiloxane), DB-17 is the best column. Generally the more polar the stationary phase is, better is the separation. Film thickness ranges 0.1 - 5.0 μm, and normally columns with 1.0 or 1.5μm film thickness are used in pesticide determination.

A few common GLC liquid phases used in pesticide residue determination are given in Table 1.

Table 1: Equivalent commercially available products

Basic structure	Capillary open tubular	Packed
Polysiloxane, 100% methyl	DB-1, HP-1, HP-101, BP-1	OV-101,OV-1, DC 200
Polysiloxane, 5% phenyl, 95% methyl	DB-5 (ht), HP-5,	OV-3, OV-73,
Polysiloxane, 50% phenyl, 50% methyl	DB-17 (ht), HP-17,	OV-17, OV-11,

(DB, HP and OV are commercial codes for each material are related to their manufacturer)

(Adopted from Sharma,2007)

1.2.4.1 Ideal Properties of Liquid Phases

i) Chemically inert

ii) Thermally stable

iii) Non volatile

iv) Samples must have different distribution coefficients.

v) Samples should have reasonable solubility in liquid phase.

vi) At operating temperature, liquid phase should have negligible vapour pressure.

1.2.4.2 Function of Chromatographic Column

It helps in separation of different constituents of the substance under the influence of a mobile phase.

1.2.4.3 How Separation Takes Place?

$$\text{By partition co-efficient (k)} = \frac{Cs}{Cg}$$

Cs = Concentration in stationery phase

Cg = Concentration in gaseous phase

Narrow ratio means faster separation and vice-versa.

Resolution power of chromatographic column is described in terms of theoretical plates to which it is equivalent to height equivalent to a theoretical plate (HETP), whereas the discrete plate is an artificial concept.

1.2.4.4 How to Measure Theoretical Plates?

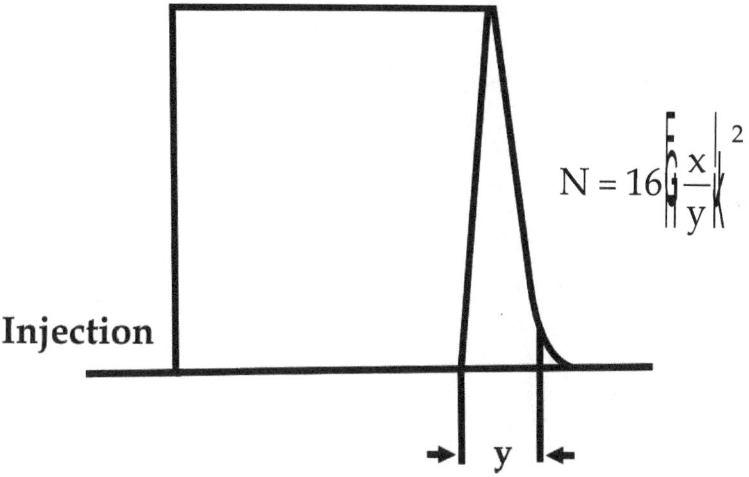

Figure 5 : Calculation of theoretical plates (Source: Mcnair and Bonelli,1967)

1.2.4.5 Calculation of Theoretical Plates

Theoretical plates can be measured from the chromatogram. Tangents are drawn to the peak at the points of inflection (about 2/3 of the height) (Figure 5). The number of theoretical plates N is given by 16 (x/y)2 where "y" is the length of baseline cut by the two tangents, and "x" is the distance from injection to peak maximum (including the dead-volume).Many factors effect column efficiency and most of these have been evaluated by their effects on N or the height equivalent to a theoretical plate, HETP. This is related to N by:

HEPT = L/N

L is the length of the column (in cm/m). Usually 1000-2000 plates are used for packed columns whereas, 100000 to 1000000 are normal for capillary columns. The more narrow the diameter, greater is the column efficiency or peak sharpness.

1.2.4.6 Resolution

Resolution is a measure of both the column and solvent efficiencies. It accounts for both the narrowness and the separation between maxima (Figure 6).

If R=1, the resolution of two equal -area peaks is approximately 98% complete.

If R=1.5, baseline separation (99.7% resolution) is achieved.

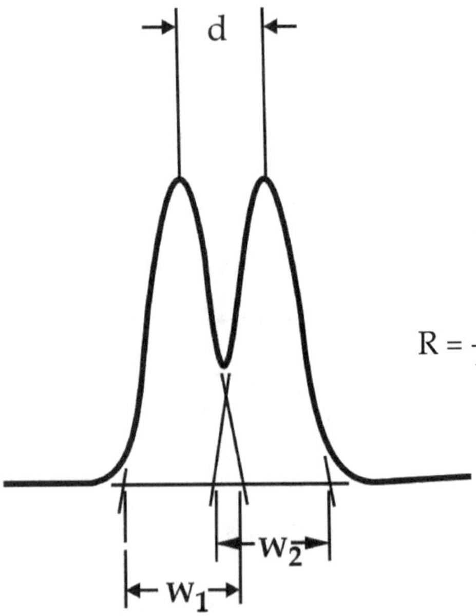

$$R = \frac{2d}{W_1 + W_2}$$

Figure 6: (Source: Mcnair and Bonelli, 1967)

1.2.5 Column Temperature

For precise work, column temperature must be controlled to within tenths of a degree. The optimum column temperature is dependant upon the boiling point of the sample. As a rule of thumb, a temperature slightly above the average boiling point of the sample results in an elution time of 2 - 30 minutes. Minimal temperatures give good resolution, but increase elution times. If a sample has a wide boiling range, then temperature programming can be useful. The column temperature is increased (either continuously or in steps) as separation proceeds.

1.2.5.1 Isothermal Conditions

In the simplest GC analysis, the column is maintained at a constant temperature, and this is termed as isothermal analysis. This temperature is selected generally when one or two compounds of similar group and with almost similar boiling points are to be analysed. In this, minimum (say 200°C) temperature is given to oven, 20-25°C higher (225°C) temperature to injector and 20-50°C higher (250°C) to detector. Comparatively, higher temperature of injector and detector avoid the condensation of injected sample. Oven temperature should not exceed the upper temperature limit of the packing material and column should be conditioned for several hours before analysis.

The dependence of GC retention on vapour pressure means that mixtures containing components with a wide range of boiling points cannot be separated satisfactorily in an isothermal run. The more volatile components may well be enough resolved, but higher boiling materials will be eluted with long retention times and very broad peaks. If the column temperature is high enough to give satisfactory peaks for less volatile compounds, the low-boiling constituents will be less well - resolved. The solution is to raise column temperature during a chromatographic run, so that for homologous series, peaks emerge at regular intervals. This is called as temperature programming.

1.2.5.2 Temperature Programming

The retention time of a compound depends not only on the type of stationary phase but also on the column temperature. As the temperature increases, compounds move faster through column and consequently, retention time decreases. Thus, analysis time can be potentially reduced by increasing column temperature. However, sufficient time has to be allowed for compounds to interact with stationary phase to provide the required separation of sample components.

1.2.6 Detectors

Detector is the device that senses the presence of components different from the carrier gas and converts that information to an electrical signal. One's choice of detector includes selectivity and sensitivity. Not all the detectors respond to all components. Selectivity is the ability of the detector to recognize and respond to the components of interest and sensitivity is the concentration level, detected. Sensitivity is defined as the change in the response with the change in detected quantity.

1.2.6.1 Characteristics of Detector

- High sensitivity
- Low noise level
- Response to all types of compounds
- Inexpensive

Following is the list of the common detectors used for pesticide residue analysis:

- Thermal conductivity detector (TCD)
- Flame ionization detector (FID)
- Electron capture detector (ECD)
- Nitrogen phosphorus detector (NPD)
- Alkali flame ionization detector (AIFD)
- Flame photometric detector (FPD)
- Photo ionization detector (PID)
- Mass selective detector (MSD)

1.2.6.2 Thermal Conductivity Detector (TCD)

It is the oldest type of GC detector used, also known as katharometer or hot-wire detector. Most organic vapors have lower conductivity than H_2, He or N_2 and their presence in one of these carrier gases reduces the amount of heat conducted from the filament. In this, filament temperature increases as analytes present in the carrier gas pass over it, causing the resistance to increase. Although its sensitivity level has dramatically improved through the years, it is still the least sensitive of the detectors. Its popularity has endured because of its ability to respond to any type of constituent that is different from the carrier gas (for example, if helium is the carrier gas, any other component would be detected as long as its concentration is in the nanogram or parts-per-million level, 10^{-9} to 10^{-12}).

1.2.6.3 Flame Ionization Detector (FID)

FID is the most commonly used detector in GC. The desirable characteristics of detector which contributed to its status include:

i) Its high and nearly uniform sensitivity (1pg/sec) to most organic compounds

ii) insensitivity to the common impurities such as carbon dioxide and moisture in the carrier gas

iii) minimal fluctuations due to changes in the flow, pressure or temperature

Organic components burn in a flame, producing ions that are collected and converted into a current. This is most widely used detector because it offers low detection limits (pico gram or parts per billion). The basic features of FID are shown in Figure 7. The FID consists of hydrogen air burner and two electrodes at a potential difference of 100-200 volts. Burner jet is usually negative electrode. The other positive electrode is often a ring, a wire mesh or a cylinder held about a centimeter above the tip of the jet. Carrier gas emerges from the tip of the jet and mix with hydrogen which is burnt in air. Air is introduced into the flame chamber independent of carrier gas and hydrogen. Organic components burn in a flame, producing ions that are collected and converted into a current. This is the most widely used detector because it offers fairly low detection limits picogram or parts-per-billion concentrations) and responds to any type of hydrocarbon component.

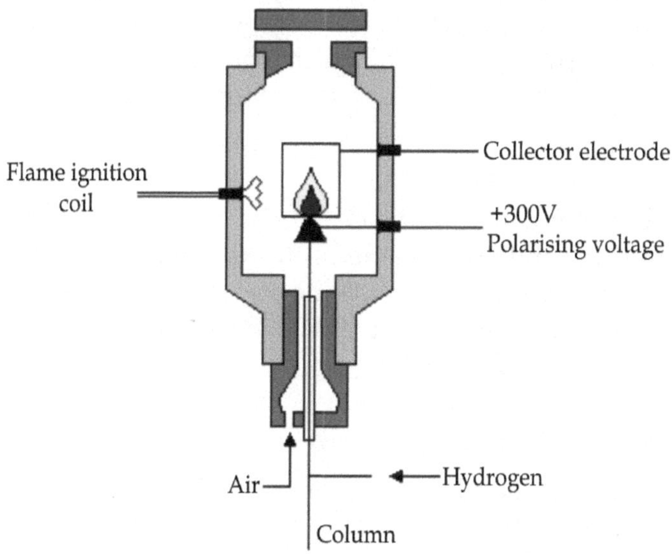

Figure 7 : Flame Ionization Detector (Source: http://www.shu.ac.uk)

At high temperature of the flame, ionization of hydrogen occurs and current is produced between the electrode. When organic compound containing C-H bonds enters the flame, positively charged ions and electrons are formed.

Organic compound + H_2 + $O_2 \rightarrow CO_2$ + H_2O + +ve ions + -ve ions + e-

The formation of ions is detected by the passage of current across two electrodes. Increase in current is proportional to the concentration of a compound which is recorded after amplification.

1.2.6.4 Electron Capture Detector (ECD)

The electron-capture detector (ECD) is a highly sensitive detector capable of detecting picogram (pg) amounts of specific types of compounds. The high selectivity of this detector can be a great advantage in certain applications. Its response can vary significantly with temperature, pressure and flow rate. Its principle is to capture of electrons by certain atom of the solute. The most common source of electron is a radio active material which emits low energy beta particles. The detector contains a sealed radioactive source, ^{63}Ni. The radiation ionizes the carrier gas into ion pairs of positive ion and electrons. As the electronegative species pass through the detector, they capture low-energy electrons, causing a decrease in cell current. It is suitable for the very low levels of halogenated pesticides. As its name indicates, its principle is to capture of electrons by certain atom of the solute. The most common source of electron is a radio active material which emits low energy beta particles. The radiation ionizes the carrier gas into ion pairs of positive ion and electrons.

$$N_2 + \beta \rightarrow N_2^+ + e^-$$

This detector contains two electrodes across which a potential is applied to collect and measure the electrons (Figure 8). The molecule of the solute capture the electrons, become negatively charged and combine with the positive ion causing decrease in the measurable current which is deflected in the ECD signal.

$$e^- + AB \rightarrow AB^- \text{ or } e^- + AB = A+B^-$$

The continued popularity of EC detector (using a ^{63}Ni source) results from its high sensitivity to halogen and certain other moieties as well as its ruggedness and low maintenance needs. Its sensitivity makes it applicable to determination of residues at the ppb and even ppt level. Its wide dynamic response range facilitates its use with automatic data system and its high operating temperature (~400°C) minimizes detector contamination by sample co-extractives and column bleed.

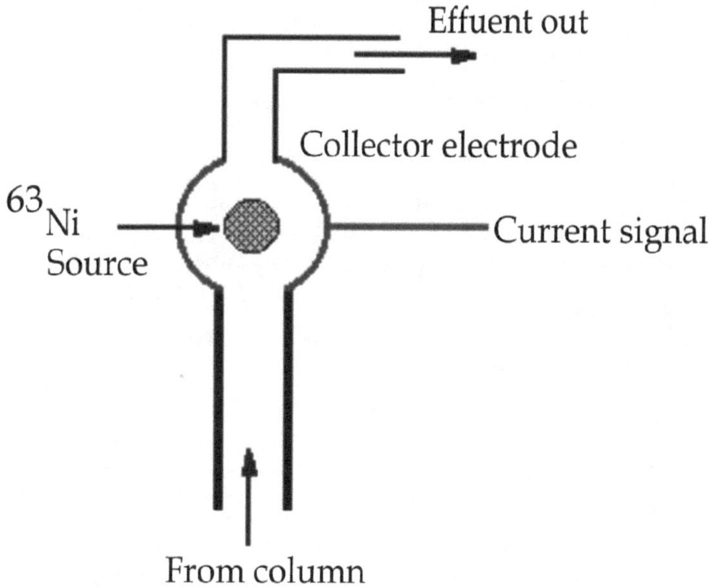

Figure 8 : Electron Capture Detector (Source: Sharma, 2007)

1.2.6.5 Nitrogen Phosphorus Detector

This is a detector of choice for analysis of organophosphorous pesticides and pharmaceuticals. Most N-P detectors (NPD) are more responsive to phosphorus than to nitrogen. It is also known as nitrogen-selective detector, because FPD-P is referred for phosphorus residues. Nitrogen and phosphorous compounds produce increased currents in a flame enriched by vaporized alkali metal salt. Response of the N and P detector also provides complementary evidence about element (s) present in the residue; information is often needed for confirmation of identity. In NPD, column effluent impose on to the surface of an electrically heated and polarized alkali source in the presence of air or hydrogen plasma, ionization occurs, and the flow of ions between plasma and an ion collector is amplified and recorded. Detector response to analytes results from increased ionization that occurs when compounds containing nitrogen or phosphorus elute from column. At gas flow rates used for N or P operations, the degree of ionization of compounds containing N or P is > 10,000 times higher than hydrocarbons. All N - P detectors provide electronic heating of alkali source to 600-800°C. This detector can detect as low as 5 -10 pg nitrogen-containing compounds and 1-5 pg phosphorus-containing compounds. Sensitivity of N-P detector is mainly dependent on hydrogen flow and the magnitude of current supplied to the alkali bead. With decrease in hydrogen flow, response to nitrogen increases.

1.2.6.6 Alkali Flame Ionization Detector (AFID) or Thermionic Detector (TID)

Alkali flame ionization detector (AFID) is also known as thermionic detector (TID) or nitrogen phosphorus detector. It is essentially the flame ionization detector modified by introduction of an alkali matter vapour into the system so that its response to nitrogen and phosphorus containing compounds is enhanced by several orders of magnitude. The modification is achieved by fusing an alkali metal salt to negatively charged electrode ring. Such a detector is 600 times more sensitive to phosphorus and 300 times more sensitive to nitrogen containing compounds relative to the unmodified FID .The sensitivity may be further enhanced by putting a second flame detector above the first but separated from it by a salt coated platinum wire mesh. In this design, the first flame burns the organic compounds and vapourizes the alkali metal salt and the vapours are then transferred to the second flame where the ionization occurs.

In the more recent versions the flame is often substituted by electrical heating of the volatile alkali metal activator and hence the name thermionic detector. Several explorations have been offered for the increased sensitivity of the AFID to phosphorus and nitrogen but none very satisfactorily. It appears that the ionization is facilitated by collision of PO, OPO, CN and halogen radicals with the excited alkali metal ions or radicals formed by the application of thermal energy to the metal. Detection limit of AFID is up to pico gram level.

1.2.6.7 Flame Photometric Detector (FPD)

In this detector, sulphur and phosphorus compounds burn in a flame, producing chemiluminescent species that are monitored at selective wave lengths.

1.2.6.8 Photo Ionisation Detector (PID)

This detector contains an ultra violet (UV) lamp. Molecules are ionized by excitation with photons from an UV radiation of sufficient energy (10-12 ev).

$R+ hv \rightarrow R+ +e-$

Where R+ is the ionized species and hv is a photon having energy equal or more than the ionization potential of the species. The charged particles are then collected in the ionization chamber, producing current and positive current is measured. The PID may be operated in a universal or selective mode by changing the photon energy of the ionization source. Major uses of this detector are for detecting small amounts of aromatics in hydrocarbon mixtures.

1.2.6.9 Mass Selective Detector (MSD)

Molecules are bombarded with electrons, producing ion fragments that pass into the mass filter. The ions are filtered based on their mass/charge ratio. The very popular GC-MS technique yields excellent qualitative identification of components by matching the compounds mass spectrum with spectra included in the libraries that are part of the system.

1.3 Data Acquisition or Identification and Quantification of the Components of Solutes

The last major component of GC system is the data acquisition system. The signal from the detector after amplification is recorded as peak called chromatogram. The area of the peak is representative of the concentration of the component, the larger the concentration, the larger the detector signal and larger will be peak area. The most commonly used devices for generating chromatogram and report are the integrator and the personal computer. Each eluate/solute is identified by its retention time which is defined as the time lapse between the injection of the sample and recording of peak maximum. Rt value is a function of column, temperature, carrier gas flow rate, affinity of the solute for mobile and stationery phase. Under given parameter, the Rt value of the compound remains constant. Quantification of a solute is done by comparing the height or area of each peak of unknown solutes with that of peaks of known standards.

1.3.1 Factors for Ensuring Good Chromatography

- Careful injection of sample into injection port gives good reproducibility.
- Proper rinsing of the syringe to avoid contamination of column and noise peaks.
- Septum should be checked frequently for leakage.
- Temperature of injection port should be at least ten degree higher than column temperature to ensure rapid vaporization of the sample.

- Increase in temperature improve peak shape but tends to bunch rapidly eluting peaks.
- Lowering of temperature leads to increase of retention time and broadening of peak, so the effect on resolution tends to be small, but in case peaks are partially resolved a change of temperature can sometimes have a beneficial effect.

1.3.2 Common Applications

- Separation of compounds in a mixture via a column and detection for qualitative and quantitative identification by any of several types of detectors
- Performs quantitative and qualitative determination of compounds in mixtres
- Used in most types of manufacturing industry for analyses of raw materials, intermediates, or final product
- Used in environmental, forensic, pharmaceutical, and clinical/ medical applications

2. MULTIRESIDUE METHODS FOR ESTIMATION OF PESTICIDE RESIDUES

For effective control of insects pests, a number of pesticides including insecticides, fungicides, herbicides and nematicides are to be applied. Estimation technique for single pesticide only can not be followed for detection of residues of all categories of pesticides. Hence, this constantly expanding use of pesticides on food crops accentuates the need for rapid, precise and sensitive method for determination of pesticide residues of all the major groups of pesticides. In such situation, multi-residue analytical technique can be efficiently followed for detection and estimation of multiresidues of intra and inter class xenobiotics .

The purpose of multiresidue analyses is to determine the residues of as many pesticides as possible within a short period of time even if the recoveries of some compounds are low. In multiresidue methods the recoveries up to 70% are accepted. The recoveries less than 70% have to be mentioned specifically. Multiresidue analysis is relatively more sensitive and requires extreme vigilance on the part of the analyst during processing of samples. Additionally, instrument, GLC, need to be equipped with capillary columns for high resolution and sophisticated detectors like ECD, NPD and MSD for detecting multi-residues of the pesticides.

2.1 Estimation of Pesticide Residues in Vegetables and Fruits (Kumari *et al.*, 2001; 2006)

2.1.1 Extraction

i) Take bulk sample (1-2 kg) of vegetable/fruit.

ii) Chop it into small pieces and mix properly.

iii) After quartering take 20g representative sample.

iv) Macerate it with 4-5g anhydrous sodium sulphate.

v) Add 100 ml acetone and extract by shaking on mechanical shaker for 1 hour.

vi) Filter the extract through 2-3 cm layer of anhydrous sodium sulphate.

vii) Concentrate the extract to 40 ml on rotary flash evaporator after adding a drop of mineral oil.

viii) Dilute the extract 4-5 times with 10% NaCl aqueous solution.

ix) Partition it thrice with ethyl acetate (50, 30, 30 ml) in a separatory funnel by shaking vigorously for one minute.

x) Combine the organic (ethyl acetate) phases and filter through anhydrous sodium sulphate.

xi) Concentrate the organic phase up to 5 ml on rotary flash evaporator.

xii) Divide the concentrated extract into two equal parts (one for organochlorines and synthetic pyrethroids and other for organophosphates and carbamates).

2.1.2 Clean-up

For Organochlorines and Synthetic Pyrethroids

i) Pack the glass column (60 cm x 22 mm i.d) with adsorbent mixture (5g) Florisil: activated charcoal (5:1 w/w) in between two layers of anhydrous sodium sulphate.

ii) Tap the column gently to ensure uniform and compact packing.

iii) Prewett the column with 50 ml hexane and transfer the concentrated extract to the column.

iv) Elute the column with 125 ml solution of ethyl acetate: hexane (3:7 v/v).

v) Concentrate the eluate to near dryness using rotary flash evaporator followed by gas manifold evaporator after adding one drop of oil.

vi) Make the final volume to 2 ml in ethyl acetate: n-hexane (3:7 v/v).

For Organophosphates and Carbamates

i) Pack the glass column (60 x 22 mm i.d.) with adsorbent mixture containing 5g silica gel (60-120 mesh): activated charcoal (5:1 w/w) in between 3-4 cm layers of anhydrous sodium sulphate. Ensure the compact packing of the column by taping gently.

ii) Prewett the column with 50 ml hexane and load the concentrated extract to the column.

iii) Elute the column with 125 ml mixture of acetone: hexane (3:7 v/v).

iv) Concentrate to near dryness using rotary flash evaporator followed by gas manifold evaporator.

v) Make the final volume to 2 ml in acetone: n-hexane (3:7 v/v).

Flow diagram of extraction of multiresidues from vegetable and fruits is shown below:

Extraction

Take bulk sample (1-2 kg) of vegetable/fruit

↓

Chop it into small pieces and mix properly

↓

After quartering take 20g representative sample

↓

Macerate it with 4-5g anhydrous sodium sulphate

↓

Add 100 ml acetone and extract by shaking on mechanical shaker for 1 hour

↓

Filter the extract through 2-3 cm layer of anhydrous sodium sulphate

↓

Concentrate the extract to 40 ml on rotary flash evaporator after adding a drop of mineral oil

↓

Dilute the extract 4-5 times with 10% NaCl aqueous solution

↓

Partition it thrice with ethyl acetate (50, 30, 30 ml) in a separatory funnel by shaking vigorously for one minute

↓

Combine the organic (ethyl acetate) phases and filter through anhydrous sodium sulphate

↓

↓

Concentrate the organic phase up to 5 ml on rotary vacuum evaporator

↓

Divide the concentrated extract into two equal parts (one for organochlorines and synthetic pyrethroids and other for organophosphates and carbamates)

Clean-up

For Organochlorines and Synthetic Pyrethroids

Pack the glass column (60 cm x 22 mm i.d) with adsorbent mixture

↓

(5g) Florisil : activated charcoal (5:1 w/w) in between two layers of anhydrous sodium sulphate

↓

Tap the column gently to ensure uniform and compact packing

↓

Prewett the column with 50 ml hexane and transfer the concentrated extract to the column

↓

Elute the column with 125 ml solution of ethyl acetate: hexane (3:7 v/v)

↓

Concentrate the eluate to near dryness using rotary vacuum evaporator followed by gas manifold evaporator after adding one drop of mineral oil

↓

Make the final volume to 2 ml in ethyl acetate: n-hexane (3:7 v/v)

For Organophosphates and Carbamates

↓

Pack the glass column (60 x 22 mm i.d.) with adsorbent mixture containing 5g silica gel (60-120 mesh): activated charcoal (5:1 w/w) in between 3-4 cm layers of anhydrous sodium sulphate

↓

Ensure the compact packing of the column by taping gently

↓

Prewett the column with 50-60 ml hexane and load the concentrated extract to the column

↓

Elute the column with 125 ml mixture of acetone: hexane (3:7 v/v)

↓

Concentrate to near dryness using rotary flash evaporator followed by gas manifold evaporator

↓

Make the final volume to 2 ml in acetone: n-hexane (3:7 v/v)

2.1.3 GLC Parameters for Estimation of Multiresidues

GLC parameters used for the above developed multiresidue methods are given below:

For Organochlorines and Synthetic Pyrethroids

GLC model	:	Hewlett Packard (HP) 5890A or equivalent
Detector	:	Electron capture detector (ECD) Ni^{63}
Column	:	Capillary column, HP-5, 30 m x 0.32 mm id x 0.25 μm film thickness of 5% diphenyl and 95% dimethyl polysilioxane

Temperature Programming

	8°/min.	15°min.
Oven 150°	→ 190°	→ 280°C
(5 min.)	(2 min.)	(10 min.)

Injection port temperature	:	280°C
Detector temperature	:	300°C

Gas flow rate : Total flow 60 ml/min. with split ratio 1:10 (Through column 2ml)

For Organophosphates (OP) and Carbamates

GLC model : HP 5890A or equivalent

Detector : Nitrogen phosphorous detector (NPD)

Column : Capillary column, HP-1, 10 m x 0.53 mm i.d. x 0.65 μm film thickness of 100% diphenyl polysilioxane.

Temperature Programming

	10°/min.	25°min.
Oven 100°	→ 190°	→ 240°C
(1 min.)	(0 min.)	(3 min.)

Injection port temperature : 250°C

Detector temperature : 275°C

Gas flow rate

H_2 : 1.8 ml/min.

O_2 : 130 ml/min.

N_2 : 18 ml/min

The chromatograms of organochlorines and synthetic pyrethroids are shown in Figure 9 whereas of organophosphates and carbamates in Figure 10. Retention times of different pesticides observed for above GC conditions are given in Table 2.

2.1.4 Final Analysis and Calculation

A stock solution of 1-2 ppm analytical grade of standard is injected and retention time is recorded. Then 1-2 μl of cleaned test samples is injected into the GC. The residues are identified by comparing the retention time of sample peaks with that of standard and the amount of residues recorded in the integrator chart. The amount of residues in ppm (μg/g) is calculated as follows:

Residues in ppm =

Where,

A_1 = Area of sample in the Chromatogram

A_2 = Area of standard in the Chromatograph

V_1 = Total volume of sample in mL

C = Concentration of analytical standard in ppm ($\mu g/mL$)

W = Weight of the sample in g

Rf = Recovery factor

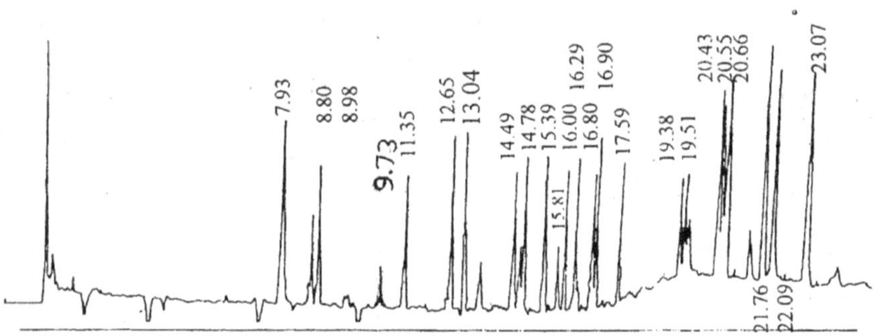

Figure 9: GC-ECD Chromatogram of organochlorines and synthetic pyrethroids

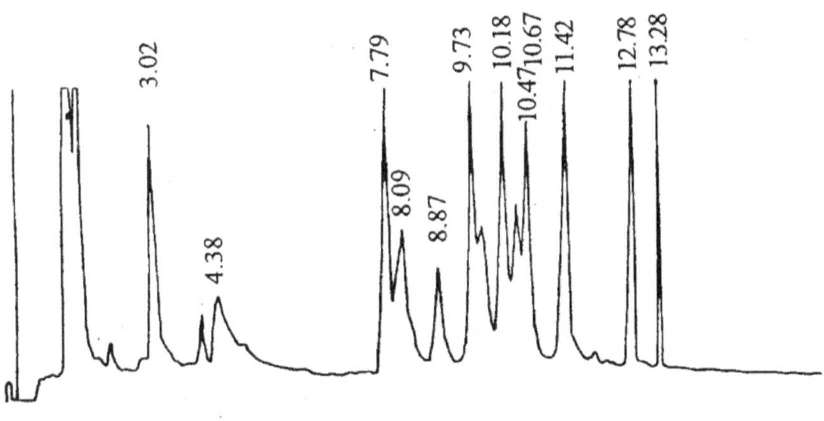

Figure 10 : GC-NPD Chromatogram of organophosphates and carbamates

Table 2 : Retention times (Rts) of different pesticides

Sr. No.	Pesticide	Retention time (Rt)
Organochlorines & Synthetic Pyrethroids		
1.	α-HCH	7.93
2.	β-HCH	8.80
3.	χ-HCH	8.98
4.	δ-HCH	9.73
5.	Heptachlor	11.35
6.	Aldrin	12.65
7.	Chlorpyriphos	13.04
8.	o, p-DDE	14.49
9.	α -Endosulfan	14.78
10.	p, p'-DDE	15.39
11.	p,p'-DDD	15.98
12.	β - Endosulfan	16.00
13.	o, p-DDT	16.29
14.	Endosulfan sulphate	16.80
15.	p, p'-DDT	16.90
16.	Fluvalinate	19.38
		19.51
17.	Cypermethrin	20.43
		20.55
		20.66
18.	Fenvalerate	21.76
		22.09
19.	Deltamethrin	23.07
Organophosphates and Carbamates		
20.	DDVP	3.02
21.	Acephate	4.38
22.	Phorate	7.79
23.	Monocrotophos	8.09
24.	Carbofuran	8.87
25.	Phosphamidon	9.73
26.	Carbaryl	10.18
27.	Fenitrothion	10.47

Contd.

28.	Malathion	10.67
29.	Chlorpyriphos	11.42
30.	Quinalphos	12.78
31.	Triazophos	13.28

References

Agnihotri, N.P. 1980. *Gas chromatography* pp: 130-140. In: Residue analysis of insecticide (ed. Gupta, D.S.), Department of Entomology, HAU, Hisar:

Dean, J.R. 1998. *Extraction Methods for Environmental Analysis.* John Willey & Sons. Ltd. West Sussex, England.

Handa, S.K., Agnihotri, N.P. and Kulshrestha, G. 1999. *Pesticide Residues: Significance, Management and Analysis.* Research Periodicals and Book Publishing House, New Delhi.

Kumari, Beena, Kumar, R. and Kathpal, T.S. 2001.An improved multiresidue procedure for determination of 30 pesticides in vegetables. *Pestic. Res. J.,* **13**(1): 32-35.

Kumari, Beena, Madan, V.K. and Kathpal T. S., 2006. Monitoring of pesticide residues in fruits. *Environ. Monit. and Assess.,* **123:** 407-412.

McNair,H.M. and Bonelli,E.J. 1967. *Basic chromatography.* Consolidated Printers, Oakland California, USA.

Pavia, Donald L., Gary, M. Lampman, George S. Kritz, Randall G. Engel 2006. *Introduction to Organic Laboratory Techniques* (4th Ed.). Thomson Brooks/Cole. pp. 797-817. ISBN 978-0-495-28069-9.

Ravinderanath, B. 1989. *Principles and Practice of Chromatography.* Pub. Ellis Horward Ltd. Chickester, England.

Sant, M.J.V. 1997. *Gas chromatography.* In: Hand Book of Instrumental techniques for analytical chemistry. (ed.Settle, F.A.), Prentice Hall, Inc. New Jersey, USA.

Sharma, K. K. 2007. *Pesticide Residue Analysis Method.* Directorate of Information and Publications of Agriculture, New Delhi.

Chapter-16

Applications of Computer Based HPLC in Horticulture Crops

Ramesh Kumari Mehta, Ashok Yadav and Vinod Kumar Madan*
Department of Agronomy
*Medicinal and Aromatic Plants Section
Chaudhary Charan Singh Haryana Agricultural University, Hisar

Computer played a distinctive role in the growth of most advanced, precise and highly sensitive instruments especially in medical and analytical instrumentation technology. Due to the development of different kinds of software the instrumentation analysis has become easy, rapid and result oriented. Auto sampling, development of new methods for detection analysis and the validity of new developed method have been possible with no trouble. Once the standardization of method is over and saved, it can be applied at any time to integrate the unknown samples and their data calculation. Thus, it avoids repetition in calculation and the end result is time saving technique due to the development of computer-organized system in analysis.

The results produced are more reliable and reproducible through various computer based instruments e.g. scanning machines, blood analyzer, gas chromatography (GC), amino acid analyzer (AAA), atomic absorption spectrophotometer (AAS) and high performance liquid chromatography (HPLC) etc.

HPLC is a technology or a system that works on the basis of software. There have been many definitions of computer shielded HPLC, such as high performance liquid chromatography (HPLC), high-pressure liquid chromatography (HPLC), high-speed liquid chromatography (HSLC), high efficiency liquid chromatography (HELC), and simply liquid chromatography (LC).

1. Background

Only a few years back, bioassays were most frequently used to study the different behaviour of pesticides:

1. Persistence
2. Movement (leaching, dissipation) in soil
3. Effect of various soil properties and environmental conditions on phytotoxicity
4. Absorption, movement and degradation of pesticides in the plant
5. Dose response curves

Bioassay is the measurement of biological response by a living organism to determine the presence and/or concentration of a chemical in a substrate. Bioassays are generally used as a means of quantitatively measuring biologically active concentrations of a pesticide known to be present. However, sometimes bioassays may be used to determine the presence or absence of a particular herbicide. No doubt the bioassays are the only suitable methods for recording the biological activity of pesticides but physical and chemical procedures are immensely needed to isolate, identify and quantitate the substances present in traces.

Differences between bioassays and chemical-physical assays

Bioassays	Chemical-physical assays
Only biologically active/phytotoxic portion of the pesticide is measured	Whole of the pesticide portion present in different forms is measured
Extraction of the pesticide from the substrate is not required and hence, circumventing the problems involved in it	Extraction of the pesticide or its residues is required and hence, it becomes troublesome
More economical, less difficult to perform, and do not require much expensive equipments	It is little bit cumbersome, costly and expensive equipments are required
Bioassay procedure for a specific pesticide is less difficult to develop	It is somewhat difficult to develop required procedure for a specific pesticide

Problems Involved in Bioassay

According to Behrens (1970) and Santelmann (1972) bioassays and chemical-physical assays, each contributes valuable information and appears to be useful in certain circumstances. Combination of these two is possible as an effort to increase the accuracy or rapidity of standard methods because of following reasons:

- Plant growth is directly influenced by environmental conditions and edaphic factors and hence must be very closely controlled.
- Untreated check plants and plants growing in known concentrations of pesticide must be included in every experiment.
- Plant uniformity is a must to increase reproducibility and reliability.
- Soil used in bioassay may absorb a portion of pesticide and either remove it from the sphere of action or allow subsequent desorption and release of phytotoxic activity.
- The time required for completion of the assay, during which the pesticide degradation may occur, may also be considered.

In the present era, there is consistent and serious concern regarding food contamination and environmental pollution. Underground water contamination due to ever-increased use of pesticides is another serious concern. So, it was realized to develop a sophisticated tool for the isolation, identification and quantitation of the compounds present in micro-level to ensure that the commodity is free from the pesticides. Therefore, based on Tswett (1954) approaches, a highly sensitive and computer-based tool namely "High Performance Liquid Chromatography (HPLC)" came into existence.

Computer Based High Performance Liquid Chromatography (HPLC)

It is an instrument that represents the most reasonable and useful example of significant role of computers in the agricultural technology. An analyst can do isolation, identification and quantification of desired analyte from the mixture compound by using classical methods i.e. distillation, crystallization etc., if it is available in sufficient amount. But problem is formidable when desired analyte is present in traces or in micro quantity in mixture compound e.g. pesticides residues persisting in fruits, vegetables, grains and beverage's etc. The dilemma arises in pharmaceutical studies where degradation study of drugs in plasma is mandatory to set the effective dose regime for living beings. Under such situations a very sophisticated, reliable and highly sensitive technique is needed to isolate, identify and quantitate the compounds accurately. The text contains brief information on bioassay, and the development of HPLC, its various components, functions, standardization of method, its applications and future prospects.

The computer directed HPLC has become the fastest growing analytical technique in the last few years. This is obvious by the increasing number of national and international conferences, symposia and seminars etc. on HPLC and also by the rapid growth of manufacturers of HPLC equipment and supplies.

A chromatography is a technique used for separation of the components of a sample, in which the analyte is distributed between two phases, one is stationary and other is mobile phase. The stationary phase may be a solid spread as a thin layer attached to glass plates (thin layer chromatography), solid particles packed in a column (column chromatography). The mobile phase may be gaseous (gas chromatography) or liquid (liquid chromatography) (Andrea and Brown, 1997). Chromatography with a liquid mobile phase can be traced back to the work of the Russian Botanist, Tswett (1906). The basic theory behind HPLC is not new, but it was not until around 1969 that HPLC as we now know it was developed. This development introduced the effective use of small diameter packing materials and columns that allowed the analyst to perform the separations faster and with greater resolution than had previously been attained by classical methods.

Computer based HPLC is not limited to thermally stable compounds as in case of gas chromatography. Nearly all classes of organic compounds can be separated by this efficient and quicker analytical technique. Roughly one million liquid chromatographic separations are performed world wide every day.

Brief History of HPLC Development

The origins of liquid chromatography began in the early 1900's with the work of the Russian botanist, Mikhail S. Tswett. His famous studies focused on separating compounds (leaf pigments), which were extracted from plants using a solvent. Two specific materials that he found useful were powdered chalk (calcium carbonate) and alumina. He created an analytical separation of these compounds based on a chemical attraction of some compounds to the particles (stationary phase). The compounds that were attracted to the particles slowed down, while other compounds that were attracted to the solvent (mobile phase), moved faster. This process can be described as follows: sample compounds "distribute" or "partition" differently between the moving "mobile phase" and the "stationary phase", creating a separation of the compounds. Tswett named this process "chromatography" [from the Greek words "chroma", meaning, "color", and "graphy", meaning, "writing" (literally "color writing"). Today, liquid chromatography, in its various forms, has become one of the most powerful tools in analytical chemistry. Liquid chromatography can be performed in three primary approaches.

Approach 1

The sample is "spotted" on to, and then flowed through a thin layer of chromatographic particles fixed on to the surface of glass plates called **"Thin Layer Chromatography** ("TLC").

Approach 2

The sample is "spotted" on to paper to which solvent is added to create flow. This is called **"Paper Chromatography".**

Approach 3

In the most powerful approach, the sample passes through a column or a device containing appropriate particles. These particles are called the chromatographic packing material, stationary phase or "adsorbent". Solvent flows continuously through the column. At a point in time, an "injection" of the sample solution is made into the solvent stream, which then carries the sample through the column. As in Tswett's experiment, the compounds in the sample can then be separated by traveling at different individual speeds through the device.

When the column or cartridge format is utilized, there are several ways to achieve flow. Gravity or vacuum can be used for columns that are not designed to handle pressure. Typically, the particles are larger in size (>50 microns) to allow flow to be generated more easily, as with open glass columns (Tswett's experiment). In addition, plastic "columns", typically in the shape of syringe barrels, can be filled with packing material particles and used to perform sample preparation. This is called **"Solid Phase Extraction"** (SPE). Here, the chromatographic device, called a "cartridge", is used usually with vacuum to clean up a very complex sample before it is analyzed further.

For improved separation power, smaller particle sizes (<10 microns) are required. These cause greater resistance to flow, resulting in higher pressures needed to create the solvent flow required. Columns and pumps, which are designed to withstand high pressure, are necessary. When moderate to high pressure is used to flow the solvent through the chromatographic column, this is called **"HPLC"**.

What is High Performance Liquid Chromatography (HPLC)?

The name "HPLC" originally referred to the fact that high pressure was needed to generate the flow required for liquid chromatography in packed columns. In the beginning, instrument components only had the capability of generating pressures of 500 psi (35 bar). This was called high pressure liquid chromatography (HPLC). The early 1970's saw a tremendous leap in technology. These new "HPLC" instruments could develop up to 6,000 psi (400 bar) of pressure, and included improved detectors and columns. HPLC really began to take hold in the mid to late 1970's. With continued advances in performance, the name was changed to high performance liquid chromatography (HPLC). It is now one of the most powerful tools in analytical chemistry, with the ability to separate, identify and quantitate the compounds that are present in any sample. Today, trace concentrations of compounds, as low as "parts per trillion" (ppt), are easily quantified. HPLC can be applied to just about any sample, such as pharmaceuticals, food, nutraceuticals, cosmetics, environmental matrices, forensic samples, and industrial chemicals.

Components of Computer Based HPLC

The HPLC is a system which typically consists of solvent reservoir, pumps, injector port, column, detector and a recorder and the collector is inter-connected by the suitable solvent resistant teflon or stainless steel capillary tubing that can withstand high pressure.

Solvent Reservoir

The solvent reservoir is very sophisticated solvent storage system for the pulse free computer controlled delivery of solvent to the column. The solvent reservoir may be equipped with magnetic stirrer to keep the composition of mobile phase constant. It must be equipped with 2-micron pore size inline filter to prevent the particulate matter from being drawn into the pump from mobile phase. Temperature sensor and a heater are important to maintain the pre-determined temperature of the mobile phase. The vacuum connection for degassing is essential especially in gradient elution where the concentration of the dissolved air in the two solvents differs or when one of the solvents is water (which has a high concentration of dissolved gases relative to other solvents).

Pump/Solvent Delivery System

The computer directs the expected, pulse free and exact supply of the solvent to the column. In reality, much of the hindrance in developing the high efficiency column packing may be accredited to the non-availability of appropriate pump that could deliver specific, resettable, reproducible and pulse free flow of solvent through the column at pressure that may range from 500 to 6000 psi, resistant to chemical action by the mobile phases and have a low hold up volume for speedy solvent changes as needed especially in gradient elution. The most modern and sophisticated is reciprocating pump (Poole and Poole, 1991) which consists of very small volume chambers (35-400 µl capacity) into which the solvent can be drawn and then pumped into the column by backward and forward movement of computer directed piston and admired in view of their exactness, continuous solvent flow over a wide range of flow rates and easy adaptableness to gradient elution (Snyder and Kirkland, 1979).

Injector/Sample Introduction Device

The computer handles the automatic injector or auto-sampler, which is very convenient for analyzing large number of samples. The sample is introduced into the chromatographic column through six port injection valve. The injector has two positions i.e. the load and the inject position. The theory of function of these sampling valve is that the valve is first switched to the fill or load position, then sample is introduced into the loop by deep syringe with no interruption of the eluent flow and then valve must be swiftly turn to the inject position to prevent the dispersal and expansion of the peaks.

Column

The column is a cylindrical tube packed with the stationary phase and complete with the end fittings so called the heart of the chromatographic system. The tube material normally used for HPLC columns is made of stainless steel in view of its potency and inertness. Where the inertness of the stainless steel is insufficient then tantalum or titanium columns may be used. The dimensions of the columns are generally determined by the instrument configuration, the convenience of packing and the size of packing particles (Knox and Kauer, 1989). During the beginning of era, long (1-2 m) and narrow columns with pellicular packing of size 30-50 μm that permitted high permeability for the mobile phase through the column were in use. But such long columns were found not permeable enough at moderate pressures. Even 25-30 cm columns packed with 10 μm particles are also not recommended. The more recent trend has been to use even shorter columns with a length in the range of 3-10 cm and packed with 3-5 μm particles. Different kinds of columns are used (Bakalyar and Spruce, 1987) for the complete resolutions of the solute because of its specification in stationary phase but frequently the normal phase and reversed phase stationary packing are in use for the analysis of different commodities. These are followings:

(i) The Normal Phase Packing

The normal phase packing is generally of silica. It is used for the separation of neutral species on the basis of their polarity such as amines and sulfoxides, etc. The solvents used for extraction from normal phase packing are non-polar hydrocarbons e.g. hexane or octane. Solvent selectivity is controlled by the nature of the added solvent. If the desired separation does not achieve then concentration of the mobile phase has to change.

(ii) Reverse-Phase (RP) Packing

The reversed phase packing is the reverse of normal-phase and extensively used. In RP, the stationary phase is non-polar and mobile phase is polar such as acetonitrile or methanol as per nature of the species. In general, the choice of the stationary phase should be such that it brings out the discrepancy that is real in the compounds to be separated. For efficient performance of the column regular care is important for its maintenance as mentioned underneath.

a) Column Safeguards

Certain elementary precautions can prolong the life of the column considerably;

- All the solvents must be filtered through micro porous sintered-glass funnels before use.
- Inline filters are also used as an extra precaution to remove the particulate matter from the mobile phase.

- The guard column of about 3 cm length is inserted between the injection port and the analytical column to prevent the clogging of the column.
- For the preliminary clean up and to get rid off much of the unwanted portion of the sample, the sep-pak cartridges packed with silica, alumina and a variety of bonded phases are used.
- Sample should be dissolved in the same solvent that is to be used as the mobile phase. In case of gradient elution, it should be the initial solvent composition to avoid the possible sample precipitation in injection loop, column bed and in the connecting tubes.
- The sample should be filtered through a 0.45-µm filtration kit attached with syringe prior to injection of the processed sample.

b) Column Regeneration and Repair

- When the performance of column has fallen below acceptable limit then it can be brought into use by cleaning with suitable solvents series.
- For silica and most normal-phase packing (Unger and Trudinger, 1989) the sequence is heptane, chloroform, ethylacetate (acetone), ethanol (methanol), water, methanol, chloroform, and heptane.
- For the reversed-phase columns, the sequence may be water, methanol, dichloromethane, methanol and water.
- In injector, inlet tubing or frits can be partially clogged. For repair, remove column and guard, and wash with 6N- nitric acid.
- The columns are generally stored in appropriate solvent that is usually marked in the manufacturer's literature; normally, it is hexane for silica gel and other normal phase packing and acetonitrile / methanol for the reversed phase packing.

Detectors

The computer control detector exchanges the column effluent into an electrical signal that is computed by the data station and presented as chromatogram on the chart. Various kinds of detectors are available (Scott, 1986) in the market. However, commonly used in the laboratories are:

i) Refractive Index Detector (RI)

It is used for all the solutes, which possess a refractive index and do not absorb UV light or don't have UV-chromophore e.g. sugars, polymers, etc can be studied.

ii) Photo Diode Array Detector (PDA)

It measures the elements lying between the ranges of 190-800 nm, hence known as UV/VIS detector. The UV or UV/VIS detector is most extensively used not only because of the relative insensitivity of the detector to temperature and gradient changes, but also because of the reason that great number of compounds absorbs radiations in UV range.

iii) Fluorescence Detector (FD)

A fluorescence detector can monitor substances those emit light in the visible region after being excited by UV radiation. Fluorescence detector is very specific and its use requires more precautions than any other detector, hence, its use is very limited.

Working and Utility

Standardization of the Method

A particular standard of technical grade (100% pure) must be in hand to standardize the method. The method is standardized by recovery experiment. The calibration curve is drawn at different concentrations (minimum five concentrations) up to the detection limit of the particular standard under particular conditions of the system. Since the results of the substances to be analyzed are predicted vis -a -vis standard hence specification of the standard is important to analyze the specific substance.

The sample to be analyzed is extracted and processed by following the accurate method for that particular compound. The processed sample is cleaned by passing through disposable Sep Pak Cartridge and in the end filtered through 0.45 μm-filtration kit attached with syringe and then sample is introduced into the stationary phase or chromatographic bed by deep syringe through injector. Allowing the mobile phase to flow through the column and the detector attains the isolation. The elution of samples for analytical separation is further categorized into two types:

(a) Isocratic Elution

The mobile phase and flow rate remains same throughout the elution of the analyte from the column and hence separation achieved, very rapidly.

(b) Gradient Elution

The composition of running mobile phase continuously changes in the column throughout the separation of sample analyte, which, have varying affinities for the stationary phase. Gradient of the mobile phase may be binary, tertiary and quaternary depends upon the flow rate.

The output of the detector, which is in the form of an electrical signal, is fed to a potentiometric recorder, which records the chromatogram on a strip chart. The detection of the desired analyte is subjected to two types of studies i.e. qualitative analysis and quantitative analysis.

i) Qualitative Analysis

It can be obtained by measuring the retention time of the peak of the substance with respect to the retention time of the peak of the respective standard. Hence, the utility of chromatography in qualitative analysis is of two folds: determination of the number of compounds and their identification in a given sample.

ii) Quantitative Analysis

It can be obtained by measuring the peak area or height of the desired peak. The peak area is directly proportional to the concentration of the substance. The response of HPLC detectors is concentration specific. The accuracy of quantification by chromatography depends on several factors such as the method of sampling and sample introduction, efficiency of the column, linearity of the detector and the method used to convert the detector response into the amount of the compound.

Precautions

- Keep the reservoir above the solvent delivery system. This assures a good siphon feed to the pump and avoids starving of the pumping system.
- The tubing from the reservoir must be of stainless steel or Teflon so that system can bear with the high pressure.

- The solvent system must be miscible with the previously used mobile phase.
- The pump should be flushed with HPLC-Grade water after using mobile phase containing salts or acids.
- The storage of the column should not be in pure water because the pure water can cause microbial growth and hence storage of the column should be in the organic solvent.
- The check valves of the pump should be periodically cleaned with 6-Nnitric acid.
- As working of the HPLC is carried out at an ambient temperature, hence, air conditioning of the laboratory should be adequate.

Applications

The development of highly sensitive computer foundation detectors have made it possible to use smaller and smaller quantity of samples for detection thus making chromatography a powerful analytical tool. This feature greatly enhances its capability and applications (Bidlingmeyer,1992) with respect to the three aspects of analytical chemistry; namely separation, identification and quantification of the components like drugs and biogenic amines in biological samples (rumen liquor and blood) Singh *et al.*, 1989), chiral pharmaceutical analysis, food (organic acids in rice straw; Kumari *et al.*, 2008) and nutritional analysis, analysis of surfactants, polymers, ions and inorganic species, physicochemical measurement, in the preservation of our cultural and historical legacy by identification of natural and early synthetic dyestuffs and by studying the degradation and polymerization and in residual and multi-residual studies of organic pollutants (pesticides, herbicides and fungicides) in different matrices i.e. in soil straw and grains up to ppb level is possible to detect (see practical example given below).

HPLC Application in Sample Processing and Estimation of Organic Pollutants Herbicides in Different Matrices:

Protocol for Analysis of Oxadiargyl (Top Star) Residues in Paddy Flow Diagram (Kumari *et al.*, 2007)

Take 20 g representative sample of grinded straw, seeds in coarse form and soil (grinded & sieved through 2 mm sieve separately in conical flask for the extraction of each sort of sample

↓

Extract each sample with 150 ml acetonitrile (GR) by shaking on mechanical shaker for 90 min

↓

Decant off each sample through anhydrous Na_2SO_4 pad (Keep the flask undisturbed and decant off. Sometimes emulsion occurs then centrifuged and decant off)

↓

Reduce the volume of above extract of each sample upto 20 ml by using Rotary Flash Evaporator

↓

The above concentrated extract of each sample is subjected to clean up process by using Column Chromatography technique

↓

Pack the column for each sample with homogenized adsorbent mixture of silica gel : activated charcoal at the ratio 10 : 1 g (w/w) between two layers of anhydrous Na_2SO_4 of 2 inches each

↓

Pre-wash the every adsorbent column prepared, with 70 ml of hexane and after that collect the clean and colourless fraction of each sample by eluting with 150 ml of hexane: toluene mixture in the ratio 7: 3

↓

Concentrate the each, clean and colourless eluate on Rotary Flash Evaporator and then dry over Manifold Evaporator till last drop

↓

Make the volume of above eluate of each sample to 2 ml in mobile phase i. e. pure Acetonitrile (HPLC grade)

↓

Prior to injection of sample extract into HPLC column, filter it through filtration kit containing 0.45 -µm pore size filter attached with syringe

HPLC parameters/conditions for analysis of oxadiargyl

High performance liquid chromatography (HPLC) from Waters Pvt. Ltd.

Detector	:	Photodiode array detector (PDA) Model 996
Column	:	XTerra^{Tm} RP$_{18}$ 5 μm (4.6 x 250 mm)
Temperature	:	Ambient (fully air conditioned)
Mobile phase (Isogradient elution)	:	Acetonitrile (100% pure)
Volume injected	:	20 μl
Flow rate	:	1 ml/min
Retention time	:	2.93 min. oxadiargyl residues herbicide
Wave length	:	202 nm
Recovery found	:	85-90%

Analysis of Indole Alkaloids in *Catharanthus roseus* Leaves

Flow Diagram (Uniyal, *et al.*, 2001)

Take 5 g of *Catharanthus roseus* leaves powder and added 30 mL of 90% ethanol, left over night and filtered.

↓

The residue was again extracted with 90% ethanol (3 x 30 mL) at room temperature (27°C), and the pooled alcoholic extract was filtered and concentrated *in vacuo* at 40°C.

↓

The dried residue was re-dissolved in ethanol (10 mL), diluted with water (10 mL) and then acidified with 3% hydrochloric acid (10 mL).

↓

This was then extracted with hexane (3 x 30 mL), the hexane extract discarded and the aqueous portion cooled to 10°C, basified with ammonium hydroxide to pH 8.5 and extracted with chloroform (3 x 30 mL).

↓

The combined chloroform extract was washed with water, evaporated to dryness, re-dissolved in 1 mL chloroform.

↓

This was then passed through a silica Sep-Pak cartridge (Waters) pre-saturated with chloroform, washed successively with 5 mL each of chloroform and chloroform: methanol (9:1, v/v), and dried over anhydrous sodium sulphate before being evaporated to dryness.

↓

The residue obtained was dried to constant weight in order to determine the total alkaloid content.

↓

An aliquot (10 mg) of the crude alkaloid was dissolved in 1.0 mL of methanol, and 10 mL was subjected to HPLC analysis.

HPLC Parameters/Conditions for Analysis of Indole Alkaloids in *Catharanthus roseus* Leaves

High performance liquid chromatography (HPLC) from Waters Pvt. Ltd.

Detector	:	Photodiode array detector (PDA) Model 996
Column	:	Symmetry C_{18} column (150 x 4 mm i.d.; 5 mm; Waters)
Temperature	:	25°C
Mobile phase (binary gradient) (Naaranlahti *et al.*, 1987)	:	Solvent A: [methanol : acetonitrile : 0.025M ammonium acetate: triethylamine (13:32:55:0.2 v/v)] Solvent B: [methanol : acetonitrile : 0.025 M ammonium acetate : triethyl amine (19:46:35:0.2 v/v)]
Linear gradient profile	:	Initially 45% B changing to 50% B at 5.0 min, isocratic to 10.0 min, changing to 57% B at 17.0 min, changing to 65% B at 25.0 min and then isocratic for another 5 min.
Flow rate	:	Initially 1 ml/min untill 17 min and then increased to 1.5 ml/min until 30 min.
Volume injected	:	10 µl
Wavelength	:	220 nm, re-plotted at 254 and 280 nm.

Multi-residue analysis

Simultaneous estimation of Aldicarb, Carbofuran and Carbaryl in water by HPLC Flow Diagram (Mukherjee *et al.*, 2007)

The water sample (2 L) was spiked with pesticide mixture at 0.5, 1.0 2.0 5.0 mg L^{-1} level and diluted with 150 ml saturated NaCl solution

↓

Counter current extraction procedure was followed

↓

The fortified samples were divided into four parts each and quantitatively transferred to a 1 L capacity separatory funnel and arranged in a row

The sample in the first separating funnel was extracted with 50 ml dichloromethane

↓

The dichloromethane extract was collected and transferred to the second separating funnel in the row and the extraction process repeated. The same solvent was used for extraction till the last separating funnel in the row (step 1). The organic phase was collected in conical flask

↓

In the second stage of extraction fresh 50 ml of dichloromethane was added to the last separating funnel and the extraction process repeated in the reverse order as in step 1

↓

In the third step 25 ml of dichloromethane was added to each of the middle separating funnels and extraction was carried out as in step 2 till the end of the row

↓

The combined extract was passed through anhydrous sodium sulphate and concentrated to dryness using rotary vacuum evaporator

↓

Traces of dichloromethane were removed completely by re-dissolving the residue in acetonitrile and again evaporating it to dryness in a rotary vacuum evaporator

↓

The samples were made upto required volume in acetonitrile and analyzed by HPLC

HPLC Parameters/Conditions for Analysis of Aldicarb, Carbofuran and Carbaryl

High performance liquid chromatography (HPLC) from Schimadzu

Detector	:	Photodiode array detector (PDA) Model 996
Column	:	RPC-80 (Stainless steel 30 cm, 4 mm, 5 m)
Temperature	:	25°C
Mobile phase	:	Acetonitrile: water (50:50)
Volume injected	:	10 µl
Flow rate	:	1 ml/min
Retention time & Carbaryl)	:	6.10, 8.61 & 9.68 min. (Aldicarb, Carbofuran

Future Prospects

Beginning in 2004, further advances in instrumentation and column technology were made to achieve very significant increases in Resolution, Speed and Sensitivity in Liquid Chromatography. Columns with smaller particle sizes (1.7 micron) and instrumentation designed to deliver 15,000 psi (1,000 bar) along with specialized capabilities were needed to achieve a new level of performance. This new technology is called ULTRA PERFORMANCE LIQUID CHROMATOGRAPHY (UPLC™Technology). A glimpse at what may be the future is crystal clear as basic research being carried out today by scientists working with columns containing 1 micron particles, and instrumentation capable of performing at 100,000psi (6,800 bar).

HPLC is computer-based precise tool hence it can be conveniently combined with ultra-violet (UV), infrared (IR) nuclear magnetic resonance (NMR) and mass spectrometry (MS). All these techniques require very small quantities of the sample and the analysis is also carried out in very short time. Moreover the HPLC results are rapid, reliable and reproducible. It is the useful example of computer applications.

Conclusion

Computer maintained high performance liquid chromatography (HPLC) is very sophisticated, reliable, highly sensitive, efficient and rapid technique, which is used to isolate, identify and quantitate the components present in mixture compounds accurately when it is present in micro-fractions i.e. up to 0.001-ppm level. Since the works of HPLC at ambient temperature, hence, compounds degrade at above than room temperature, can be analyzed.

References

Andrea, Weston and Brown, Phyllis R. 1997. *"HPLC and CE Principles and Practice"* Academic Press, California, USA.

Bakalyar, S.R. and Spruce, B. 1987. *How to use column selection valves for HPLC. Technical Notes* 8: Feb.

Behrens, R. 1970. Quantitative determination of triazine herbicides in soil by bioassay. *Residue Rev.,* **32:** 355 - 369.

Bidlingmeyer, B. A. 1992. *Practical HPLC Methodology and Applications.* New York: John Wiley & Sons.

Knox, J.H. and Kauer, B.1989. In: *High Performance Liquid Chromatography* (eds. P.R. Brown and R.A. Hartwick,), Chapter 4. Wiley, New York.

Kumari, A., Kapoor, K. K., Kundu, B.S., Mehta, R. K. 2008. Identification of organic produced during rice straw decomposition and their role in rock phosphate solubilization. *Plant Soil. Environ.* **54** (2) : 72-77

Kumari, Ramesh, Kumari, Beena and Punia, S. S. 2007. High performance liquid chromatographic method for estimation of oxadiargyl residues in paddy grains, straw and soil. *Ann. of Biol.,* **23:** 33.

Mukherjee, I. Gupta, S., Kulshrestha, A., Pant, S., Singh, A., Kumar, A., Gajbhaiye, V.T., Dikshit, A.K. and Kulshrestha, G. 2007. Simultaneous estimation of aldicarb, carbofuran and carbaryl in water by HPLC. *Pestic. Res. J.,* **19 (1):** 128-130.

Naaranlahti, T., Nordstrom, M., Uhtikangas, A. (1987) Determination of *Catharanthus* alkaloids by reversed-phase high-performance liquid chromatography. *J. Chromatogr.* **410:** 488-493.

Poole, C.F. and Poole, S.K., 1991. *Chromatography Today* Elsevier, New York.

Santelmann, P.W, 1972. Herbicide bioassay. pp 91-101. In: *Research Methods in Weed Science. South Weed Sci. Soc.*USA.

Scott, R. P. W. 1986. *Liquid Chromatography Detector.* Amsterdam : Elsevier.

Singh, N., Kumari. R., and Akbar, M.A. 1989. Concentration of biogenic amines in the rumen liquor and blood of healthy buffaloes and buffaloes suffering from indigestion. *Indian Vet. J.,* **13:** 84-87,1989.

Snyder, L.R. and Kirkland, J.J. 1979. *"Introduction to Modern Liquid Chromatography"* (2nd Ed.). Wiley, New York.

Tswett, M.S. Ber. dtsch. Ges. *24, 316, 384, 1906. Vgl.* Hesse, G., Weil, H., In: Woelm-Mitteilungen, Al. I, Eschwege 1954.

Unger, K.K. and Trudinger, U. 1989. In: *High performance Liquid Chromatography* (eds. P.R. Brown and R.A. Hartwick), Chapter 3. Wiley, New York.

Uniyal, G.C., Bala, S., Mathur, A.K. and Kulkarni, R.N. 2001. Symmetry C_{18} Column: a better choice for the Analysis of Indole Alkaloids of *Catharanthus roseus. Phytochem. Anal.,* **12:** 206-210.

Chapter-17

Effect of House Hold Processing on Pesticide Residues in Vegetables

Indu Chopra and Beena Kumari*
Department of Chemistry and Physics, *Department of Entomology
CCS Harayana Agricultural University, Hisar - 125 004 (Haryana)

Nature has bestowed our country with diverse agro climatic and topographic conditions ranging from temperate to humid and from sea bed to snow line. The existing climatic conditions with a plenty of sunshine, nutrient rich soil, abundant water resources and reservoir of trained manpower, enable our country to grow various types of vegetables in tropical, subtropical, temperate, alpine and desert vegetation zones in one or the other seasons during the year.

Vegetables, the cheapest and nutrient-rich food source within the economic reach of poor man, play a vital role in human diet since they provide carbohydrates, protein, fat, minerals, vitamins, fibers and phytochemicals (non- nutrient bioactive compounds), which are essential for making the body immune system strong, detoxifying carcinogens, reducing muscular degeneration and protecting the body from infectious ailments. Vegetables besides having medicinal values also play an important role in nutritional security and national economy of the country since they are quick growing and short duration, and their yield per unit area per unit time is 5-10 times higher than the cereal crops.

India has made commendable progress in vegetable production enabling it to secure 2nd position in the global vegetable production matrix, accounting for about 10 per cent of the world's production. More than 40 kinds of vegetables belonging to different groups are grown in different agro climatic conditions of the country. Currently, India's share in the world's total vegetable production is 13.6 per cent and the demand for vegetable is projected to rise 170 million tons by the year 2025 (Phogat *et al;* 2010).However, several factors limit their productivity, mainly insect pests and diseases. In order to combat the insect pest problem, lot of pesticides is used by vegetable growers. For better yield and quality, insecticides are repeatedly applied during the entire period of growth and sometimes even at the fruiting stage. It accounts for 13-14 per cent of total pesticide consumption, as against 2.6 per cent of cropped area (Sardana, 2001). Indiscriminate use of pesticide particularly at fruiting stage and non adoption of safe waiting period leads to accumulation of pesticide residues in consumable vegetables.

Most analyses of pesticide residues in foods are being performed in Raw Agricultural Commodities (RAC) for a variety of purposes, which include regulatory monitoring, risk assessment, field-application trials, organic food verification, and marketing to consumers. The levels of the positive detections in these analyses are generally being estimated on the basis of established Maximum Residue Limits (MRL's) which are set using field trial data for a particular pesticide to arrive at the highest residue levels expected under use according to Good Agricultural Practice (GAP). However, MRL's use, proved to be inadequate as a guide to pesticide residue consumption through nutrition in health risk assessment studies from residues in food of plant origin and this is mainly because a wide range of RAC's are processed before they are consumed. Storage and other post-harvest practices prior the further management of the product, as well as household and industrial food preparation processes may alter pesticide residues as compared with raw crops via chemical and biochemical reactions (hydrolysis, oxidation, microbial degradation etc.) and physicochemical processes (volatilization, absorption etc.). Although these processes usually are leading to reduction of any residues left on crops at harvest (Kaushik *et al.*, 2009; Holland *et al.*, 1994), in special cases residues may concentrate in the final product (e.g in the production of dry fruits and unrefined vegetable oil) (Amvrazi and Albanis 2008; Guardia Ruibio *et al.*, 2006; Lentza-Rizos *et al.*, 2006; Lentza-Rizos and Kokkinaki, 2002; Cabras *et al.*, 2000; Cabras *et al.*, 1998; Cabras *et al.*, 1997; Holland *et al.*, 1994; Cabras *et al.*, 1993; Leandri *et al.*, 1993; Ferriera and Tainha, 1983) and/or be formed in more toxic by-products or metabolites of the pesticide parent compound on raw crop (Holland *et al.*, 1994; WHO 1988).

Most of the vegetables for processing are subjected to different unit operations during preparation and preservation. These operations are essential for cleaning, eliminating wastes and making raw materials more palatable. The procedures used may actually reduce or remove pesticide residues, when present. Therefore, in this chapter, to assess the effect on pesticide residues of different processing methods have been discussed.

Mechanisms for Post Harvest Alteration of Residues in Vegetables

Basic processes acting on pesticide residues in the field can continue to operate after crops are harvested. These include volatilization, hydrolysis, penetration, metabolism, enzymatic transformation and oxidation. Photo degradation generally ceases or is greatly reduced once a crop is removed from the field situation.

The use of various physical unit processes on crops such as washing, trimming, peeling or juicing apportions residues between various processed food fractions. This often leads to direct reductions in the levels of residues in remaining edible portions. However, lipophilic pesticides tend to concentrate in tissues rich in lipids and thus residue levels can increase in fractions such as vegetable oils. Processes involving heat or chemicals can increase volatilization, hydrolysis or other chemical degradation and thus reduce residue levels. However, drying process may result in higher concentration of residues due to loss of moisture.

PROCESSING

I) Processing Systems

A variety of unit operations are used in commercial or domestic food processing. These are summarized in Table 1.

Table 1: Different unit operations and conditions

Process	Conditions
Washing	Cold water, hot water (blanching), chemical baths (caustic, acid, detergent, hypochlorite)
Peeling	
Husking	
Hulling	
Shelling	
Trimming	
Comminution	Blending, chopping, mincing
Cooking	Steaming, boiling, baking, frying, grilling, microwave
Pickling	
Drying	Oven, solar, spray, freeze

In many cases, a raw commodity undergoes a series of unit processes to form a food. For example, a range of specialized processes are used in canning tomatoes. The effects of various unit steps on residue levels of parent pesticides are likely to be additive although the quantitative effects on levels of any transformation products may not be so evident.

II) Effects of Processing

Washing

Washing of RAC is the preliminary step in both household and commercial food preparation and the effect of washing on the fate of the pesticide residues on RAC has been well studied and recently well reviewed (Kaushik *et al.*, 2009; Zabik *et al.*, 2000; Krol *et al.*, 2000; Holland *et al.*, 1994). The most interesting conclusions of these studies are that the rinsability of a pesticide is not always correlated with its water solubility (Cengiz *et al.*,2007; Boulaid *et al.*, 2005; Angioni *et al.*, 2004; Krol *et al.*, 2000) and that different pesticides may be rinsed from processed units of RAC by different washing procedures (Angioni *et al.*, 2004; Pugliese *et al.*, 2004; Lentza-Rizos and Kokkinaki, 2002; Cabras *et al.*, 1998). The removal of pesticides with the washing of RAC may be performed not only through the dissolution of pesticide residues in the washing water or the rinsing with chemical baths (detergents, alkaline, acid, hypochlorite, metabisulfite salt, ozonated water etc.) (Holland *et al.*, 1994) but also through the removal of dust or soil particles previously absorbed residues from the outer layer of RAC (Guardia Rubio *et al.*, 2007; Guardia Rubio *et al.*, 2006; Angioni *et al.*, 2004; Cabras *et al.*, 1997). Penetration is again the most dynamic process that may control the fate of a pesticide residue on RAC during washing.

a) Washing with water

Generally vegetables are washed before consumption with running or standing water at moderate temperatures. If necessary, several washing steps can be conducted consequently.

The effects depend on the physicochemical properties of the pesticides, such as water solubility, hydrolytic rate constant, volatility and octanol- water partition coefficient (P_{ow}), in conjunction with the actual physical location of the residues; washing processes lead to reduction of hydrophilic residues which are located on the surface of the crops.

Dikshit *et al.*, (1986) reported that the initial residues of methamidophos applied at 0.03 and 0.05% concentrations reduced to 65.71-77.67% by simply washing the leaves and curds of cauliflower, heads of cabbage , leaves and pods of Indian colza. In leafy vegetables, concentration of the pesticide residue was higher in outer leaves than in inner ones (Yoshida *et al.*, 1992). There was no pesticide contamination of inner leaves of Chinese cabbage with a well formed head. Removal rates of pesticide residues by washing with water or 0.1% liquid detergent were 8.52% and 19.67%, respectively. Awad and el-Shimi (1993) studied the effect of washing on the removal of Actellic residue from fresh and processed vegetables namely, spinach and egg plant reporting high percent removal (> 45%). Miyahara and Saito (1994) reported that during Tofu production, the pesticide levels were reduced to about 10 and 20% of the initial levels by washing of soybeans with water twice. Washing removed more residues from carrots than from tomatoes, but it did not affect the relative distribution of the residues (Burchat *et al*; 1998). Cabras *et al*; (1998) observed a significant decrease in pesticide residues during the washing treatment of prunes. In case of iprodione, the residue at harvest time was 0.68 ppm and washing for 5 minutes resulted in six times decrease. Persistence of malathion in bell peppers (*Capsicum annum var. grossum*) was studied in field experiments (Bhagirathi *et al*, 2001). Washing of treated fruits with running tap water removed 67-78% of residues from samples. Gill *et al.*, (2001) reported that washing reduced residues of alphamethrin in brinjal from 25 to 33% while reduction was 11 to 30% in tomato for the same pesticide. Lee (2001) observed the effect of different washing methods on levels of organophosphate (OP) pesticide residues in red pepper. Red pepper fruits were artificially contaminated with chlorpyrifos and fenitrothion after harvest, at levels of 4 and 10 ppm, respectively. Decrease in pesticide levels were approximately 30 to 40% after shaking/ sonicating the peppers for 5 minutes in water. Decontamination processes such as washing and steaming dislodged the cypermethrin residues in pulses by 37-49% and 63-74%, respectively. Maximum concentration of cypermethrin was found on seed coats. Treatments with cypermethrin did not affect viability of pulses (Dikshit, 2001). Chavarri *et al.,* (2005) reported that fortified levels of chlorpyrifos dropped by 38% in tomatoes while that of ethylenbisdithiocarbamates dropped by 43% after washing. In the process of making tomato paste, washing decreased endosulfan and deltamethrin residues to 30.62% and 47.58%, respectively. Kumari (2008) studied the effects of washing on residues of organochlorines, synthetic pyrethroids, organophosphates and carbamates in brinjal, cauliflower and okra. Organochlorine residues reduced by 27 -44% in brinjal, 34-36% in cauliflower and 20-38% in okra. The residues of synthetic pyrethroid insecticides in brinjal, cauliflower and okra were reduced to 26, 29 and 31%, respectively. In case of organophosphates, the maximum reduction of residues was observed which decreased to the extent of 77, 74 and 50% in brinjal, cauliflower and okra, respectively.

b) Washing with salt solution

Washing with dilute salt solution is a convenient method to lower the load of contaminants from food surfaces particularly fruits and vegetables. This method could be equally effective for reducing the pesticide residue from other commodities too. This procedure is recommended as being practical for household use. Chlorothalonil was found best removed from Chinese cabbage by 1% saline exposure for 10 min (Lee and Chou, 1995). Decontamination of cabbage treated with chlorpyriphos (0.05%) and quinalphos (0.05%) was studied by Nagesh and Verma (1997). Decontamination through different processes showed that residues were reduced to some extent by washing and cooking. Washing the samples with salt water did not differ significantly by ordinary washing with tap water. Dipping of green chillies in 2% salt solution for 10 min followed by washing in water removed 90.56 and 66.93% of residues from chillies at 0 and 5 days after final spraying, respectively (Phani-Kumar *et al.,* 2000). Dissipation and decontamination of triazophos and acephate in chillies was also studied by Phani-Kumar *et al.,* (2000) Dipping of green chillies in 2% salt solution for 10 min followed by washing in water proved effective in removing residues of both pesticides, 32.56 and 84.21% of triazophos residues were removed, respectively, at 0 and 5 days after the final application of spray, while 78.95% of acephate was removed immediately after the final application.

c) Washing with chemical solution

Chlorine water and dilute solutions of other chemicals are commonly used for disinfection of fruits and vegetables. These chemicals also play an effective role in removing the pesticide residues. The effectiveness of chlorine, chlorine dioxide, ozone and hydrogen peroxyacetic acid (HPA) treatments on the degradation of the fungicides mancozeb and ethylenethiourea (ETU) in apples was studied by Hwang *et al.,* (2001). Fresh apples were treated with 2 different levels of mancozeb (1 and 10 μg/mL). Several of the treatments were effective in reducing or removing mancozeb and ETU residues on spiked apples. Mancozeb residues decreased 56-99% with chlorine and 36-87% with chlorine dioxide treatments. ETU was completely degraded by 500 ppm of calcium hypochlorite and 10 ppm of chlorine dioxide at a 1 ppm spike level. However, at a 10 ppm spike level, the effectiveness of ETU degradation was lower than observed at the 1 ppm level. Mancozeb residues decreased 56-97% with ozone treatment. At 1 and 3 ppm ozone, no ETU residue was detected at 1 ppm of spiked mancozeb after both 3 and 30 min. HPA was also effective in degrading the mancozeb residues, with 44-99% reduction depending on treatment time and HPA concentration. ETU was completely degraded at 500 ppm of HPA after 30 min of reaction time. These treatments indicated good potential for the removal of pesticide residues on fruit and in processed products.

Pesticide residues present in potatoes obtained from a market in Minufiya, Egypt, were determined and effectiveness of various washing solutions for removal of pesticides from the potatoes was investigated (Zohair, 2001). Potatoes were washed for 10 min in tap water, an acid solution (5 or 10% solutions of an aqueous extract of radish leaves, acetic acid, citric acid, ascorbic acid or H_2O_2), a neutral solution (5 or 10% NaCl) or an alkaline solution (5 or 10% $NaHCO_3$). Concentration of individual organochlorine and organophosphorus pesticides in the potatoes were measured before and after washing. The residues, pirimphos-methyl, malathion, profenofos, endrin, lindane, aldrin, heptachlor-epoxide, o,p'- and p,p'-DDE, and o,p'-DDD were detected in the potatoes while p,p'-DDD, and o, p'- and p, p-DDT were not. Results showed that acidic solutions were more effective for extraction of organochlorine pesticides from potatoes than the other washing solutions used; however, in general, removal of organophosphorus pesticides was more complete. Soliman (2001) reported that washing potatoes with tap water or aqueous solutions of acetic acid and/or NaCl and blanching or frying of potatoes removed most of organochlorine and organophosphorus residues.

Use of ozone to prolong the shelf life of fruits and vegetables is discussed in relation to its ability to remove ethylene, inhibit respiration activity and inactivate microorganisms (Fanchun *et al* ; 2003). They also considered the potential for ozone to reduce pesticide levels in fruits and vegetables and the possible negative impacts of ozone treatment on fruit and vegetable quality.

Oxidizing agents for decreasing pesticide levels in foods are discussed, together with their mechanisms of action and the effectiveness of various treatments (Hwang and Cash, 2003). Aspects covered include: developments in pesticide use in recent decades; fate of pesticides in the environment; principal pathways for pesticide degradation in the environment; and degradation of pesticides using chemical oxidation with chlorine, chlorine dioxide, ozone or H_2O_2.

Efficiency of non-toxic aqueous solutions for removal of the pesticides, chlorpyrifos, fenarimol, iprodione, malathion, methidathion, myclobutanil, parathion and pirimicarb from nectarines was determined for various concentration of non-toxic compounds and washing times (Pugliese *et al.,* 2004). Aqueous solutions of potassium permanganate, citric acid, glycerol, ethanol, hydrogen peroxide, sodium metabisulfite, sodium laurylsulfate, urea and sodium hypochlorite were prepared and compared to washing with tap water. Residues present on nectarines were determined by analysing ethyl acetate and anhydrous sodium sulfate extracts of washings. Formation of possible non-toxic by-products (such as paraozon-methyl, chlorpyrifos oxon, malaoxon or methidaoxon) as a result of interactions between different aqueous solutions used to wash nectarines and organophosphorus pesticides (chlorpyrifos oxon, malaoxon, methidaoxon and paraoxon methyl) were also determined. Washings from nectarines using aqueous solutions containing large amounts of hypochlorite, hydrogen peroxide and potassium permanganate contained large amounts of toxic by-products (oxons). With the exception of sodium lauryl sulfate, glycerol and ethanol (which removed 50% of pesticide residues), washing solutions prepared from the remaining compounds were no more effective than washing with tap water. It is concluded that the octanol-water partition coefficient and water solubility of a given pesticide were factors determining the amount of pesticide removed from nectarines by washing with aqueous solutions.

Effects of ozonation on growth and pesticide residues in soybean sprouts during cultivation were evaluated by Kim *et al.*, (2000). Soybeans were treated with 0.3 ppm ozone water for 30 min during the soaking period. Sprouts from ozonated beans were heavier and longer than control sprouts, with lower root weight. Residues of carbendazim, captan, diazinon, fenthion, dichlorvos and chlorpyrifos as affected by various soaking/ozonation treatments were examined. Ozone treatments destroyed more pesticide than soaking in pure water, captan being the most susceptible and chlorpyrifos least. Use of ozone at 15-20 ppm to reduce levels of organophosphorus insecticides, including pirimiphos-methyl, chlorpyrifos-methyl, malathion, fenitrothion and DDVP, in wheat and corn was investigated by Zhanggui *et al.*,(2003). Pesticide residue levels were reduced by ozone treatment, with a maximum degradation rate of 12.3% per day.

d) Trimming, washing, peeling and cooking

The removal of the outer part of RAC by peeling, husking, hulling, shelling, or trimming is the most effective food preparation process for pesticide residues removal from RAC. Numerous studies report the elimination of different pesticide residues on different RAC through peeling to range from 70 to 100% (Cengiz *et al.*, 2007; Boulaid *et al.*, 2005; Fernández-Cruz *et al.*, 2004; Rasmusssen *et al.*, 2003; Burchat *et al.*, 1998; Clavijo *et al.*, 1996; Celik *et al.*, 1995; Holland *et al.*, 1994; Rouchaud *et al.*, 1991). The systemic action of a pesticide residue in this case is not always correlated with decreased reduction of pesticide residues through peeling. Thus, although, residues of the systemic organophosphorus phorate were only reduced by 50% through peeling of potatoes (JMPR, 1992) and similarly disyston residues in potatoes were only reduced by 35% after peeling (Holland *et al.*,1994), quinalphos residues on apples and procymidone residues on tomatoes were reduced by "73% (Cengiz *et al.*, 2007). Furthermore, the reported data of the reduction of dichlorvos and diazinon on cucumber by 57.2% and 67.3% respectively (Cengiz *et al.*, 2007), the low reduction of pyridaben (~70%) in tomatoes (Boulaid *et al.*, 2005).

Pesticide residues may vaporize, hydrolyse and /or thermally degrade during cooking. However, the processes and conditions used in food cooking are highly varied. The details of time, temperature, degree of moisture loss, whether the system is open or closed and whether water is added or not in the process are important for the estimation of the fate of a residue level. In general, rates of degradation and volatilization of residues are increased by the heat involved in cooking or pasteurization and rates of hydrolysis may also be increased by the water addition and the increase of temperature. In different studies of organophosphorus (OP) insecticides (fenitrothion, fenitrothion oxon, 3-methyl-4-nitrophenol) on RACs (cauliflower, kaki fruits) it was reported that OPs are stable through heating without the addition of water for 10-15 min (Fernández-Cruz *et al.*, 2006; Fernández-Cruz *et al.*, 2004) and unstable to heating in aqueous solution. Coulibaly and Smith (1993) studied famphur, fenthion, parathion, stirofos, chlorpyrifos, and ronnel in aqueous solutions that were unheated, heated to 70 °C for 1 or 2 h, and heated to 80 °C for 1 h. Stirofos and famphur were largely unaffected by heating whereas fenthion, parathion, chlorpyrifos and ronnel were hydrolyzed by 53.2-80% in unheated water and further heating did not result in further degradation. Other studies confirm the reduction of OPs (32% reduction of fenitrothion and 89.5% reduction of triazophos) after boiling (Rasmussen *et al.*, 2003; Holden *et al.*, 2001), but also reported chlorpyrifos and acephate reduction (50.5-68%) after cooking of peppers, asparagus and peaches without the addition of water (at 100- 110ºC for 20-80 min) (Chavarri *et al.*, 2005).

Numerous investigations have been carried out to study the effect of kitchen-type processing on pesticide residues in various fruits and vegetables. Kitchen-type processing techniques could remove 40-77% of diazinon residues and 37-82% of dimethoate residue in green beans and cauliflower (Bognar, 1977). It has further been reported that washing alone reduced the levels of dimethoate residues by approximately 25-80% whereas washing and cooking of cauliflower curds reduced the level of residues by 52-91% (Khaire and Dethe, 1983). Sugibayashi *et al.*, (1996) compared the effects of washing, peeling and cooking on the residue levels of dichlorovos, dimethoate and other pesticides in white potatoes and carrots. Chlorfenvinphos-E, S-benzyl diisopropyl phosphorothiolate and monocrotophos-E were efficiently removed by washing alone. However, peeling was found to be the most effective way to remove the pesticide from the vegetables followed by frying. Boiling was effective in reducing the level of water-soluble pesticides. The residues of dimethoate were reduced by approximately 50% by boiling in water for 10 min in vegetables of the *Brassica rapa* type (Watanabe *et al.*, 1988). Sharma *et al.*, (1994) studied the effect of washing on the removal of mancozeb (Dithane M-45) residues on cabbage, knol-khol, tomato, okra and brinjal and found that plain washing dislodged 20-52% mancozeb residues while washing coupled with cooking led to 53-79% decontamination.

Nagesh and Verma (1997) studied the decontamination of cabbage treated with chlorpyriphos and quinalphos. Cabbage was treated with chlorpyriphos (0.05%) and quinalphos (0.05 %) in separate plots to control important pests. Decontamination through different processes showed that the residues were reduced to some extent by various home processing methods like washing and cooking. Washing the samples with salt water did not differ significantly from that of ordinary washing with tap water. Cooking also did not help much in reducing the residue below the MRLs of 0.25 and 0.05 mg/kg for quinalphos and chlorpyriphos respectively.

Lalitha *et al.,* (1998) studied the dissipation of quinalphos and the effect of processing in cauliflower. It was concluded that residues were considerably reduced from curds roasted in oil after boiling (36.3-68.6%) followed by simple boiling (10.1-60.0%) compared to the curds soaked in 2% salt water (14.7-43.2%) and soaked in tap water (13.0-40.5%).

The results of the study conducted by Schattenberg *et al.,* (1996) indicated that residue levels in most commodities were substantially reduced after typical household preparation. Washing and cooking treatments resulted in considerable reduction of lindane residues in brinjal and okra (Patel *et al.,* 2001). Kadian *et al.,* (2001) reported that cypermethrin residues declined in tomato, okra, bottlegourd and ridge gourd after all processing steps i.e about 5-14% by washing, 6-26% by blanching, 6-19% by washing in brine solution and 15-33% by cooking. They suggested that cooking was more effective in reducing the cypermethrin residues in vegetables. Kang and Lee (2005) investigated the effects of brining, blanching and cooking on 8 pesticide residues in Chinese cabbage and spinach. After harvest cabbage samples were brined in 8% brine for 4 h and boiled in water for 20 min, whereas spinach samples were blanched by dipping in water for 2 min and cooked in the same way as cabbage. Brining had little impact on pesticide levels in cabbage apart from those of diazinon and dichlorvos, decreased by about 20%. Residues undergoing greatest concentration decreased during cooking were diazinon and dichlorvos (80-90%) whereas levels of cypermethrin and deltamethrin and fenvalerate slightly increased. During blanching of spinach 72% decrease in dichlorvos concentration took place whereas decreases in levels of other pesticides ranged from 0-17%. During spinach cooking, decreases in pesticide concentration ranged from 0% for fenvalerate to 81% for dichlorvos.

Randhawa *et al.,* (2007) studied the fate of chlorpyrifos and its degradation product 3, 5, 6- trichloro-2-pyridinol during boiling of different vegetables with water and the decrease of chlorpyrifos ranged from 12 to 48%. The effect was more obvious in spinach (38%) followed by cauliflower (29%). 3,5,6-trichloro-2-pyridinol was substantially increased during the course of cooking consisting that metabolites of toxicological concern should also be measured in cooked fruits and vegetables (and in the boiled water that might be used in further cook).

Peeling-off the fruit skin was reported to dislodge the residues to varying degree depending on constitution of the fruit, chemical nature of the pesticide and environmental conditions (Nath *et al.*, 1975; Awasthi 1986). Awasthi (1993) studied the effects of washing and peeling of insecticide-treated mangoes on residues of dimethoate, fenthion, cypermethrin and fenvalerate. Washing reduced levels of dimethoate and fenthion on mangoes to 66-68% and levels of fenvalerate and cypermethrin to 21-27%. Peeling removed 100% of the residues in all cases. This reflected the accumulation of residues in the fruit pericarp only and no further movement to fruit pulp. Thus, the fruit pulp was free from any toxic residues at any stage of residue persistence after peeling-off the treated fruits.

Effects of washing (water or detergent), peeling, blanching-boiling, milling and processing on efficacy of removal of organophosphorus pesticide (OP) residues from various foods were studied by Lee and Lee (1997). It was revealed that 45% of OP residues were eliminated when foods were washed in water, 56% with detergent washing, 91% with peeling, 51% with blanching-boiling and 90% in milling and processing.

Persistence of propham and chlorpropham residues in stored potatoes was studied by Singh *et al.*, (2000). Potatoes were treated with a 29% (1:6) mixture of propham and chlorpropham at a concentration of 60 ml/ton prior to storage at 12°C. At 2 days after treatment, propham was not detected in samples, while chlorpropham was detected in peel, whole tubers and pulp at concentration of 4.69, 0.5 and 0.085 μg/g, respectively; these concentrations fell to 2.61, 0.15 and 0.032 μg/g at 21 days after treatment. Peeling of potatoes caused a substantial decrease in chlorpropham residues.

Organochlorine and organophosphorus pesticide residues in potatoes, fries and chips were evaluated and effects of washing, peeling and cooking on removal of residues were assessed (Soliman, 2001). From the analysis of potatoes and potato products collected from various regions of Egypt, it was found that malathion, lindane, HCB and p,p DDD were present universally; highest levels were present in raw potatoes while chips had the lowest levels. Potato skins contained higher amounts of pesticides residues than potato pulp, and peeling markedly decreased pesticide content of potatoes. Washing potatoes with tap water or aqueous solutions of acetic acid and/or NaCl, and blanching or frying of potatoes also removed most residues.

Gill *et al.*, (2001) studied the brinjals and tomatoes sprayed with recommended dose (0.005 %) of alphamethrin up to drenching level and stored at ambient (40°C) and refrigerated conditions (5°C). Dissipation of alphamethrin was observed faster at room temperature as compared to cold conditions in both the vegetables. Washing and cooking were not found to be very effective in lowering residues in brinjals. However, reduction of residues was more due to cooking than simple washing (25-33%). In tomato both the processes reduced the residues almost to the same extent (11-30%). Effect of household processing (washing, peeling and cooking) and unit-to-unit variability of pyrifenox, pyridaben and tralomethrin residues in tomatoes were carried out by Boulaid *et al.*, (2005). Levels of pyrifenox, pyridaben and tralomethrin residues were determined in unprocessed and processed tomatoes. The washing processing factor results were 0.9 ± 0.3 for pyridaben, 1.1 ± 0.3 for pyrifenox and 1.2 ± 0.5 for tralomethrin, whereas the peeling processing factors were 0.3 ± 0.2 for pyridaben and 0.0 ± 0.0 for both pyrifenox and tralomethrin. The average loss of water in the tomato samples during cooking was approximately 50%; the cooking processing factors were 2.1 ± 0.8 for pyridaben, 3.0 ± 1.1 for pyrifenox and 1.9 ± 0.8 for tralomethrin. The unit-to-unit variability factor results were within the range 1.3-2.2.

Steeping in water followed by dehulling reduced residues of deltamethrin in chick peas (*Cicer arietinum L.*) by around 70-73%; however, this decontamination method was not sufficient to reduce deltamethrin concentration to the prescribed maximum residue level of 0.1 mg/kg. Washing and steaming removed from 40 to 60% of deltamethrin residues remaining on chick peas after storage (Kumar and Dikshit, 2000). Reddy *et al.* (2001) studied the dissipation and decontamination of triazophos and lindane in brinjal. Washing with water followed by steam cooking removed triazophos to an extent of 64-88 %. Lindane residues were reduced to an extent of 42-56%.

The effect of microwave and oven baking on residues of the post-harvest fungicide thiabendazole (E-233) in potatoes was investigated by comparing amounts present in raw, microwave and oven baked tubers by Friar and Reynolds (1991). It was found that thiabendazole was predominantly retained in the peelings (96.3-98.8%) and not lost during the two types of processing treatments.Brinjal fruits were sprayed with recommended dose (0.001%) of cypermethrin by Walia *et al.* (2010). Effects of processing viz. washing, cooking in water, cooking in oil, microwave cooking, and grilling was studied to dislodge cypermethrin residues on brinjal. Cypermethrin residues remaining in the control and processed samples were analyzed by gas chromatography equipped with ECD. Dislodging of cypermethrin residues was observed more in grilling (50.12%), followed by cooking in oil (45.2%), cooking in water (41.4%), and microwave cooking (40.89%) after 1st day of the treatment. Reduction of residues after washing treatment was minimal.

Residues of cypermethrin and decamethrin and effect of processing were assessed in brinjal fruits by gas liquid chromatography following single application of Cymbush 25 EC@43.75 and 87.50 g a.i./ha and of Decis 2.8 EC@11.20 and 22.40 g a.i./ ha at fruiting stage by Kaur *et al.* (2011). Washing and washing followed by boiling/cooking processes were found to be effective in reducing the residues of both the insecticides in brinjal fruits. Maximum reduction (31-42%) and (26-37%) was observed by washing followed by boiling/cooking for cypermethrin and decamethrin, respectively. Samriti *et al.* (2011) investigated the effect of processing (washing and washing +boiling) on okra fruits and found very effective in reducing the levels of chlorpyriphos residues in okra fruits. Maximum reduction (64-77%) was observed by washing + boiling followed by washing (13-35%).Chauhan *et al.* (2011) reported that the process of washing followed by boiling reduced the residues effectively (74-84%) in tomato fruits whereas by washing only, residues could be reduced in the range of 37-40%. In samples under refrigerated condition, residues decreased slightly less than the samples stored under room temperature. Washing followed by boiling reduced the residues from 72 to 80% whereas only washing reduced the residues from 35 to 36%.

e) Comminution

Comminution of RAC through chopping, blending, crushing and similar processes usually do not affect pesticide residues in RAC since most pesticides are relatively stable in acidic plant tissue homogenates for the moderate periods of time involved in food preparation. However, comminution leads to release of enzymes and acids which may increase the rate of hydrolytic and other degradative processes on residues that were previously isolated by cuticular layers. Special concern on the fate of residues should be paid on acid sensitive pesticide compounds (e.g. ethylene bis dithiocarbamate (EBDC), carbosulfan, benfuracarb, pymetrozine, dioxacarb, thiodicarb and others) that readily are hydrolyzed in the presence of trace amounts of acids and the most toxic metabolites formed should be studied. A typical example is the rapid degradation of EBDC fungicide residues to the formation of the toxic ETU, carbon disulphide and ethylenediamine in slightly acidic media, similar to the pH of the tomato homogenates (4.0-4.2) (Kontou *et al.*, 2004; Kontou *et al.*, 2004b; Holland, 1994; Howard and Yip, 1971).

Conclusion

It is obvious from the foregoing review that pesticide residues are left in almost all the food commodities resulting from the pre harvest or post harvest treatments. The level of pesticide residues is affected by washing, preparatory steps, heating or cooking, processing during product manufacturing, post harvesting handling and storage. The extent of reduction varies with nature of pesticide molecule, type of commodity, processing steps and product prepared. The washing of raw materials is the simplest way to reduce the pesticide residue in the final product. The more effective and convenient alternative could be washing with chlorine water or with dilute solutions of other chemicals depending upon food commodity. Special precautions should be taken to dislodge the residues from raw materials to be used for preparation of concentrated and dehydrated products. Judicious and systematic approach should be followed to adopt pre harvest practices and post harvest handling treatments to minimize the residue levels in finished products. There is urgent need to monitor the pesticide residues to standardize the application doses. Equally important is to develop or find new pesticide molecules with high effectiveness and fast degrading capabilities. Because pest management requires one of the major inputs in agricultural productivity, this area needs great attention to economize the production, to provide safe foods, to improve human/animal health and lower the medical expenses for treatment of ailments. The alternative ways of pest management should also be explored. In the present scenario of globally competitive trade, all concerted efforts should be done to ensure food safety for boosting the export.

References

Amvrazi, E. and Albanis, T. 2008. Multiclass pesticide determination in olives and their processing factors in olive oil: Comparison of different olive oil extraction systems. *J. Agric. Fd. Chem.*, **56:** 5700–5709.

Angioni, A. Garau, V.L. Aguilera Del Real, A. Melis, M. Minelli, E.V. Tuberoso, C. and Cabras, P. 2004. GC-ITMS determination and degradation of captan during winemaking, *J. Agric. Fd. Chem.*, **51:** 6761–6766.

Awad, OM and el-Shimi, N.M. 1993. Influence of different means of washing and processing on removing of Actellic residue from spinach and eggplant and its in-vivo action on mice hepatic biochemical targets. *J Egypt Public Health Assoc.*, **68:** 671–86.

Awasthi, M.D. 1993. Decontamination of insecticide residues on mango by washing and peeling. *J. Fd. Sci. Technol.*, **30(2):** 132-133.

Awasthi, M.D. 1986. Chemical treatments for the decontamination on brinjal fruits from the residues of synthetic pyrethroids. *Pestic. Sci.*, **17:** 89.

Bhagirathi, D. Kapoor, S.K. and Singh, B. 2001. Persistence of malathion residues on/in bell pepper (*Capsicum annuum Linn*) *Pestic. Res. J.*, **13(1):** 99-102.

Bognar, A. 1977. Studies to reduce or eliminate residues of crop protection products in vegetal foods through kitcen processing. *Dtsch Lebensm Rundsch,* **73(5):** 149-157.

Boulaid, M. Aguilera, A. Camacho, F. Soussi M and Valverde A. 2005. Effect of household processing and unit-to-unit variability of pyrifenox, pyridaben, and tralomethrin residues in tomatoes. *J. Agric. Fd. Chem.*, **53(10):** 4054-4058.

Burchat, C.S. Ripley, B.D. Leishman, P.D. Ritcey, G.M. Kakuda, Y. and Stephenson, G.R. 1998. The distribution of nine pesticides between the juice and pulp of carrots and tomatoes after home processing. *Fd. Addit. and Contamn.*, **15(1):** 61-71.

Cabras, P. Angioni, A. Garau, V.L. Pirisi, F.M. Brandolini, V. Cabitza, F and Cubeddu, M. 1998. Pesticide residues in prune processing. *J. Agric. Fd. Chem.*, **46:** 3772-3774.

Cabras, P. Angioni, A. Garau, V.L. Pirisi, F.M. Cabitza, F. and Pala, M. 2000. Acephate and buprofezin residues in olives and olive oil, *Fd Addit. Contamn.*, **17:** 855-858.

Cabras, P. Angioni, A. Garau, V.L. Melis, M. Pirisi FM, Karim M and Minelli EV. 1997. Persistence of insecticide residues in olives and olive oil. *J. Agric. Fd. Chem.*, **45:** 2244-2247.

Cabras, P. Garau, V.L. Melis, M. Pirisi, F.M. and Spanedda, L. 1993. Persistence and fate of Fenthion in olives and olive by products. *J. Agric. Fd. Chem.*, **41:** 2431-2433.

Celik, S. Kunc, S. and Asan, T. 1995. Degradation of some pesticides in the field and effect of processing. *Analyst.*, **120:** 1739–1743.

Cengiz, M.F. Certel, M. Karakas, B. and Göçmen, H. 2007. Residue contents of captan and procymidone applied on tomatoes grown in greenhouses and their reduction by duration of a pre-harvest interval and post-harvest culinary applications. *Fd Chem.*, **100:** 1611–1619.

Chauhan, Reena, Samriti and Kumari Beena. 2011. Effect of processing on reduction of λ cyhalothrin residues in tomato fruits. *Bull Environ Contamn Toxicol.*, DOI 10.1007/s00128-011-0483-9.

Chavarri, M.J. Herrera, A. and Arino, A. 2005. The decrease in pesticides in fruit and vegetables during commercial processing. *Int. J. Fd. Sci. Technol.*, **40(2):** 205-211.

Clavijo, M.P. Medina, M.P. Asensio, J.S. and Bernal, J.G. 1996. Decay study of pesticide residues in apple samples. *J.of Chromatog., A.* 740: 146–150.

Coulibaly, K and Smith, J.S. 1993. Thermostability of organophosphate pesticides and some of their major metabolites in water and beef muscle. *J. Agric. Fd. Chem.*, **41:** 1719-1723.

Dikshit, A.K. Handa, S.K. and Verma, S. 1986. Residues of methamidophos and effect of washing and cooking in cauliflower, cabbage and Indian Colza. *Indian J. Agric Sci.*, **56:** 661.

Dikshit, A.K. 2001. Persistence of cypermethrin on stored pulses and its decontamination. *Pestic. Res. J.*, **13(2):** 141-146.

Fanchun, K. Shengmin, L. and Qun, W. 2003. Application of ozone for pesticide degradation and preservation of fruits and vegetables. *Fd Machinery,* **5:** 24-26.

Fernández-Cruz, M.L. Barreda, M. Villarroya, M. Peruga, A. Llanos, S and García-Baudín, J.M. 2006. Captan and fenitrothion dissipation in field-treated cauliflowers and effect of household processing. *Pest Managt. Sci.*, **62:** 637–645.

Fernández-Cruz, M.L. Villarroya, M. Llanos, S. Alonso-Prados, J.L. and García-Baudín, J.M. 2004. Field incurred fenitrothion residues in kakis: comparison of individual fruits, composite samples, peeled and cooked fruits. *J. Agric. Fd. Chem.*, **52:** 860–863.

Ferreira, J.R. and Tainha, A.M. 1983. Organophosphorous insecticide residues in olives and olive oil. *Pestic.Sci.,* **14:**167–172.

Friar, P.M. and Reynolds, S.L. 1991. The effects of microwave baking and oven baking on thiabendazole residues in potatoes. *Fd Addit.Contamn.,* **8(5):** 617-26.

Gill, K. Kumari, B. and Kathpal, T.S. 2001. Dissipation of alphamethrin residues in/on brinjal and tomato during storage and processing conditions. *J. Fd. Sci. Technol.,* **38 (1):** 43-46.

Guardia Rubio, M. Ruiz Medina, A. Molina Diaz, A. and Cañada Ayora, M.J. 2006. Influence of harvesting method and washing on the presence of pesticide residues in olives and olive oil/ *J. Agric. Fd. Chem.,* **54:** 8538-8544.

Guardia Rubio, M. Ruiz Medina, A. Molina Diaz, A and Cañada, Ayora M.J. 2007. Multiresidue analysis of three groups of pesticides in washing waters from olive processing by solid-phase extraction-gas chromatograhy with electron capture and thermionic specific detection. *Microchemi.J.,* **85:** 257-264.

Holden, A.J. Chen, L. and Shaw, I.C. 2001. Thermal stability of organophosphorus pesticide triazophos and its relevance in the assessment of risk to the consumer of triazophos residues in food. *J.Agric. Fd. Chem.,* **49:** 103- 106.

Holland, P.T. Hamilton, D. Ohlin, B and Skidmore, M.W. 1994. Effects of storage and processing on pesticide residues in plant products. *Pure and Applied Chem.,* **66:** 335-356.

Howard, S.F. and Yip, G. 1971. Stability of metallic ethylene bisdithiocarbamates in chopped kale. *J Assoc. Off. Anal. Chem.,* **54(6):** 1371-72.

Hwang, E.S. and Cash, J.N. 2003. Control of pesticide residues by oxidizing agents to ensure food safety. *Fd. Sci. and Biotechnol.,* **12(5):** 581-587.

Hwang, E.S. Cash, J.N. and Zabik, M.J. 2001. Postharvest treatments for the reduction of mancozeb in fresh apples. *J Agric. Fd. Chem.,* **49(6):** 3127-3132.

JMPR 1992. Refers to the FAO Plant Production and Production Paper series "Pesticide Residues in Food-Evaluations Part 1" published annually by FAO, Rome. issue summarises residue data submitted to the previous JMPR meeting and is organised by pesticide in alphabetical order.

Kadian, S. Kumar, R. Grewal, R.B. and Srivastava, S.P. 2001. Effect of household processing on cypermethrin residues in some commonly used vegetables. *Pestology.,* **25:** 10-13.

Kang, S.M. and Lee, M.G. 2005. Fate of some pesticides during brining and cooking of Chinese cabbage and spinach. *Fd. Sci. Biotech.,* **14(1):** 77-81.

Kaur, Prabhjot, Yadav, G.S. Chauhan, Reena and Kumari Beena. 2011. Persistence of cypermethrin and decamethrin residues in/on brinjal fruits. *Bull. Environ. Contamn. Toxicol.,* **87:** 693–698.

Kaushik, G. Satya, S and Naik, S.N. 2009. Food processing a tool to pesticide residue dissipation - A review. *Fd. Res. Intl.,* **42:** 26–40.

Khaire, J.T. and Dethe, M.D. 1983. Effects of washing and cooking on the dimethoate level in cauliflower curds. *J Maharashtra Agric. Univ.,* **8(1):** 61-62.

Kim, S.D. Kim, I.D. Park, M.Z. and Lee, Y.G. 2000. Effect of ozone water on pesticide-residual contents of soybean sprouts during cultivation. *Korean J. Fd. Sci. Technol.* **32(2):** 277-283.

Krol, W.J. Arsenault, T.L. Pylypiw, H.M. Jr. and Mattina, MJI. 2000. Reduction of Pesticide Residues on Produce by Rinsing. *J. Agric. Fd Chem.,* **48:** 4666-4670.

Kumar Lal, A. and Dikshit, A.K. 2000. Persistence of deltamethrin on chickpea and its decontamination. *Pestic. Res. J.*, **12(1):** 74-79.

Kumari, Beena. 2008. Effects of household processing on reduction of pesticide residues in vegetables. *ARPN J. Agric. Biologi. Sci* **3(4):** 46-48.

Lalitha, P. Uma Reddy, M. Narsimha Rao, B and Reddy, DDR. 1998. Dissipation of quinalphos and the effect of processing in cauliflower. *Indian J. Nutr. Dietetics.,* **35:** 129-32.

Leandri, A. Pompi, V. Pucci, C and Spanedda, A.F. 1993. Residues on olives, oil and processing wastewaters of pesticides used for control of Dacus oleae (Gmel.) (Dipt., Tephritidae), *Anz. Schädlingskde., Pflanzenschutz, Umweltschutz.*, **66:** 48-51.

Lee, M.G. and Lee, S.R. 1997. Reduction factors and risk assessment of organophosphorus pesticides in Korean foods. *Korean. J Fd. Sci. Technol.,* **29(2):** 240-248.

Lee, M.G. 2001. Reduction of chlorpyrifos and fenitrothion residues in red pepper peel by washing and drying. *Fd. Sci. Biotech.*, **10(4):** 429-437.

Lee, Y.S. and Chou, S.S. 1995. Reduced pesticide residues in vegetables by various methods of washing. *Conf Proc IFT Annual Meeting* 1995. Food Science Program, Univ. of the District of Columbia, Washington, DC 20008, USA

Lentza-Rizos, C. Avramides, E.J. and Kokkinaki K. 2006. Residues of azoxystrobin from grapes to raisins, *J. Agric. Fd Chem.*, **54:**138- 141.

Lentza-Rizos, C. and Kokkinaki, K. 2002. Residues of cypermethrin in field-treated grapes and raisins produced after various treatments. *Fd. Addit. and Contamn.,* **19:**1162–1168.

Miyahara, M. and Saito, Y. 1994. Effect of the processing steps in toffee production on pesticide residues. *J. Agric. Fd. Chem.,* **42(2):** 369-373.

Nagesh, M. and Verma, S. 1997. Decontamination of cabbage treated with chlorpyriphos and quinalphos. *Indian J. Entomol.,* **59:** 404-10.

Nath, G. Jat R.N. and Srivastava BP 1975. Effect of washing, cooking and dehydration on the removal of some insecticides from okra (*Abelmoschus esculentus* Moench.). *J. Fd. Sci. Technol.,* **12:** 127-130.

Patel, B.A. Shah, P.G. Raj, M.F. Patel, B.K. and Patel, J.A. 2001. Dissipation of lindane in/on brinjal and okra fruits. *Pestic. Res. J.,* **13(1):** 58-61.

Phani Kumar, K. Jagdishwar Reddy, D. Narasimha Reddy, K. Ramesh Babu, T. and Narendranath, VV. 2000. Dissipation and decontamination of triazophos and acephate residues in chilli (*Capsicum annum* Linn). *Pestic. Res. J.,* **12(1):** 26-29.

Phogat, V. Sharma, S.K. Kumar, S. Satyavan and Gupta, S.K. 2010. Vegetable cultivation with poor quality water. *Tech.Bull.* Department of Soil Sciences, CCSHAU, Hisar, pp72.

Pugliese, P. Molto, J.C. Damiani, P. Marin, R. Cossignani, L. and Manes, J. 2004. Gas chromatographic evaluation of pesticide residue contents in nectarines after non-toxic washing treatments. *J Chromatog.-A.* **1050(2)**: 185-191.

Randhawa, M.A. Muhammad, F. Anjum, F.M. Ahmed, A. and Randhawa, M.S. 2007. Field incurred chlorpyrifos and 3, 5, 6-trichloro-2-pyridinol residues in fresh and processed vegetables. *Fd. Chem.,* **103:** 1016–1023.

Rasmusssen, R.R. Poulsen, M.E. and Hansen, H.C.B. 2003. Distribution of multiple pesticide residues in apple segments after home processing. *Fd Addit. and Contamn.,* **20:** 1044–1063.

Reddy, N.K. Sultan, M.A. Reddy, D.J. and Babu, T.R. 2001. Dissipation and decontamination of triazophos and lindane in brinjal. *Pestology.,* **725:** 51-54.

Rouchaud, J. Gustin, F. Creemers, P. Goffings, G. and Herrgods, M. 1991. Fate of the fungicide tolylfluanid in pear cold stored in controlled or non controlled atmosphere. *Bull. Environ. Contamn. and Toxicol.*, **46:** 499- 506.

Samriti, Chauhan, Reena and Kumari, Beena. 2011. Persistence and effect of processing on reduction of chlorpyriphos residues in okra fruits. *Bull Environ Contamn Toxicol.*, **87:**198–201.

Sardana, H.R. 2001. Integrated pest management in vegetables. In: Training Manual-2; Training on IPM for Zonal Agricultural Research Stations, from May 21-26.pp105-118.

Schattenberg, H.J. III, Geno, P.W. Hsu, J.P. Fry, W.G. and Parker, R.P. 1996. Effect of household preparation on levels of pesticide residues in produce. *JAOAC Int.*, **79(6):** 1447-1453.

Sharma, I.D. Nath, A. and Dubey, J.K. 1994. Persistence of mancozeb (Dithane M 45) in some vegetables and efficacy of decontamination processes. *J. Fd. Sci. Technol.*, **31:** 215-18.

Singh, R. Madan, V.K. Singh, B. and Kathpal, T.S. 2000. Dissipation of propham and chlorpropham residues in potato tubers, peel and pulp. *Pestic. Res. J.*, **12**(1): 133-136.

Soliman, K.M. 2001. Changes in concentration of pesticide residues in potatoes during washing and home preparation. *Fd. Chem. Toxicol.*, **39**(8): 887-891

Sugibayashi, S. Hamada, I. Mishima, I. Yoshikawa, N. Kataoka, J. Kawaguchi, Y. Fujimoto, Y. Semma, M and Ito Y 1996. Reduction in the residual levels of dichlorvos and 19 other agricultural chemicals by washing and cooking in experimental models of white potato and carrot. *Nippon shokuhin Kagaku Gakkaishi* **2**(2):97-101.

Walia, Shweta, Boora, Pinky and Kumari, Beena. 2010. Effect of processing on dislodging of cypermethrin residues on Brinjal. *Bull. Environ. Contam. Toxicol*, **84:**465-468.

Watanabe, S. Watanabe, S. and Ito, K. 1988. Residue of synthetic pyrethroid insecticide fenvalerate in vegetables and its fate in the process of cooking. *Kanagawa-ken Eisei Kenkyusho Kenkyu Hokoku*, **18:** 43-45.

WHO. 1988. Environmental Health Criteria No. 78, Dithiocarbamate pesticides, ethylenethiourea and propylene thiourea: A general introduction. *World Health Organisation*, Geneva 1988.

Yoshida, S. Murata, H and Imaida, M. 1992. Distribution of pesticide residues in vegetables and fruits and removal by washing. Nippon Nogeikagaku Kaishi. *J. Agric. Chem. Soc. of Japan.*, **66(6):** 1007-1011.

Zabik, M.E. Booren, A. Zabik, M.J. Welch, R. Humphrey, H. 1996. Pesticide residues, PCBs and PAHs in baked, charbroiled, salt boiled and smoked Great Lakes lake trout. *Fd Chem.*, **55(3):** 231-239.

Zabik, M.J. El-Hadidi, MFAJ, Cash, N. Zabik, M.E. and Jones, A.L. 2000. Reduction of azinphos-methyl, chlorpyrifos, esfenvalerate, and methomyl residues in processed apples. *J. Agric. and Fd. Chem.*, **48:** 4199-4203.

Zhanggui, Q. Xiaoping, Y and Xia, W. 2003. Trials of ozone reducing pesticide residues in grain. *Grain-Storage*, **32(3):** 10-13.

Zohair, A. 2001. Behaviour of some organophosphorus and organochlorine pesticides in potatoes during soaking in different solutions. *Fd. Chem. Toxic.*, **39(7):** 751-755.

Papier | Fördert gute Waldnutzung
FSC® C083411

Zeitfracht Medien GmbH
Ferdinand-Jühlke-Straße 7
99095 Erfurt, Deutschland
produktsicherheit@kolibri360.de